The Atom in the History of
Human Thought

The Atom
in the
History of
Human
Thought

Bernard Pullman

Late Professor of Quantum Chemistry
at the Sorbonne

translated by
Axel Reisinger

OXFORD
UNIVERSITY PRESS

OXFORD
UNIVERSITY PRESS

Oxford   New York
Athens   Auckland   Bangkok   Bogotá   Buenos Aires
Cape Town   Chennai   Dar es Salaam   Delhi   Florence   Hong Kong   Istanbul
Karachi   Kolkata   Kuala Lumpur   Madrid   Melbourne   Mexico City   Mumbai
Nairobi   Paris   São Paulo   Shanghai   Singapore   Taipei   Tokyo   Toronto   Warsaw

and associated companies in

Berlin   Ibadan

Copyright © 1998 by Oxford University Press
Originally published by Librarie Arthème Fayard,
75 rue des Saints Péres,
75278 Paris Cedex 06, France
Original French title: L'atome dans l'histoire de la pensée humaine

First published in 1998 by Oxford University Press, Inc.

First issued as an Oxford University Press paperback, 2001
198 Madison Avenue, New York, New York 10016

Oxford is a registered trademark of Oxford University Press

All rights reserved. No part of this publication may be reproduced,
stored in a retrieval system, or transmitted, in any form or by any means,
electronic, mechanical, photocopying, recording, or otherwise,
without the prior permission of Oxford University Press.

Library of Congress Cataloging-in-Publication Data
Pullman, Bernard, 1919–1996
[Atome dans l'histoire de la pensée humaine. English]
The atom in the history of human thought / by Bernard Pullman,
late professor of quantum chemistry at the Sorbonne; translated by Axel Reisinger.
p. cm.
Includes bibliographical references and index.
ISBN 0-19-511447-7 (Cloth); ISBN 0-19-515040-6 (Pbk.)
1. Atomic theory—History. I. Title.
QC171.2.P85   1998   97-36040
539'.14'09—dc21

Oxford University Press would like to acknowledge a
generous subsidy from the French Ministry of Culture
toward the cost of translating this work into English.

1 3 5 7 9 8 6 4 2
Printed in the United States of America
on acid-free paper

If all scientific knowledge were lost in a cataclysm,
what single statement would preserve
the most information for the next
generations of creatures? How could we
best pass on our understanding of the
world? [I might propose:] "All things are
made of atoms—little particles that
move around in perpetual motion,
attracting each other when they are a
little distance apart, but repelling upon
being squeezed into one another." In that
one sentence, you will see, there is an
enormous amount of information about
the world, if just a little imagination
and thinking are applied.

—Richard P. Feynman

 CONTENTS

PREFACE

This book endeavors to describe the turbulent relationship between atomic theory and philosophy and religion over a period of twenty-five centuries. I am not a philosopher, as the experts will immediately realize. Nor am I a man of religion, as God is my witness. I am—but then so is everybody—made of molecules, themselves made of atoms, which in turn means essentially vacuum, even though the modern vacuum is not nearly as empty as it once was thought to be. My only credential for writing this book is to have, by virtue of my profession, long lived side by side with atoms. This has given me a chance to ask them many questions and to glean from them a few answers. Asking questions presents no difficulty at all—atoms have been queried endlessly over the last twenty-five centuries. But extracting answers is another matter altogether. Yet I was fortunate to live in the twentieth century, a time when atoms finally became accessible to scientific study and began grudgingly to surrender a few of their secrets.

Born twenty-five hundred years ago "on the shore of the heavenly sea," the atomic hypothesis is the most important and enduring legacy bequeathed by antiquity. For all its scientific implications, it was at first primarily a philosophical idea—just one link in the chain of re-flections on the part of Greek thinkers in search of things essential and universal. On this abstract level it remained almost until the nineteenth century. As such, its scientific epoch is relatively recent, while its philosophical heritage is ancient; ancient and prestigious, one might add, because of its association with the most illustrious names in this great intellectual epic of mankind. On the other hand, by encompassing in its doctrine a body of cosmogenic and cosmological propositions, by concerning itself as much with the soul as with the body, by professing a particular system of ethics, by adopting toward God (or the gods, to be more precise) an equivocal attitude, the atomic theory could not—and indeed never did—leave theologians and the religious authorities indifferent.

Because of its amazing durability, the import of the debates and controversies it elicited, the originality and variety of the arguments invoked—which combined scientific, philosophical, and religious concepts in an often passionate and sometimes rancorous rhetoric—the atomic theory provided a battleground for a clash of ideas spanning twenty-five centuries of the history of human thought. The intense philosophical discussions about its deeper meaning showed no sign of abating with the advent of scientific atomism. On the contrary, they have experienced a renewed impetus.

It is an overwhelming feeling to realize that we owe this extraordinary adventure of the human spirit to the creative imagination of a handful of Greek thinkers from the distant past. Even though the atom of modern science bears only a vague resemblance to the kind of atom envisioned by these early thinkers, the concept they handed down to future generations proved to be one of the most important gifts ever bestowed by man or heaven. None could have been more weighted with momentous consequences for mankind's own existence and future. I shiver at the thought of what might have been lost had fate dictated that someone like Auguste Comte, rather than Democritus, be born in Abdera during the fifth century B.C.

I would have dearly loved to gather together in some transhistoric symposium—capped by a grand banquet in deference to the etymological sense of the word—all those who at one time or another were part of this long quest for the ultimate truth, so that they may confront various points of view and, in light of what we know today, get a sense of their frailty and frequent temerity. Since this is not to be, I will instead let them speak in turn, often quoting them verbatim, since I am firmly convinced that in order to fully appreciate the richness and depth of this remarkable debate one has to rub elbows with the protagonists, understand their arguments from their own perspective, hear them express their views in the language of their time—in short, to relive history.

I wish to thank my many colleagues and friends who helped me create this book. My special gratitude goes to Henri Berthod for carefully reviewing the entire manuscript, and to Gaston Berthier for critiquing the section dealing with the period of scientific discoveries. I benefited immensely from their comments and feedback. I am greatly indebted to Isabelle Lépine, my secretary, for her constant, diligent, and courageous efforts in bringing the manuscript to completion.

Last but not least, I want to take this opportunity to express my deep appreciation to my sons, Michel and Bertrand, who sacrificed a good portion of their vacation time to submit the manuscript to a rigorous regimen of constructive criticism, and to my wife, Alberte, for her invaluable, untiring, and patient help throughout the conception, preparation, and final editing of the book. Without her continuous presence at my side, her encouragement and sound advice, this project would never have come to fruition.

The Atom in the History of
Human Thought

 **PART ONE**

The Birth of
the Atomic Theory

I

The Backdrop

The Greek Miracle

The European spirit spent its youth in Greece.
—G. W. F. Hegel

Greece, cradle of the arts and of mistakes.
—Voltaire

The concept of atoms, which was to ultimately triumph after a tortuous history, was born almost twenty-five centuries ago in ancient Greece. Although not entirely impossible, it seems difficult to view this birth as purely fortuitous. In order to fully understand the origins and intellectual impact of this development, it is useful to consider both the geographical locations and historical circumstances involved: Why in this particular place and at this particular time? Such questions are, of course, not easy to answer, and there is the ever-present danger of straying down blind alleys. Nevertheless, this period marks a turning point in the history of mankind that has fascinated a great many historians, and it would be foolish to dismiss their efforts as frivolous.

The starting point of this remarkable and unique experiment was the "Greek miracle." At a time when most people believed the observable world to be ruled by a plethora of gods with extraordinary powers but strangely human weaknesses and appetites, a handful of Hellenic thinkers set out to provide instead a rational explanation involving only natural causes and effects, without invoking transcendent powers. That is not to say they denied the existence of such powers, but they were determined not to get things mixed up. To use Farrington's phrase, they "left the gods out of" the picture and barred them from the world, relegating them to the "interworld" or, in the context of our theatrical parlance, to the background.[1] Although we are not dealing with a genuine scientific inquiry as we understand it today, with its mix of hypotheses, experiments, and proofs, we are, nevertheless, witnessing the birth of a natural philosophy that recognizes only what reason can conceive of and approve.[2] Indeed, the principal ingredients of a scientific approach are beginning to take form: the drive to explore the universal and the essential; the belief that nature, under its complexity and astonishing diversity, hides an order that can be articulated in terms of simple elements and their interactions; the hope that, in the best of cases, a unifying reason might even preside over the extraordinary variety and endless changes of the elements of nature; and, above all, the conviction that, in this grand cosmic puzzle, only rational elements and events intervene—in other words, that there is no place for supernatural mediation.

To illustrate this intellectual process, it suffices to review briefly the contribution of what was probably the first school of rational thought in ancient Greece: the Milesian school in Ionia (on the shore of the Aegean Sea, in Asia Minor), in the sixth and fifth centuries B.C. Its most prominent representatives are Thales, Anaximenes, and Anaximander.[3] All three were dedicated to a cosmogonic quest for the "rational" and "natural" cause of things, which they envisaged in terms of a "primary reality" or "primordial substance." They differed, however, in their choice of what that substance might be—water for Thales, air for Anaximenes, and *apeiron* (the "boundless" or the "indeterminate," an undifferentiated and indefinite substance) for Anaximander. They also disagreed on the mechanisms by which these simple elements gave rise to the remarkable multiplicity of observable things: an unknown mechanism for Thales, a process of rarefaction and condensation for Anaximenes, and a growth, or "push," for Anaximander.[4]

The development of such different theories about a common theme could not have taken root and flourished without the two essential components of a burgeoning scientific method. The first is based on freedom of expression, an acceptance of divergent opinions, an openness toward debate, and the absence of dogmatism or claims of infallibility within the scholarly community. The second is a relative tolerance on the part of the public authorities: They refrained from intervening in the debate, at least as long as it did not raise seditious issues or openly clash with established beliefs.

We will later return to these crucial aspects of the Greek scientific adventure. But for now, let us focus on the central topic of this chapter: Why in this part of the world? More precisely, why Miletus in Ionia? Several arguments can be put forth to answer this question.

Geographic argument. Ionia was at the crossroads of numerous trade routes between East and West, particularly those linking Greece with the Near East and Egypt. Located on the southern end of Greece's Aegean coast (in present-day Turkey), Miletus found itself exposed to many intellectual influences, which undoubtedly contributed much to its advanced cultural level. Indeed, some claim that Thales himself traveled to Mesopotamia and Egypt, from where he is said to have brought back the seeds of astronomy and mathematics. Anaximander is often mentioned as the first to have drawn a geographical map. In fact, it appears that all pre-Socratic philosophers traveled extensively.

Ethnographic argument. Some, like Robert Cohen, believe in the stimulating virtues of cultural miscegenation: "The Aeolians and the Ionians, who had previously migrated from the Phocis, the Peloponnesus, and Attica, had long populated the gulfs of Smyrna, Ephesus, and Miletus. Their cities, framed to the north and south by other urban centers, were in the process of forming a powerful confederation. A new Greece was sprouting everywhere along the coast of Asia Minor. The mixing between the Hellenes and indigenous populations could only produce the most favorable results, not the least of which was an extraordinary movement of thought."[5]

Likewise, Nietzsche wrote: "Nothing is more preposterous than to attribute to the Greeks a civilization of their own; in fact, they were simply assimilated into numerous other cultures. If they succeeded in progressing as far as they did, it is only because they picked up the javelin left on the ground by other peoples and threw it farther."[6]

Economic argument. This argument is closely linked to the previous one. From the eighth to the sixth century B.C. (at least until its destruction by the Persians in 497 B.C.), Miletus was the most important Asian city on the coast of the Aegean Sea.[7] It was a leading and prosperous commercial center with large avenues, magnificent plazas, and sumptuous temples. It had successfully founded a number of colonies.[8] Its zenith coincided, incidentally, with a number of inventions (the most important of which is probably the introduction, in the seventh century B.C., of a state-guaranteed currency) that only accelerated the rise of its economic power. The argument certainly has merits: Did Aristotle not state that the birth and development of science requires men of leisure?[9]

Sociopolitical argument. This argument is related to the establishment— quite unique in the world at the time—of city-states all across Greece and the attendant debates about what constituted the best political structure and form of government. This phenomenon promoted a culture of open debate and divergent ideologies, and a willingness to accept the possibility of more than one solution. That this atmosphere of openness and tolerance fostered the rise of critical philosophical and scientific thinking seems plausible enough. Moreover, these favorable circumstances appear to have remained immune from official control or interference on the part of civilian or religious authorities. As Lenoble observed: "Because they were free, the Greeks were able to found cities, get rid of tyrants, and recognize that there can be no real freedom without laws. That was the beginning of the miracle. They dramatically exemplify the belief of many sociologists that a coherent description of the world requires an organized society."[10] A. Pichot expressed a similar opinion: "Democracy and constitutional laws give man a status in society that molds his conception of his own place in nature and, by extension, of his overall view of the world. By promulgating societal laws (as opposed to natural or divine laws), democracy can only encourage a quest for a specific natural order, an order that is no longer divine but directly accessible to man."[11] E. Schrödinger stressed the favorable circumstances prevailing in Hellenic society; unlike the situation in Mesopotamia or Egypt, power was not concentrated in the hands of dominant priestly castes opposed to novel ideas out of fear that any change in the way the world is perceived might jeopardize their own privileges.[12]

Of course, it is virtually impossible to ascertain if these arguments reflect accurate historical reality or if they are merely *a posteriori* attempts at rationalizing otherwise random events.

Under these circumstances, it might be easier to attack the problem from a different angle and explore, instead, the reasons for the lack of a similar "miracle" among another ancient people whose intellectual achievements were no less momentous than those of the Greeks. Indeed, of all the concepts born in antiquity, none had a more profound impact on mankind than those originating either with the Greeks or with the Hebrews. Both peoples were clearly gifted with exceptional spiritual qualities and a yearning toward things universal (which explains the permanence of their legacy). Yet their timeless contributions are in decidedly different realms. In particular, one may wonder why, despite their long political existence and enduring spiritual influence, the deeply religious Hebrews, unlike the Greek pagans, contributed virtually nothing of significance to the birth or development of science. Understanding the reasons for this shortcoming might in turn help us better pinpoint what fueled the advent and pursuit of natural philosophy among the Greeks.

I believe that the best explanation for what the Hebrews failed to give us is to be found in the nature of what they did contribute. Some twelve or thirteen centuries before the philosophers of Miletus, in the nineteenth century B.C., Abraham, who was himself searching for the primordial cause and principle of all things, found Yahweh, a unique creator and ruler of the universe. Not water, or air, or some indefinite substance, or numbers, but God. As a result, Abraham was to mark the future of his people with an indelible seal. As the Milesian school was reaching its peak in the sixth century B.C., the Hebrew view of the world had already been in the process of codification for three centuries. The Pentateuch was "canonized," as it were, around the middle of the fifth century B.C. It provided a definitive, indisputable, and unquestionable explanation of the origin and evolution of the world (including man) that was set in a unique framework of thought and placed in the custody of religious and political authorities.

Truth having thus been bequeathed from on high and frozen, no allowance was made for scientific scrutiny or independent philosophical developments short of challenging the established authorities. Nor was any room afforded for disagreements on the fundamental issues of knowledge. The certitude of an absolute truth obviated any need to question its validity. The Hebrews experienced no urge to ask the questions that so preoccupied the Greek philosophers. "God created . . ." fully satisfied their curiosity, and their steadfast belief in the immutable truth of the Revelation rendered any inquiry about the origin and purpose of things, or about man's role in this world, superfluous. Flavius Joseph, a Jewish historian of the first century of the common era, loudly proclaimed the glory of his people blessed with "not a multitude of works full of contradictions," but with "a unique source, the Book of Books" to which he could constantly refer "with reverence."[13] There was no doubt in his mind, or in that of his coreligionists, that its meaning was literal and that it described the permanent reality of the world. Centuries had to pass before certain, mostly scientific, developments would even suggest the possibility of an allegorical interpretation of the Bible, particularly of those passages that mirrored issues of concern to the Greek philosophers.

What a contrast with the Milesian intellectual universe! For the Hebrews, everything flowed from the starting divine premise. God, the primary cause, creator of the world and its laws, constituted a ready-made answer to questions about the origin of things, their diversity, and their changes—issues that were the essential focus of the early scientific thinkers. By contrast, the promotion of water to primordial substance by Thales almost inevitably invited counterproposals: that of air, by Anaximenes, or of fire, by Heraclitus of Ephesus (Ephesus was the second most important Ionian city). Since none of these substances could, of course, provide a completely satisfactory answer, the stage was set for more elaborate efforts: two-element theories of the type advocated by Parmenides of Elea, and, later, four-element theories (water, air, fire, and earth) advanced by Empedocles of Agrigentum and subsequently adopted by Aristotle (who even added a fifth—ether), or the theory of an infinitude of substances proposed by Anaxagoras of Clazomenae, another native of Ionia. Out of this fertile intellectual breeding ground sprang yet another approach, one based on the concept of atoms. It met with mixed success because, as we will see, it raised a host of thorny problems centered around the discontinuous, the void, the infinite, and other issues. Thus, like a snowball, a propitious start down a false path fostered the development of contradictory or complementary propositions that, through a dialectical process of restructuring and refining, ensured an intellectual dynamism that was the key to the birth of science. The Greek miracle is truly phenomenal, although less for its actual content than for the horizons it opened up. As M. A. Rey wrote in *Jeunesse de la science grecque* (The youth of Greek science): "That is the great turning point. . . . That is the entrance of Science onto the stage, conceived in its universality, in its logical and rational aspects. The legacy of the Milesian school in terms of concrete results adds up to very little—some might even argue nothing. But what it started and bequeathed in terms of spirit, method, and thought is everything; Ionia founded a science that came to be our Western science, our entire intellectual civilization. It constitutes the first realization of the Greek miracle, and it is to be viewed as its key."[14]

One can also quote Daniel Rops: "Athens and Jerusalem are the epitome of two contradictory attitudes of the spirit: one calls only on the intellect for an explanation of the world, of life, and of man, while the other relies exclusively on faith to reach the same ultimate goal. In the fifth century B.C., these two paths are pursued independently, totally oblivious to each other. They will eventually collide . . . ; the ultimate showdown was to build up through a lengthy journey across history."[15]

As George Minois wrote: "The question of the relationship between science and faith, science and the Church, science and reason, would never have come up had the sacred Book, viewed as the expression of divine thought itself, written by authors 'inspired' directly by God, not existed. Every battle waged between the Church and science until modern times is traceable to the fact that the Scriptures provide a comprehensive explicative account of the structure of the world that seemingly freezes scientific knowledge into a static framework. Any theory that appears to depart from this framework will be judged *a priori* false."[16] Although the comment spoke to much more recent events, it is no less relevant to the underlying conflict taking shape in antiquity.

And so a nascent antagonism between science and religion began to take root in these early times on two separate stages. But the play had only begun. It would unfold throughout the history of our civilization with highs and lows, ruptures and more or less sincere attempts at reconciliation, if not outright synthesis.

Finally, it must be acknowledged that, for many Greek philosophers, science was an antidote to religious fright, to the fear inspired by unknown gods with unpredictable moods (the God of Israel was constraining in other ways), and the obstacles put in the path of scientific development by the cosmologic and cosmogonical concepts in the Bible in no way diminished the universal and moral impact of the Book of Books, whose message on that level remains eternal. Nor should one infer that moral issues were absent from the minds of the Milesian philosophers. The fact that Thales is included among the "seven sages" of pre-Socratic Greece (he is even one of the four mentioned in all references) attests to the inverse.[17] We might quote a few of his apothegms as they were passed on to us by Demetrios of Phaleron: "Remember your friends, whether they are absent or present. Do not grace your appearance; it is through your lifestyle that you can grace yourself. Do not enrich yourself dishonestly. Beware of making yourself obnoxious by your words to those linked to you by oath. Reject all that is dishonest. Indolence is troublesome. Intemperance is a disease. Ignorance is a heavy burden. Learn and teach what is most valued. Reject indolence even if you are rich. Show leadership. If you are in a position of power, show self-discipline." Simple truths, to be sure, but important ones nonetheless.

This comparison between the circumstances leading to the rise of scientific philosophy in Greece and those that resisted the emergence of any similar trend among the Hebrews highlights what I consider to be the two most fundamental differences between these peoples: the climate of their political and social structures and, perhaps more important, the degree of their religious engagement.

An alternative point of view would be simply to accept the explanation of the Greek enigma proposed by Emil Ludwig: "He who lives on the sea shore and constantly gazes out toward far-away horizons will be inspired to have a more profound view of human models, God, love, music. . . . The ancients themselves claimed that the spirit of the Eleatics, which had driven them to the western sea, had also guided them over the ocean of pure thought."[18] Although this assessment referred specifically to the Eleatics, it applies to Greece as a whole and to the Milesians in particular.

2

The Foreground

Archè, the Primordial Substance

The learned man is not the one who provides the right answers, but the one who asks the right questions.
—C. Lévi-Strauss

From the perspective of its relation to philosophy and religion, the theory of atoms was only one foray of a multi-pronged Greek intellectual offensive aimed at elucidating the structure of matter, its origin, its polymorphism, and its transformations. The goal was to understand the significance of change in the midst of the immutable, to forge explanations on a universal scale, applicable to both the inanimate and the living worlds, encompassing heavens and earth. Atomism was but one of many links in that quest, with no more *a priori* significance than any other theory. As a matter of fact, it is more abstract than most other doctrines proposed at about the same time. Nevertheless—but only history will ultimately demonstrate this—it is a most remarkable and fertile example of intuitive imagination, and one with enormous scientific impact.

To truly understand the originality and significance of the atomic theory, it is essential to measure it against competing doctrines that had been advanced for the same purpose. The relevant period stretches across the sixth and fifth centuries B.C. The foundations of the atomic theory began to be laid in the fifth century. In the foreground are the various theories of primordial substances, the first of which appeared during the sixth century, although some were concurrent with the beginnings of the atomic doctrine and developed in parallel with it. It is difficult to pinpoint exactly when the main actors participating in these developments lived in such a distant past. We seldom know with any certainty the dates of their birth or death. Worse, the discrepancies among various sources can even be considerable. I will, therefore, simply give in Figure 1 a schematic chronological chart showing the approximate periods when the main protagonists in this story were living.

In Figure 1, four names are marked with an asterisk. These are the four most famous atomists of antiquity (the last one, Lucretius, was Roman and lived much later than the others, who were Greek). The other philosophers all conceived of or adopted the idea of an originative substance (or several) as the "first of all things," the primordial stuff of all the matter present in the cosmos, as diverse as it may appear. It is to this specific aspect of their natural philosophy that we will confine our discussion. Furthermore, given the frequent overlap of their lives, we will not always rigorously respect the chronological order of the various theories, but we will instead attempt to group them according to their affinities in order to extract the essential thrust of their ideas.

On the leftmost side of Figure 1, the names of two individuals stand out. They are Thales and Anaximander, both appearing in the seventh century B.C. Thales was the founder of the Milesian school, and Anaximander was his pupil and successor at the helm of the school.

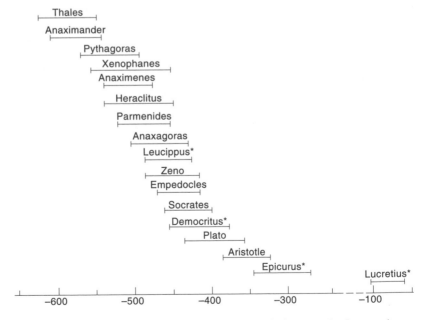

Figure 1. *Schematic diagram of the periods during which the main Greek natural philosophers were living.*

THE MILESIAN SCHOOL

Thales

"Most of the early philosophers believed that the essence of all things could be reduced to material principles. That from which all things come to be, the original stage of their generation and the ultimate stage of their degradation—while the state of the substance can change, the substance itself remains—that is what they hold to be the element and the principle of things; they believe, furthermore, that nothing is created and nothing is destroyed, since nature is forever conserved." These are Aristotle's own words, and they define the problem well. The great adventure of the scientific spirit apparently began when man undertook to explain the complex visible world with a simple and natural visible cause, or sometimes with an equally simple invisible cause. Between these two approaches are some similarities and some profound differences. They are best illustrated by comparing the doctrines of Thales and Anaximander.

An explanation based on a simple visible cause is undoubtedly the most pragmatic approach. Thales, the originator of this philosophical concept, considered that the primordial substance, the "material cause" of things (to use Aristotle's expression), was water. We have from him no direct written document, and all our information about him comes from the accounts of later authors. According to Aristotle and Simplicius, Thales was led to his theory by the observation that "all foods are moist and that warmth itself draws generation and life from moisture (that which begets generation is the principle of all things); and also the fact that the seeds of all things have a moist character," while, by contrast, "cadavers undergoing necrosis dry out." According to other sources, he was guided by the observation of the polymorphism of water, a liquid that can change into solid ice or gaseous vapor (perhaps identified with air). It was, in those days, the only substance for which such a remarkable behavior was known. Still others suggest that he was influenced during his visits to Egypt by the indigenous belief, linked to the periodic floodings of the Nile, that life germinates in water; some even maintain that he had revived the ancient teachings of the Sumerians, whose mythology considered water the primordial stuff, the element of life *par excellence*, already infused with spirit. Or perhaps he was simply echoing some verses by Homer (*Iliad* XIV, 201) celebrating the ocean, which is depicted as a powerful river surrounding the world, irrigating and fertilizing it.

If water was for Thales the primal stuff, the source of all substances, we do not know how he visualized the way these substances were formed. Starting with water, he may have envisioned first the formation of three other elements: air and fire, which according to him are merely exhalations of water, and then earth, which effectively constitutes the lees or sediments. We can quote Galen, who claimed to reproduce a passage from a book titled *On the Principles* written by Thales himself: "As for the famous four elements, of which we hold that the first is water, which we view as a special element, they mix into one another by combination, solidification, and incorporation of things of the world."

This brief review summarizes the extent of our knowledge of Thales's conception of the origin and composition of matter. It is admittedly rather sparse, but, as de Crescenzo asserted: "Thales occupies a very important place in the history of philosophy, not as much for his answers to specific questions as for the questions themselves he was intent on posing."[1] In fact, it may be more accurate to say that he earned his place mainly because of the *nature* of the answers he proposed, answers that reflect his determination to substitute a rational explanation, of a physical type, for the mythical interpretations commonly accepted at the time. Quoting G. Lloyd: "As naive as these explanations may be, their significance lies not as much in what they contain as in what they omit, namely, the arbitrary whims and quasi-human motives of anthropomorphic gods."[2] The most definitive judgment is probably Nietzsche's: "Greek philosophy seems to start with this absurd notion that water would be the origin and maternal womb of all things. Should one pause and take this seriously? The answer is a resounding yes, and for three reasons: first, because it is an axiom which speaks to the origin of things, second because it deals with the issue without resorting to folklore and tales, and finally because it contains this idea, albeit in an embryonic form, that 'everything is one.' The first of these three reasons puts Thales in the category of religious and superstitious men, but the second sets him apart from this company and points to the naturalist scientist in him, while the third makes him the very first Greek philosopher." He goes on later: "The mediocre, inconsistent, and completely empirical observations which he made about the properties and physical changes of water, or, more precisely, of moisture, could in no way have permitted, or even suggested, this bold generalization; what led him to make that jump was a philosophical axiom, the basis of which is a certain mystical intuition typical of all philosophies, with diligent efforts to articulate it ever more clearly; it is the principle that *all is one*."[3]

Since we have just mentioned the mystical, one may ask what becomes of God in this matter. Is there a place for the notion of the divine in Thales's "rational" conception of the world? We stated in chapter 1 that the Greek philosophers left the gods in the background without necessarily denying their existence. We will return repeatedly in this chapter to the changing attitude of the Greek philosophers vis-à-vis the gods or God. Interestingly, Thales's position is as ambiguous as that of many other philosophers of the day. Aristotle's affirmation that Thales believed "all things to be filled with gods" is often cited in this context. While it is difficult to gauge the precise meaning of this apparent polytheism, it is doubtful that Thales could have attributed to these multiple deities an effectual role in the course of the universe, and it appears all but ruled out that they would correspond in his mind to the traditional mythological gods. In fact, he seems to have conceived of a God that was rather more abstract but at the same time authoritarian and engaged. In the words of Aëtius: "Thales said that God is the intellect of the world, that the whole is made alive and full of demons; and also that the elementary moisture is the seat of a divine force that moves it." Cicero commented: "Thales of Miletus, who was the first to undertake studies on this subject, said that water is the principle of things and that God is the intellect that fashions all things from water."[4]

Thus, although Thales sublimated the idea of a supreme divinity, the rupture with traditional descriptions of the world was not yet complete. In his work *From Religion to Philosophy*, F. M. Cornford strongly emphasizes this particular aspect of Ionian philosophy. As summarized by Vernant and Vidal-Naquet: "Ionian physics transposes into a laicized form and on a more abstract intellectual level the description of the world elaborated by religion. . . . By becoming 'nature,' the 'elements' have shed their trait of individualized gods; they nonetheless remain active forces, living and imperishable, and perceived as divine." Perhaps this judgment, which belittles the conceptual innovation we owe to the Ionians, is too radical. Vernant writes again, this time expressing his personal view: "The elements of the Milesians are both divine and natural 'powers.' . . . They confine themselves to producing a physical effect that is completely natural."[5] Compared to the world of Homer, another Milesian, progress is appreciable.

The notion of water as primordial substance would be exploited again later by other thinkers, notably Hippon of Samos (at least according to Simplicius, John Philoponus, and Aëtius) in the middle of the fifth century B.C.; according to others (notably Hippolytus and Sextus Empiricus), Hippon accepted two principles: water and fire. Indeed, Thales's ideas would reverberate until the dawn of modern science and would reemerge, for instance, with Jean-Baptiste Helmont, a Belgian physician (1579–1644).

One had to wait until the end of the eighteenth century, twenty-three centuries later, for the compound nature of water to be established, thanks mostly to the studies of Priestley (1733–1804), Lavoisier (1743–1794), and Cavendish (1731–1810). Water as an element was finally defeated. The French chemist Guyton de Morveau exclaimed in 1789: "Water is a compound! It is difficult to avoid a feeling of surprise the first time one hears a proposition so contrary to the tradition of so many centuries and to the principles taught until now in every school. . . . A new generation is approaching, which, not having been biased by the weight of a long tradition, will need only a simple explanation in order to decide."

Anaximander

After Thales's provocative proposition bestowing on water the supreme dignity of "primordial substance," one might expect that his colleagues, hence competitors, would try to dethrone the first claimant to this enviable position and substitute their own candidates. There was no shortage of plausible candidates, both visible and simple, and indeed many attempts to install them took place, as we will see shortly. Thus, it is all the more surprising that Anaximander, the pupil of Thales and his immediate successor as head of the Milesian school, did not follow this natural path. Instead, he raised the level of the debate by adopting as element and prime principle a simple *invisible* fabric: the *apeiron*, the "unlimited," an infinite, eternal, and immutable medium, out of which all elements, first among them water, air, fire, and earth, come into being through "the effect of an eternal movement."[6] The process takes place "by dissociation of opposites," such as warm and cold, while the harmonious interaction between opposites is indispensable to "the justice" (or equilibrium) of the world.[7] In this model, "the unlimited is the universal cause of all generation [coming to be] and corruption [passing away], from which the heavens and all the worlds, which are innumerable, 'separated' themselves off." He invoked both generation *and* corruption, because "it is to the unlimited that all things return."

The reasons that led Anaximander to make this jump from a primordial substance as simple and visible as water, postulated by his master, to an invisible medium with no definite boundaries, such as the *apeiron*, is not obvious. Had he simply become aware of the difficulties raised by the notion that fire, for example, could emanate from water, when every ordinary observation suggests that these two elements are rather incompatible? By designating as *archè* an undetermined and undifferentiated medium, even one that could conceivably contain in its midst all that is needed to produce all the substances of the world, he freed himself, formally at least, from this difficulty and proposed the kernel of a solution to the problem of change, the nature of which was one of the central issues of interest to the pre-Socratics.

At the same time, he displayed much greater originality than many advocates of a more natural *archè*, whether they proposed a single one or several.

Remarkably, he managed to substitute for the visible *archè* an invisible one without, in the process, forgoing the concept of the material or natural. To be sure, like virtually all pre-Socratics, he did not exclude the gods from his universe. Quoting Aëtius: "Anaximander declared that the limitless heavens are gods." Similarly, Cicero wrote: "The opinion of Anaximander is that the gods are subject to generation, since they are born and die at long intervals and constitute innumerable worlds," which prompted the following question from a Roman writer: "As for us, how could we conceive of a god who is not eternal?" The gods of Anaximander are not even in the background any longer; they appear almost banished to the coatroom.

Anaximenes

With Anaximenes, Anaximander's successor at Miletus, we return to Thales's way of thinking. The fundamental proposition of this young disciple is, at first sight, nothing more than a variation of the axiom of the school's founder: Air replaces water as the primordial substance. Had Anaximenes's contribution been limited to such a simple substitution, it would evidently be of rather trivial significance. In fact, the importance and originality of Anaximenes rest with his explicit proposition concerning the manner in which other natural material bodies issue from this primordial substance. To that end, he envisioned two basic processes: rarefaction and condensation, each associated with an increase or decrease in temperature. For instance, through rarefaction, air would turn to fire, while by condensation, it would successively become wind, water, earth, and stone. All other objects in turn stemmed from these. Underlying this doctrine was a belief that the various natural elements differ from one another in a quantitative, rather than qualitative, way, since all of them, including the soul, are fundamentally made of the same stuff. The mechanical character of the universe thus appears reaffirmed.

Compared with Anaximander's *apeiron*, Anaximenes's proposition that air is the primordial stuff may seem like a step backward, a retreat from abstraction. Yet an elegant interpretation by André Pichot suggests that it may not be so at all.[8] Noting that Anaximenes added to the principle of air the epithet *apeiron*—so that his principle then became "undetermined air" or "boundless air"—and since the corporeality of air was not yet established at the time, Pichot argues that the ability of immobile, invisible, impalpable, and shapeless air to become perceptible under certain conditions (in motion, warm, cold, etc.) represents, in a manner of speaking, the *apeiron* of Anaximander *together with* its potential for "concretization."

At the same time, considering the preeminence Anaximenes granted to air, and given the prevailing beliefs of the time, it is hardly surprising that he equated this element with God. Thus, Aëtius stated: "Anaximenes said that air is God." Cicero confirmed this assimilation, even ridiculing it somewhat: "Then Anaximenes established that air is God, that it is engendered, that it is immense, infinite, and always in motion. But how could shapeless air be God, especially since not only must God have a shape, but it must be the most beautiful of all? And how could something engendered be God when all such things are destined to die?"

Anaximenes's premise that air is the primordial substance was adopted by several other pre-Socratics: Idaeus of Himera (around 450 B.C.), Archelaus of Miletus (also around 450 B.C.), and Diogenes of Apollonia (about 450–400 B.C.). A few fragments of Diogenes's writings are extant, and selected excerpts are reproduced here to illustrate the reasoning of the Greek substantialists: "In my opinion, and to say it all succintly, all things come from transformations of a single primordial substance and are this same substance. And this is easy to see. Indeed, if what exists in our actual world—earth, water, air, and fire, and everything that appears to our eyes in this world—if, I repeat, these things differed from one another by their nature, and if, despite changes and multiple modifications, they did not remain the same primordial essence, they could neither mix together nor exert an action on one another, whether favorable or injurious. No plant could be born of the earth, no animal or anything else could be created, were it not for a law of nature that there be a substance unique and singular. . . . In my opinion, the primordial substance that harbors intelligence is what men call air; that is what governs and rules everything; and that is what, in my view, is God; it is ubiquitous, commands everything, and exists in all things. And there exists nothing, absolutely nothing, which does not participate of air; but this participation is different for various things, and there are many degrees for air just as there are for intelligence. It is, in truth, very diverse, sometimes warmer, sometimes colder, sometimes dryer, sometimes more humid, sometimes calmer, sometimes more violently agitated. From it arise many modifications and infinite degrees of sensations of taste and color. In all living beings, the soul is composed of the same substance, namely, air, which is warmer than the outside air spread around us, but much colder than the air found near the sun. But in no two living beings is the degree of warmth of the air identical."[9]

As was the case with water, one would have to wait twenty-four centuries before the British chemist Joseph Priestley (1733–1804) demonstrated that air is a composite substance and that therefore it could not be considered an "element."

HERACLITUS

Heraclitus of Ephesus belongs in the category of advocates of a single primordial substance. According to him, this substance is fire, which, by condensation, generates water and, in turn, through a more complete condensation, produces earth. The process can be reversed, regenerating fire. The entire world is based on these three elements, to which Heraclitus appears later to have added air.[10]

The pre-Socratics are generally viewed as both physicists and philosophers. Heraclitus leaned more toward the latter. While he proclaimed that "all is one and one is all," that the world is eternal and "was not created by any god," his primary interest was to philosophize on the issue of continual change. He saw a world undergoing endless transformations. A number of his aphorisms have become classics, such as "One cannot twice enter the same river" or "Every day sees a new sun." He believed that the existing order of the world is the result of opposites and of balance between contrasts: "What is contrary is useful, and out of conflict emerges the most beautiful harmony; everything stems from strife." Moreover, Heraclitus's particular conception of fire takes on a meaning that transcends that of a mere primordial substance; interestingly, fire is identified with *logos*, the universal harmony and intelligence governing all things. *Logos* roams freely throughout the world; part of it resides in man for his lifetime (it is his soul, which is, in effect, derived from fire), and its interplay with the outer *logos* is the source of our knowledge of the world.

According to Simplicius and other doxographers (including Aëtius, Clement of Alexandria, and Aristotle), the Pythagorean Hippasus of Metapontum also made fire the principle out of which everything came into being through rarefaction and condensation.

Finally, in a more general vein, there is a famous aphorism attributed to Heraclitus worthy of mention: "A great knowledge does not teach one to be sensible. Otherwise, it would have taught Hesiod and Pythagoras, as well as Xenophanes and Hecataeus." Democritus would later express a similar opinion: "Many individuals full of knowledge possess no reason."

THE ELEATIC SCHOOL

Xenophanes of Colophon

Xenophanes of Colophon is often considered the founder of the Eleatic school. He is of interest to us here primarily as a champion of earth as primordial substance. Theodoretus attributes to him the phrase: "Everything is born of earth and everything returns to earth."

Both Stobaeus and Olympiodorus recount that Xenophanes conferred a special status on earth, although Galen questioned that conclusion. Perhaps inspired by the Milesians, who made the position of earth in space dependent on which substance they considered primordial, Xenophanes declared that "the lower part of earth is unlimited" or "plunges its roots into the unlimited." These statements, related by Aristotle and other doxographers, tend to corroborate the opinion that Xenophanes did indeed choose earth as primordial substance.

According to some sources (e.g., John Philoponus, Sextus Empiricus), Xenophanes took the view that the world was derived from earth and water. Underscoring how unusual it was for Xenophanes to pick a solid substance—earth—as primordial matter, Pichot speculates that adding water to the list may have served to make earth more fluid and malleable.[11] Be that as it may, giving primary status to *two* substances places Xenophanes somewhere between the early pre-Socratics and later philosophers who would explicitly postulate the existence of four such primordial substances. Although Xenophanes is not unique, advocates of two substances are relatively few.[12]

At any rate, Xenophanes saw a remarkable unity in the cosmos. For instance, Cicero attributed to him the proposition that "all things are one, and this One is not subject to change and is God. It has no beginning and is eternal." Unfortunately, the solemnity of this proclamation is somewhat spoiled by the next sentence: "Its [God's] shape is spherical," although in all likelihood this satisfied a demand of esthetics, the sphere being considered the most perfect of shapes. Likewise, Galen wrote: "Xenophanes expressed doubts about many things, but there is one belief he proclaimed dogmatically: 'All is one, and that One is God, beauty, reason, and permanence.'"

These postulates can be considered a preview of what was to become one of the essential tenets of the Eleatic school, namely, the denial of change.

Parmenides

Parmenides is commonly considered the most prominent member of the Eleatic school. Indeed, he plays a particularly significant role in the history of ancient philosophy, having been the first to formulate the concept of "being," necessary, eternal, without beginning, perfectly homogenous, indivisible and indestructible, excluding all movement and all becoming. The precise meaning of this concept is evidently not easy to articulate. According to Pichot, Parmenides's "being" represents "the hidden face of nature," "the other side of appearances," "the manifestation of perceptible nature," "what exists beyond observable qualities" and "can be grasped only by the pure intellect," "a sort of matter-space." For our immediate purpose, it is particularly important to recognize that this concept implies the rejection of the divisibility and discontinuity of matter as well as the denial of void, the existence of which is to be, by contrast, the backbone of the atomic doctrine, as we will explore in depth later.

Parmenides's successors, Zeno of Elea (known primarily for his famous paradoxes, which tended to repudiate plurality and change, and thus motion), and Melissus of Samos, share similar views: The universe is uncreated and eternal, filled with indestructible matter, and does not contain void. Movement, change, indeed all impressions perceived by the senses are only illusions. Their beliefs will bring onto them the wrath of Aristotle, as told by Sextus Empiricus: "The disciples of Parmenides and Melissus have rejected the existence of movement. Aristotle labeled them immobilists and antiphysicists: immobilists because of paralysis, and antiphysicists because movement is the very essence of nature, and to pretend, as they did, that nothing moves is tantamount to abolishing nature."

In the context of this chapter, it is interesting to note that Parmenides, like Xenophanes, advocated two primordial substances, fire and earth, which, according to Clement of Alexandria, he viewed as gods. Water and air represented two secondary, intermediate elements, resulting from the mixing of the primary ones. The two prime principles appear, incidentally, not to play equal roles in his doctrine, earth being considered more like matter and fire more like cause and agent. According to Aristotle, "He postulates two causes and two principles, warmth and cold, in other words fire and earth. . . . Since, according to them [the Eleatics], warmth has a natural separating function and cold a reassembling one, and since each of the other elements is naturally subject to change and evolution, it is from them and through them that they explain the generation and corruption of all other things."

Zeno of Elea and Melissus shared Parmenides's main ideas, although they appear to accord almost the same importance to four fundamental substances. They also rejected the concept of void. Diogenes Laertius relates Zeno's opinion in these words: "There exist several worlds, but void does not exist. All things derive their essence from warmth, cold, dryness, and moisture, which transform mutually one into another. . . . The soul is composed of a mixture of the four elements without any one of them dominating over another." According to Galen, Melissus would have envisioned the existence of a single substance that served as support for the four elements and was nonengendered and indestructible, a substance he called the One and the All.

THE PLURISUBSTANTIALISTS

Empedocles of Agrigentum

With Empedocles, the theory of primordial substances reached, in the fifth century B.C., the pinnacle of its expression. It was to survive for the next twenty-five centuries with few changes and scant opposition until the advent of the theories of modern chemistry. In fact, certain aspects of the ancient propositions constitute the outline of a primitive chemical view of the structure of existing matter, while others literally usher us into the antechamber of atomic theory.

The notion that it is possible to distinguish in nature several—more specifically, four—essential constituents was promoted by numerous ancient philosophers, even though some viewed only one or two of them as truly primordial. What made Empedocles's theory original and compelling, and what ensured its remarkable durability, was to propose *four primordial substances of equal importance*, from which all other substances in the universe are derived.[13] These four primordial elements are fire, air, water, and earth.[14] They are eternal, uncreated, and irreducible, in keeping with Empedocles's contention that nothing can come into being from nonbeing. To emphasize their primordial character, Empedocles called them "roots." This multiplicity of original substances, as opposed to a single one claimed by Parmenides and others, reinstates the notion of change. In this doctrine, change results essentially from a process of mixing and separating of the primordial materials, which occurs under the influence of two external agents: love and hatred, which together represent necessity (these are *material* agents, not to be equated with attractive and repulsive *forces*, although they may well constitute a premonition of these concepts). What becomes is explained as something taking place *within* being, rather than as a characteristic *of* being. There is neither generation nor destruction, only changes in the composition of things.

Three characteristics of this process of assembling and dismantling, which is the source of the organization of the visible universe, epitomize the advance made by Empedocles over prior ideas and even hint at theories yet to be born.

Every primordial substance is considered made of small particles that are homogeneous, changeless, indivisible, and eternal, separated by pores (but not by void, which Empedocles specifically rejected). All observable material bodies are formed by particles of one substance penetrating into the interstices between the particles of another, the mixing being promoted when these interstices have the proper size. Although Empedocles's concepts are quite distinct from those of the atomists, as we will amplify later, they can rightfully be regarded as the harbinger of the notion of elementary corpuscle and the precursor of the atomic theory.

Empedocles probably did not view the primordial substances as pure "chemical" materials in the modern sense of the term. Hence, for him, the word *earth* applied to a variety of solid substances, *water* to different liquids, and *air* to any gas. It must be appreciated that Empedocles, like Anaxagoras at about the same time, seems to have had a clear vision of the corporeality of air. That was not necessarily true of all proponents of air as a primordial element.

To recapitulate, Empedocles posited that all existing substances result from a combination of four primordial substances. A complementary proposition, obviously added to account for the practically unlimited variety of material bodies, claimed that the four "roots" are capable of mixing in different proportions, a particular proportion, fixed and defi- nite, being associated with a specific body. Empedocles even ventured so far as to specify these proportions for some ordinary substances; for instance, he asserted that bones are made of water, earth, and fire, in the ratio of two parts of water and earth to four parts of fire. How could any- one fail to see in this elegant idea a prescience of the law of constant pro- portions, which was to become, twenty-three centuries later, one of the foundations of modern chemistry?

Moreover, it should be noted that the four primordial substances are intimately linked to, although not identified with, two pairs of opposites: cold/warm, and humid/dry. Fire and air are warm and dry, earth and water are cold and humid. The association of these elements with "quali- ties" must be pointed out here insofar as these notions will be later revived by a number of natural philosophers, notably Aristotle (see chap- ter 5). According to Pichot, the "roots" are probably to be understood as "qualities materialized into substance."

Two more general remarks are in order. Several authors have speculat- ed about the possibility that foreign influences may have inspired Emped- ocles to adopt the four primordial substances. Such influences, which would have originated in Egypt, Mesopotamia, Persia, or even India, are indeed evident with all natural philosophers of pre-Socratic Greece. While the Greeks unquestionably did draw from these sources much information about, and knowledge of, the world, they were the first to have tried to sys- tematize the body of empirical evidence in a rational effort to devise explanatory principles about nature and its manisfestations.

In conformity with other pre-Socratic thinkers, Empedocles's doc- trine is not averse to a supernatural or divine presence, although none plays any explicit role; for instance, he introduced his "roots" under the names of deities (Zeus for water, Hera for earth, Aidoneus for air, and Nestis for fire). In addition, according to Aëtius, "he believes that the four elements are gods. . . . He believes that souls are divine."

The theory of the four primordial substances would be embraced by several other pre-Socratic philosophers, in particular Philistion of Locris (427–347 B.C.) and Diocles of Carystos (400–350 B.C.). Later, Plato and Aristotle will also make it one of the foundations of their natural philoso- phy (see chapters 4 and 5). Furthermore, it is significant that the concept spilled even into the field of medicine. The doctrine of the "four humors," the cornerstone of the teachings of Hippocrates (460–377 B.C.) and of Galen (A.D. 129–201), two of the foremost physicians of antiquity, bears a direct kinship to the theory of the four elements and four "qualities." Similarly, Pliny the Elder (A.D. 23–79), whose *Natural History* exerted an enormous intellectual influence for many centuries, was a fervent proponent of the doctrine of the four elements, even though he shrouded it in mystical speculations.

In summary, the Greek thinkers offer us a choice of one (the Milesian school), two (the Eleatics), or four (Empedocles) primordial substances. The number three is largely missing. It is apparently found only in the teachings of Ion of Chios (452–421 B.C.), a later representative of the Pythagorean school. John Philoponus puts him on record as choosing fire, earth, and air, although Harpocration interpreted that to mean intelligence, power, and fortune.

Anaxagoras of Clazomenae

Among the champions of primordial substances, Anaxagoras of Clazomenae stands diametrically opposite the Milesians, monists who consistently ascribed this quality to a single medium. Starting from the tenets that nothing can come into being from nonbeing and that nothing perishes (which were accepted by the vast majority of pre-Socratic philosophers, as we have seen), he declared that all existing things contain infinitely small particles harboring the qualities of all other existing things: "In every thing is a small part of all things." He named these particles *omoiomeres*. Thus, the prime substances were, to him, quantitatively and qualitatively infinite. The unique substance of the Milesians is replaced by a medley of qualities contained within infinitesimal particles, and change results from their mixing and separation, which obviate the need for their generation and disappearance.

All objects in nature contain all possible *omoiomeres*, but in variable proportions. What a given object appears to be depends on which particular type of *omoiomere* is dominant. For instance, what we perceive as gold is indeed mostly gold, but it also contains small amounts of all other substances. In the words of Simplicius: "In every thing are included parts of all other things, and every unique thing is and was made of those which, being the most abundant, become then the most visible." Elementary particles no longer have a common structure, since, as stated by Aëtius, "*omoiomeres* assume multiple aspects." Furthermore, they are infinitely divisible, and although they are separated by pores, void does not exist. The difference with the atomic theory, which we will examine later, will become evident.[15]

In the opinion of some, Anaxagoras's main contribution was his formulation of the idea of an intellect (called *nous*), the architect of the world; not its creator, but the primary cause of universal order and arrangement. Direct intervention of this intellect is responsible for the making of the world and its contents out of the initial chaos of *omoiomeres*. Commenting on Anaxagoras, Diogenes Laertius wrote: "He is the first to impose an intellect over matter, writing at the beginning of his work, in an elegant and lofty style: 'All things were lumped together. Then came the intellect which restored order to them.'" Let us again quote Aëtius: "Anaxagoras states that, in the beginning, material bodies ignored all movement, and it is the divine intellect that brought order to them and was the agent of the generation of all things." He later reiterated: "Anaxagoras says that God is the intellect, agent of order in the world."

Interpreting these citations demands some wariness, however. Voilquin, for instance, cautions: "This intelligence, although conceived as a more subtle substance, could not be endowed with personality. It would be a mistake to try to view it as a providence or a spirit ruling the world. From the very first moment, everything is amenable to a mechanistic explanation." Similarly, according to J.-P. Dumont: "The action of the universal intellect takes on primarily the form of an initial intervention. . . . It is clear that the intellect is a purely mechanistic cause and is in no way final."[16]

In the end, *nous* is quite reminiscent of Heraclitus's *logos*. Quoting Pichot: "As was the case with *logos*, *nous* would govern the world. But part of *nous* (in which it would remain whole) would be hosted (if not imprisoned) in living beings, particularly in man; this inner presence would afford a 'participation' of living beings in this *nous*, making our comprehension and knowledge of the world possible."

PYTHAGORAS AND THE PYTHAGOREAN SCHOOL

None of the writings of Pythagoras himself, who was a contemporary of the first Milesians, have survived. Although it might appear irrelevant to our present concern, the doctrine of his eponymous school, which flourished over the centuries, deserves nonetheless to be outlined. The reasons are twofold: first, because of the possible connection between his ideas and the atomic doctrine or related concepts of philosophers such as Plato, and, second, because it was the first attempt, albeit a rather peculiar one, at a mathematical description of the world. J.-P. Dumont has likened this doctrine to an "arithmetic theology."[17] Indeed, the Pythagoreans held that numbers are the essence of all things. Numbers are the source of what is real; they themselves constitute the things of the world. Put in another way, the number is the *archè*, the primordial element, the principle from which all things come to be. In the words of Aristotle: "Those who are called Pythagoreans were the first to have an interest in mathematics, which they perfected. Because they were steeped in this science, they believed that its principles were the principles of all things; and since, by their nature, numbers are the first of the mathematical principles, it is in these numbers that they came to see many similarities with eternal beings as well as with creatures subject to becoming; they believed in numbers far more than they did in fire, earth, and water."

In addition to being the essence of things, numbers possessed, from the Pythagoreans' vantage point, a concrete reality as well. They were identified with space and were related to specific figures: The number one is a point, two is a line, three a plane, four a volume, and "since everything has a shape, this shape can always be resolved into a ensemble of points, lines, or surfaces, hence into numbers."[18] The Pythagoreans pushed their fantasies even farther. Again, quoting Aristotle: "Evidently, the number is, for the Pythagoreans, principle, not simply as the essence of beings but also insofar as it constitutes their properties and their behavior." Thus they identified the number eight with love (as well as with cunning). With a small leap of imagination, one can detect in such a Pythagorean physics a premonition of an atomism of sorts. Such is the opinion of Pichot, who writes: "Thus, proceeding down this line, units (which make up numbers as well as things) are endowed with spatial extent and matter (two concepts still poorly differentiated then). In this respect, the Pythagoreans are seen to have evolved a concept in which the unit (and the numbers it constitutes) was not just an abstraction, but a real particle, with thickness and consistency. While it does not qualify as genuine atomism, the notion that things are made of particles could be construed as presaging it. . . . The only concrete aspect of Pythagorean physics would be the outline of a physics of particles, in which the units (which constitute numbers and things) have a thickness and a consistency, and are indivisible; hence, they are atoms (this view is attributed to Ekphantos, reputed to be a Pythagorean of the end of the fifth century B.C., but whose existence has sometimes been disputed)."[19]

Note that even though the Pythagoreans assigned a primordial role to numbers, they still attributed an intermediate but significant purpose to the four elements (fire, water, earth, and air) in their practical description of the universe. Diogenes Laertius wrote: " In his *Succession of Philosophers*, Alexander [a reference to Alexander Polyhistor, who lived during the first half of the first century B.C.] claims to have made this other discovery in some Pythagoreans' writings: that the monad is the principle of all things; derived from the monad, the indefinite dyad exists as a material substrate for the monad, which is the cause; it is the monad and the indefinite dyad that [beget] numbers, then numbers [beget] points, then points [beget] lines. In turn, these produce planar figures, which produce three-dimensional figures, which themselves produce the perceptible bodies, made of precisely four elements: fire, water, earth, and air, changing and transforming completely one into the other; and they give rise to an animate world, intelligent and spherical, at the center of which is Earth, itself spherical and inhabited over its entire surface."

Moreover, according to Aristotle, the Pythagoreans accepted the concept of void, and their physics of change was based on the dualism of opposites (odd/even, clear/obscure, etc.), the constraints of which governed, in their view, the affairs of the world. This duality is also found in their division of the universe into two regions: the sublunary region, where change, birth, and death prevail, and the celestial region, where objects are immutable. This dichotomy was to be adopted by Aristotle and would enjoy an extended career (see chapter 5).

The Pythagorean Philolaus of Tarentum (circa 480–unknown B.C.) is often given credit for proposing the following relations between the four elements and regular polyhedra: Earth equals cube, fire equals tetrahedron, air equals octahedron, and water equals icosahedron. A fifth relation completes the list: Ether equals dodecahedron, an association that will be exploited and amplified by Plato (see chapter 4). A few statements relating to Philolaus are worth mentioning: "According to Philolaus, geometry is the principle and the spring of all sciences" (after Plutarch); "according to Philolaus, the number is the all-powerful link, and begets that which eternally unites all objects of the world" (after Jamblique).

SYNOPSIS OF THE PROPONENTS OF PRIMORDIAL SUBSTANCES

At this point in our discussion, before embarking on the advent of the theory of atomism, it might be useful to summarize how the Greek natural philosophers defined the principle or principles on which they believed the material structure of the universe to be founded. A schematic representation of the various theories advocating one or several primordial substances as such principles is sketched in Figure 2.

In this figure, arrows pointing toward the four elements identify the philosophers who proposed them as primordial substances. A number in parentheses next to a name—for instance, (+3) next to Anaximander—signifies that, in addition to air, which this philosopher considered the primordial substance, he also recognized the other three substances (water, earth, and fire) as privileged. Moreover, for completeness, we have added at the bottom of the figure the three philosophers whose principles were not substances in the classical sense of the term. The number (+4) next to Pythagoras's name is there to remind us that, while his principles were, first and foremost, numbers, he nonetheless attributed to the four elements an intermediate but important role in the making of the world.

What conclusions can we draw at this stage of our analysis of the significance (historical rather than physical, of course) of the various

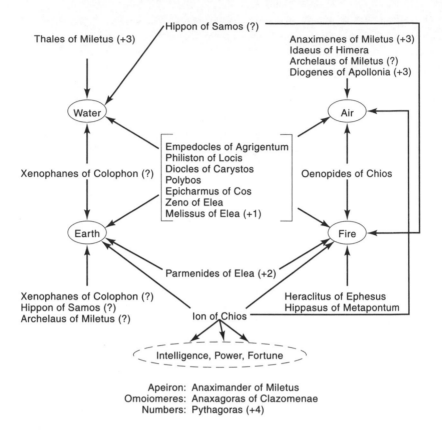

Figure 2. *Primordial substances and their advocates.*

propositions put forth? Aside from the fact that a particular doctrine may often merely reflect the personal preferences of its author, my own view is that the most important contributions, at least when it comes to primordial substances, are those of Thales, Anaximander, and Empedocles: Thales, because he was the first to envision the existence of such substances, and also because of the symbolic value of his conception of the unity of matter; Anaximander, because while he too recognized the unity of matter, he placed its essence beyond the level of an observable material substance; finally Empedocles, because his proposal of four primordial substances, while departing from this notion of unity, would be accepted by virtually every natural philosopher for nearly two millennia. He also deserves credit for having envisioned a corpuscular structure of matter.

A separate place in this brief synopsis must be reserved for Pythagoras. As we have seen, a little bit of imagination reveals in his doctrine a glimpse of what amounts to a corpuscular physics, a sort of arithmetic atomism (he even seems to have accepted the existence of void). The main failing of his doctrine, however, was to have incorporated a definite theological, if not mystical, component. Also most unfortunate was his division of the world into two regions: celestial and sublunar. Revived later by Aristotle, this differentiation would prove to have a decidedly negative effect on the development of physics.

Finally, although God or the gods are not shut out from the world conceived by the Greek philosophers, they are, generally speaking, involved neither in its creation nor in its forward march. Anaxagoras's view on the subject deserves to be singled out: For him, God, the intellect, may not be the creator of the world, but he is, nevertheless, its architect and agent of order.

3

The Atomists' Entry
onto the Stage

Democritus had found the most philosophical way to
enjoy nature and men; he studied one and laughed at
the other.

—D'Alembert

LEUCIPPUS AND DEMOCRITUS

In the overall perspective of the development of natural philosophy, the dawn of the theory of atomism represents, at first sight, but one stage in the long and obsessive quest of the Greek thinkers for "principles," or primordial "elements." The founders of this theory are Leucippus and Democritus, who appear on the scene after the great philosophers of the Milesian school and the first Pythagoreans; they are contemporary with the philosophers of the Eleatic school, notably Empedocles and Anaxagoras (see Figure 1). Atomism constitutes an original attempt to resolve the central problem addressed by the pre-Socratic thinkers, namely, the nature of coming into being and of change. It represents an effort to synthesize sometimes fiercely antagonistic positions. Historically, Leucippus is considered the father of the theory, insofar as one accepts that he existed, which appears to have been denied only by Epicurus. Unfortunately, no original documents describing his doctrine have come down to us. On the other hand, Democritus, Leucippus's successor, has left us more substantial writings. This makes it difficult to ascertain which of the primitive ingredients of the theory is due to one or the other, particularly since they are often lumped together in ancient accounts. We too will adhere to that tradition, although in some rare cases when it appears possible to do so, we will try to identify their respective contributions.

The fundamental concepts of the theory seem to have been formulated by Leucippus. Diogenes Laertius wrote explicitly: "Leucippus was the first to elevate atoms to the level of principles."[1]

Aëtius and Cicero are more thorough and express more rigorously the essence of his teachings. Aëtius wrote, "Leucippus of Miletus says that the filled and the void are principles and elements," and Cicero stated, "Leucippus admits two principles: the filled and the void."

While the general public thinks of the atomic theory as rooted in a single fundamental reality, namely, atomic corpuscles, the doctrine actually was based from the very outset on two concepts: corpuscles and void, inseparably and wholly integrated into one proposition. Although the introduction of the concept of void forces any reference to primordial substances to be jettisoned, the notion of principle or element does survive, and the atomic theory of Leucippus and Democritus clearly postulates two of them. It is important to stress that the primordial character attributed to void was to play a central role in the long battle facing this doctrine before its eventual vindication in the modern age.

The philosophical significance of this dual set of principles becomes apparent when they are considered in the context of ancient Greek thought concerning the questions of being and becoming. First, it offers an alternative to the restrictive doctrine of the Eleatics, particularly of Parmenides, that being is immutable, perfectly homogeneous, indivisible, and indestructible, excluding all movement and change. The preeminence of both void and atoms—in other words, of matter—means that from the atomists' vantage point nonbeing, or nothing, has just as much reality as being, or something. Moreover, the doctrine taught by Leucippus and Democritus can be construed as a process of breaking apart the infiniteness of a unique entity (the One) into many infinitesimal parts, into a dust of atoms, which can then be assimilated with an infinite multitude of ones in the Eleatic sense. Such a view accounts simultaneously for the diversity of things and for the potential for change, while preserving the concept of a unique elementary substance. Finally, the existence of void (rejected by the Eleatics), which makes movement possible, gives the doctrine the power to interpret the reality of the observable world. In particular, it makes allowance for the separation of space and matter, whose presumed identicalness had until then prevented them from being described in strictly mechanistic terms in most philosophical systems.

Having conceded to void its status of primordial principle, we must nonetheless now focus our attention on atoms.

First, as their name implies, atoms, the elementary corpuscles of matter, are indivisible. This property is due, according to Leucippus and Democritus, to their "impassivity" (hardness, incompressibility). They are compact and full, without parts, of homogeneous composition, and exhibit no qualitative difference. They are infinite in number, and they are in constant and eternal motion. However, they differ among themselves by three geometrical characteristics: shape, arrangement (order), and position. To these one must add the two individual properties of size and weight, although weight does not appear in the original version of the theory conceived by Leucippus; opinions differ on whether it was introduced by Democritus or much later by Epicurus.

Atoms undergo a continual and endless motion, random and not pre-ordained, determined by mutual collisions. The many consequences resulting from this process are due to the various quantitative and mechanical characteristics of the atoms involved. For instance, upon colliding, atoms can either recoil "in the direction where fate throws them," or "coalesce according to the congruence of shapes, sizes, positions and arrangements, and remain bound, thereby causing the generation of aggregates," according to Simplicius's account. Clustering is promoted by the diverse shapes atoms can have—polished, rough, round, pointed, hooked, twisted, bent—and, their number being infinite, their association can produce an infinite number of objects and worlds, since, in Aristotle's words, "after all, it is the letters of the same alphabet that compose a tragedy and a comedy." This vision, striking in its simplicity and coherence, was skillfully summarized by Simplicius in this phrase: "Leucippus and Democritus hold that the worlds, which are in unlimited numbers and occupy the unlimited void, are formed by an unlimited number of atoms." The concepts of the infinitude of the universe and of the uniformity of the matter that forms it are the two essential components of the theory of atomism.

In the process of aggregating, atoms do not lose their identity; they remain in contact, clustered and juxtaposed, separated by an absolute void (unlike Anaxagoras's *omoiomeres*, which were separated by pores filled with a light and subtle medium). The individuality, impenetrability, and absolute perenniality of atoms effectively made the notion of molecule, as we understand it today, inconceivable. Aristotle sensed this difficulty quite well. According to Simplicius, he would have remarked that, in Democritus's conception, "atoms end up assembling and adhering without actually generating a new entity, whatever that may be, for it would be absurd to believe that two or several things could ever engender one" (see also chapter 4). Galen rejected atomism for the same reason, as he could not envision how the association of atoms could produce compounds with properties different from those of their constituents.

Following the tradition of almost all ancient philosophers, the atomists constructed from their principles an entire cosmogony, of which certain propositions are worth mentioning.

In this cosmogony, atoms were initially assembled into an infinite mass, from which they escaped to form aggregates and vortexes, where similar atoms clustered. In the words of Diogenes Laertius: "Here now is Democritus's doctrine. The principles of all things are the atoms and the void, and everything else exists only by convention. The worlds are unlimited and subject to generation and corruption. Nothing could come to be from nonbeing, and nothing could return by corruption to nonbeing. Atoms are unlimited in size and number, and they are the seat of a vortex motion in the universe, which results in the creation of all compounds: fire, water, air, and earth, which are simply organizations of certain atoms, themselves resistant to change and alteration by virtue of their hardness. The sun and the moon are composed of such particles, smooth and round, as is the soul, which is the same thing as the intellect."

A few points in this quotation must be underscored:

1. The appearance of the word *vortex,* describing a generative motion capable of shaping worlds out of the primordial stuff, a concept to be taken up again by other philosophers.
2. A certain individualization of the role played in the making of the worlds by the four substances mentioned by numerous pre-Socratic philosophers. Naturally, these are not primordial substances. Such substances do not exist in this doctrine, since only atoms embody the primeval stuff of matter.
3. The idea that the soul is also made of atoms, hence of matter; subtle matter perhaps, but matter nonetheless. It is degradable and decomposes at the same time as the body, at which time it returns to the universal cosmos.
4. The intriguing assertion that like atoms associate preferentially. Democritus appears to be (with Empedocles) one of the few pre-Socratic thinkers to have adopted this point of view, most others opting in favor of "the opposition of contrasts" as the driving force behind the organization of things.[2]

This last point leads us to examine more closely the mechanism responsible, from the atomists' perspective, for this fundamental process of associative collisions, which is the source of the generation of things and of the worlds. Its expression sometimes takes two different forms. Some authors (Cicero, Lactantius, Simplicius) declare that the atomists ascribed the process to "chance." Others (Aëtius, Diogenes of Oinoanda, Aristotle, and again Cicero and Simplicius on other occasions) associate it with "necessity." Still, it has been argued that the apparent contradiction may be misleading, since the two notions appear rather equivalent in the context in which they are used. For instance, J.-F. Duvernoy points out: "Atomic collisions happen randomly; but they cannot possibly happen otherwise, nor can they be other than what they are. Atomism does not imply a complementarity of chance and necessity, but a strict identity."[3] Brunschwig asserts even more explicitly that in this conception "necessity and chance are not in contradiction, but, instead, stand together in contraposition to the action of an intelligence or providence, much like mechanism and finalism; chance appears to be the name given to the cause when it is unknown or unknowable."[4] Jean d'Ormesson claims eloquently that "chance and freedom are the temporary names of necessity."[5] Without belaboring the point, let us underline the one point these two descriptions have in common: the affirmation that the world is governed by a natural law, a decidedly mechanistic physics, without intervention by the divine or by mysterious forces (such as Empedocles's love and hatred), without design or inevitability. According to Nietzsche, Democritus (who was of far greater interest to him than Leucippus) would have been the first to formulate the notion of "an anthropomorphic conception of the mythical," the "first rationalist," a proponent of causality without final purpose. Nietzsche's interpretation of Democritus's vision was that "matter moving according to the most general laws produces, via a blind mechanism, effects that *seem* like the will of a supreme wisdom."[6]

This discussion would be incomplete if we failed to mention here the attempt of these early atomists to create a theory of sensations. They distinguished between primary and secondary properties, between objective and subjective, or in the language of these early thinkers, as told by ancient writers and doxographers, "those [the properties] that exist by their nature and those that exist by convention" (after Aëtius), in other words, those that constitute "the legitimate form of knowledge and those that represent a corrupted form of it" (after Sextus Empiricus). The meaning of this distinction was magnificently analyzed by Galen: "Indeed, color is only by convention, by convention is the sweet, by convention is the bitter; and in fact, there are only atoms and void, as Democritus assures us; it is the encounter of atoms that produces all the sensory qualities for us to experience, while nothing is by nature white, black, yellow, red, bitter, or sweet. By the word 'convention' he means 'relative to custom' and 'relative to us,' and not according to the nature of things themselves; which is also what he means by the phrase 'in reality,' a term forged from the word 'real,' which designates truth. And the complete meaning of the formula is the following: people believe that there exist white, black, sweet, bitter, and other similar qualities; but, truth be told, all things are made of 'being' and of 'nothingness.' Those are his own words: 'being' is the name given to atoms, 'nothingness' that given to void."

Equally articulate is the account of Sextus Empiricus: "Democritus, for his part, abolishes the phenomena that concern the senses, and believes that no problem can appear in conformity with truth, but only in conformity with opinion, that what is true in substances consists in the reality of shapes and void: convention is the sweet, he proclaims, convention is the bitter, convention is the warm, convention is the cold, convention is color; and in reality, there exist only atoms and void."

The theory of color of the atomists is worthy of special mention. According to Alexander of Aphrodisia: "Democritus believes that seeing consists in receiving a visual perception emanating from the objects seen. . . . He himself thinks—as did Leucippus before him, and as will Epicurus's disciples subsequently—that a wave made of certain simulacra, the shape of which is the same as that of the objects emitting this wave [which makes these objects visible], strikes the eyes of those who see, and that so is seeing produced." The term *simulacra* might be translated today as "effluences" or "emanations." Thus, J.-F. Duvernoy writes: "Compounds emit small material 'images' in all directions: *eidolè*, or simulacra. These tiny emitted atoms are 'similar' to the compounds that emit them. But only the compounds are perceptible, not the isolated atoms which are 'simple' and cannot be resolved into atom-substrate or into simulacra." We will analyze the theory of sensations of the atomists in greater detail in subsequent pages devoted to Epicurus and Lucretius.

The attitude of the atomists toward the gods is ambiguous. As documented in the comments of Sextus Empiricus, Democritus was quite conscious of the "natural" origin of religious beliefs: "Some have supposed that the notion of gods was inspired in us by awesome events encountered in the world, and such appears to be Democritus's thesis: When, proclaims he, the ancients witnessed events of which the heavens are the theater, like thunder, lightning, thunderbolts, the conjunctions of celestial bodies, or the eclipses of the sun and the moon, their fright made them believe that certain gods were responsible."

Democritus's vague conception (or conceptions, to be more exact) is summarized by Cicero in a way that is at once critical and witty: "Yes, even Democritus, great man as he may be, at whose source Epicurus drew the water to irrigate his small gardens, appears to me tentative when it comes to the nature of the gods. Sometimes he thinks that the universe contains images imbued with a divine character, sometimes he claims that the gods are the principles of the intellect residing in the same universe, sometimes that they are living images which we customarily hold to be either beneficial or harmful, sometimes that they are certain immense images, so vast that they embrace the whole of the universe. All these conceptions are assuredly more worthy of the home of Democritus than of Democritus himself."

The criticism aimed at Epicurus is unmistakable. The last sentence is a reference to the reputed stupidity of the inhabitants of Abdera. It is ironic that the most remarkable philosophical concept to come out of Greek antiquity should have originated in a city considered so intellectually wanting.

In spite of this ambiguity, perhaps deliberately cultivated, there is little doubt about the limited role Democritus reserved for the gods. It is evident that they do not intervene in the march of the world, governed as it is by physical laws. According to Minois, Democritus would have admitted their existence only "out of prudence," philosophical prudence and prudence toward the authorities (let us not forget he was a contemporary of Socrates).[7]

EPICURUS AND LUCRETIUS

The atom is but the last refuge where being, reduced to
its primary elements, can pursue a sort of deaf and
blind immortality.

—Albert Camus

The man who eats is the most just of all men . . .
he nourishes both good and evil.

—Socrates

One century after Democritus (see Figure 1), the atomic theory had acquired the refinements of a more elaborate articulation and polished logical structure, which was to remain practically unchanged for the next two thousand years. The author of this overhaul was Epicurus, of whom sufficiently abundant writings have survived (notably his letters to Herodotus and to Pythocles) to give us what is probably a fairly accurate view of his doctrine. The original sources are supplemented by Lucretius's famous poem *On the Nature of Things*. Although composed two centuries later (see Figure 1), it provides an account that is in all likelihood reorganized, condensed in some aspects, amplified in others, but by and large reliable, since the contribution of the poet himself to the body of ideas appears minimal. Consequently, we will draw simultaneously from both sources.[8]

Upon reading these documents, one realizes that Epicurus, although he denied it, adopted every single fundamental proposition of Leucippus and Democritus. Such is the opinion of Cicero, who wrote in *On the Nature of the Gods*: "What is there in the physics writings of Epicurus that does not come from Democritus? In spite of a few changes, such as what I said earlier concerning the deviation of atoms, he generally reiterates the same things: the atoms, the void, the images, the infinitude of spaces and the innumerable multitude of worlds, their birth and their death; almost everything that contains the rational explanation of nature." Cicero might as well have added the fundamental principle common to all atomists, namely, that "nothing can be born of nothing and nothing returns to nothing."

All the same, Epicurus proposed a number of innovations that went beyond the teachings of his predecessors and differentiated his doctrine from theirs.[9]

In the above-cited passage, Cicero mentions the "deviation of atoms" —some will use the term "declination" *(clinamen)*, or "swerve"—as one of the most noteworthy innovations introduced by Epicurus. Indeed, it is one of the most original and significant of his personal contributions. It also turned out to be the most controversial.

In reality, at the basis of this concept of "deviation" or "swerve" are two other concepts added by Epicurus to the Leucippus-Democritus theory. The first is *weight*, considered a complementary property of atoms, on a par with their size and shape. Although there is no consensus about whether it was Democritus or Epicurus who ascribed weight to atoms, the balance appears to tilt in Epicurus's favor. Such was the opinion of Aristotle, Simplicius, and Aëtius, among others. For instance, Aëtius wrote: "Democritus believed that there were two attributes: size and shape, while Epicurus added a third: weight; for, he proclaims, it is necessary for objects to receive the impulsion of weight in order to move. . . . Democritus states that the primordial bodies have no weight, but move, instead, by mutual collisions in the unlimited."

The second of Epicurus's additions is his explicit assertion that in void, which makes movement possible, atoms move about at the same speed no matter what their weight and volume (a remarkable claim for the time, especially since it contradicted Aristotle).

These two propositions led Epicurus to his hypothesis of the *clinamen*. The reader may recall that in Democritus's theory, atoms are in constant and eternal, albeit disordered, motion. Such a view had the distinct advantage of providing an environment compatible with continual collisions that cause atoms to either fall back "in the direction in which chance throws them" or to aggregate and cluster, and, consequently, to form objects and even entire worlds.

The concept of weight, together with the assertion that in void (the second reality of the world) all atoms move about with the same speed, forced Epicurus to infer that all these particles were subject to a steady, rectilinear, and unidirectional motion "toward the bottom." However, such conditions would preclude collisions and encounters, and, as a result, the creation of substances and of the worlds would become impossible. The only way to avoid this difficulty was to postulate that atoms must occasionally deviate from their natural trajectories. That is the idea behind the *clinamen*.

The citations of some ancient authors concerning the introduction of the *clinamen* stress two aspects that made this tenet intriguing, one more physical, the other more philosophical. Cicero, in particular, relates: "Epicurus considers that atoms are carried by their own weight in a straight line toward the bottom, this motion being natural to all physical bodies. But when this penetrating mind realized that, if all atoms always moved downward along a straight line, they could never touch each other, he invoked the fantasy that they experience an imperceptible sideways declination, which enables them to embrace and cling together. And that is how the world is formed, and all its parts and everything they contain. . . . Seeing that if atoms proceeded toward the bottom under their own weight, we would be deprived of all freedom, since their movement would be preordained and necessary, Epicurus imagined, in order to free himself from this necessity, a means that had evidently escaped Democritus. He claimed that while in straight-line motion toward the bottom imposed by their weight and gravity, they deviate slightly to the side."

Diogenes of Oinoanda likewise states: "If someone were to use Democritus's doctrine to affirm, on the one hand, that atoms have, because of their mutual collisions, no freedom of movement and that, on the other hand, all things seem to be moving necessarily downward, we would say to him: Do you not know, whoever you are, that atoms do also enjoy a certain freedom of movement that Democritus evidently failed to discover, but that Epicurus has brought to light? It consists in a sideways inclination, as he shows with the help of phenomena."

Lucretius also recounts: "Although the elements tend, because of their own weight, toward the lower regions, be nevertheless cognizant, O Memmius, that they all swerve from a straight line in undetermined times and spaces; but these declinations are so small as to barely deserve the name."

This last phrase by Lucretius can serve as a lead-in to this slightly sarcastic comment made by Plutarch: "They [the Stoics and the Peripatetics] do not forgive Epicurus for having supposed, to account for the most important things, an event as small and insignificant as the minute declination of a single atom, in order to introduce with much cunning the celestial bodies, the living beings, and destiny, and for the purpose that our free will not be annihilated."

These quotations highlight the dual role of the *clinamen*: (1) a mechanism to allow for collisions between atoms and, consequently, the building of complex entities resulting from favorable impacts; and (2) the source of a certain freedom characteristic of the behavior of atoms, the fundamental and unique particles forming everything that exists. The significance, both material and philosophical, of the *clinamen* was summarized by M. Conche, for whom the deviation is simply a manifestation "of an essential spontaneity of the stuff of atoms." Through it, a miracle takes place: "Then collisions between atoms can occur, superseding the dull uniformity of the original rain, a fecund disorder fills the universe, masses form, worlds are aborted, and cohesive systems ultimately emerge.

Actually, this process has always existed (so that the 'original rain' is a mis-nomer). Because spontaneity coexists essentially and eternally with the atom, there always has been a *clinamen* (that is, the deviations have always been there). Without it, all would have been death; and without it, every-thing, even now, would surely return to death."[10]

To this poetic euphoria the author adds more in-depth considera-tions, of some complexity, the responsibility for which I leave to him: "Since the atom has no connection to anything else, the power to deviate from the direction in which gravity or a collision would push it can only come from itself. It then becomes necessary to endow the atom with spontaneity. The power to deviate becomes its eternal prerogative. The coherence of this hypothesis has been debated: It would contradict the principle of causality, since deviation would come about without cause. It is clear, however, that the principle of the deviation is to be found *ab aeter-no* in the atom itself. . . . The cause of the deviation, which is for all times inherent to the atom, acts, therefore, in a way that is completely unrelat-ed to any external condition. The effect could then not be inserted into any causal chain. . . . In brief, we are dealing with causality without laws; there is no indeterminism since there is a cause; yet there is contingency, because the effect is in no way the result of prior phenomena, and, rela-tive to them, it seems to be without cause since the cause is not to be found among them." Along the same line of thought, St. Augustine later declared that the *clinamen* is "the soul of the atom."

Much like gravity centuries later, the *clinamen* is all by itself the cause of physical phenomena. While the theory is interesting and ingenious, it never had very many proponents. In the end, the inherent disorder of Democritus's atomic collisions always seemed to describe the intrinsic behavior of these particles better. Lucretius himself adopted an equivocal attitude on this subject, as he presented first Democritus's concept of chaotic collisions without voicing any reservations or questioning its merit, and later described, with seemingly equal approval, Epicurus's theory of the *clinamen*. In fact, his discussion of the consequences of the disordered movement of atoms is particularly remarkable in its pre-science of reality: "It is appropriate to examine with greater attention these corpuscles, the disorderly motion of which can be observed in rays of sunshine: such chaotic movements attest to the underlying motion of matter, hidden and imperceptible. You will indeed observe numerous such corpuscles, shaken by invisible collisions, change path, be pushed back, retrace their steps, now here, now there, in all directions. It is clear that this to-and-fro movement is wholly due to atoms. First, the atoms move by themselves, then the smallest of the composite bodies, which are, so to speak, still within the reach of the forces of the atom, jostled by the invisible impulse from the latter, start their own movement; they themselves, in turn, shake slightly larger bodies. That is how, starting from atoms, movement spreads and reaches our senses, in such a way that it is imparted to these particles which we are able to discern in a ray of sunshine, without the collisions themselves which produce them being manifest to us."

This passage is nothing short of an excellent introduction to the concept of Brownian motion. Those familiar with the important role played by the study of this phenomenon in demonstrating the existence of atoms will find it difficult to repress a feeling of admiration for the prophetic inspiration of the poet.[11]

Is there, then, a contradiction between the apparent acceptance by Lucretius of the ideas of both Democritus and Epicurus? Not everyone thinks so. Marie Cariou writes: "It makes more sense to recognize that the hypothesis of the *clinamen* is quite consistent with numerous directions of motion. It merely favors one of the many possible directions."[12]

Be that as it may, it is logical that Epicurus should also ascribe all sensations, to which he assigns an essential role in our perception of things, to the primary properties of atoms. According to him, our knowledge comes entirely from our senses, which are never mistaken; there can be errors only at the level of interpretations, of the power of the thought to discriminate. For instance, hearing is due to corporeal particles, made of a certain number of atoms, striking the ear. Variations in sound are explained by the different shapes of the atoms making up these particles. Taste and smell involve a somewhat more complex mechanism. Here, differences in impressions are attributed not only to different shapes and arrangements of atoms, but also to the multitude of crevices and channels located in our sensory receptor organs; these cavities are extensions of the void in material bodies themselves. In this picture, a pleasant and sweet taste would be due to smooth and spherical atoms, while a bitter taste "would originate with an assemblage of crooked atoms, which, being strongly bonded, penetrate the seat of our sensations only by bruising the fabric of our sensory organs."

In this interactive process, a particularly important role is assigned to the "simulacra," material emanations emitted by the surface of objects, of which they effectively represent extremely fine and thin replicas, in monoatomic layers, and which provide us with direct information about the outer appearance of these objects, that is to say, of their contour and shape. Color, too, is linked to the arrangement of atoms on the surface of bodies, and is communicated by means of simulacra.

The role played by simulacra in mental visions and dreams is particularly fascinating. In Lucretius's own words: "First, I say this: Myriad simulacra of all different sorts roam about in all directions, on all sides, tenuous images which, when they come together, fuse effortlessly in the air one onto another, like spiderwebs or gold foils. They are of a much finer texture than the simulacra that strike the eyes and elicit vision when they penetrate through the pores of the body to stir the subtle substance of the spirit, the sensibility of which they excite. Thus are we able to see Centaurs, the body of Scylla, or the head of Cerberus, or the likenesses of the dead whose bones are covered with earth, because all manner of simulacra are carried here and there, some forming spontaneously in the air itself, some escaping from different objects, some, likewise, coming into being by the fusion of those of which I just spoke. For, quite obviously, it is not from a living being that the image of a Centaur comes, since such a beast has never existed, but it is the image of a horse and that of a man coming together by chance, instantly adhering with ease one onto the other, as we have said before, due to their subtle nature and fine texture. Any other image of this sort forms in the same manner. As they are carried about, as I have shown, with extreme mobility and lightness, any one of these subtle images easily stirs our mind at the first contact, for the mind itself is astonishingly supple and mobile."

Lucretius's description of "the subtle substance of the mind" brings us to how atomists pictured the soul. Their conception was sketched by Democritus and substantially elaborated by Epicurus, for whom the idea of the *clinamen* will have a particularly significant and original role to play in introducing the possibility of free will in human behavior.

As it was for Democritus, Epicurus's soul is material, composed of the smallest possible atoms, smooth and round, of very great mobility. It "pervades the entire aggregate constituting our body" and is actually what holds the different parts of the body together.

What, then, is the connection between the simulacra, the *clinamen* of the atoms of the soul, and the free exercise of our will? The link is not clear but is definitely theorized. Let us again quote Lucretius, followed by two interpreters who are both qualified, although of somewhat divergent opinion.

First, Lucretius: "One must therefore also recognize in the principles of matter a driving force that is different from gravity and collisions, out of which freedom arises, without which one would be forced to admit effect without cause. Gravity prevents in truth that all movements should be the effect of an external force; but, if the soul is not predetermined in all its actions by an innate necessity, and if it is not a purely passive substance, it must be the outcome of a slight declination in times and spaces that are indeterminate."

We turn next to the interpreters. First, Marcel Conche writes: "In wakefulness, will reigns. For example, we walk because we want to. Simulacra of walking, triggered by an intention (distinct from a host of others) strike our mind. It is at this instant that, having seen ourselves walking, the free will to walk asserts itself. Once the decision is made, the atomic impulse of the mind resulting from that decision is transferred to the atoms of the soul, and, from the soul, in turn, to the whole body, which begins the process of walking. It all happens mechanically, except at the very instant of the decision. It is freely that we want, because we could just as well not have wanted. The mind can shield itself from the power of simulacra. In other words, by imparting onto atoms a deviation from the motion that would otherwise result from mechanical collisions with simulacra, the mind has the ability to defeat them."[13] Lloyd, our second interpreter, states: "As a materialist, Epicurus explained mental events in terms of the physical interactions of soul-atoms, and his problem was to say what moral responsibility could mean in such a system. It has commonly been supposed that his solution was to postulate a swerve in the soul-atoms for every 'free' action. Yet there is no direct evidence that this was his view, and indeed to account for *choice* by assuming the intervention of an *uncaused* event at the moment of decision is bizarre. It is more likely that . . . Epicurus's account of responsibility, like Aristotle's, depends rather on his notion of character, and the function of the swerve is merely to introduce a discontinuity at *some* point in the motions of the soul-atoms in order to make room for the possibility of free choice. The swerve need not, indeed should not, take place at the moment of choice: all that is necessary is that a swerve occurs at some stage in the soul-atoms to provide an exception to the rule that their interactions are fully determined."[14]

Anticipating the future position of the Church toward Epicurean doctrine, it must be added that both Epicurus and Democritus considered the soul perishable, since it is corporeal, and that at the moment of death its constituent atoms are simply scattered in the universal cosmos. In Epicurus's moral philosophy, which we will explore in more detail later, this certainty was intended to free man from the fear of death, a fundamental source of unending apprehension in his existence, and from his superstitious beliefs in certain gods. Again we quote Lucretius, whose words say it best: "Death is therefore nothing to us and does not concern us at all, since it appears that the substance of the soul is perishable. And just as we felt no sorrow in the past when the Carthaginians rushed in all directions to wage war, when the world, all of it shaken by the shuddering tumult of war, trembled with fear under the towering canopy of heaven, and when it was uncertain which of two peoples would inherit the right to domination over men on earth and at sea, likewise when we will cease to be, when the separation of body and soul, whose union is the essence of our being, is consummated, it is clear that absolutely nothing will be able to reach us and awaken our sensibility, not even if earth mixes with sea, and sea with heaven."

In the process, Epicurus rejected hell. According to Lactantius, "Epicurus says that the punishments of hell are not to be feared, because souls perish after death, and hell no longer exists at all." Ironically, Epicurus was the only one of the ancient philosophers Dante put in his hell; a nonexistent hell, of course.

Besides the concepts of weight and *clinamen*, Epicurus also proposed certain complementary notions regarding two other phenomenalistic qualities of atoms, namely, their size and their shape. Concerning size, to which Democritus had imposed no upper limit, Epicurus, probably guided by Aristotle's criticisms, admitted only "a very small size, and, as such, imperceptible." This he did "so as not to contravene the testimony of phenomena." In addition to having considered that atoms can have only a finite, albeit large, number of shapes (in contradiction to Democritus's beliefs), Epicurus's most original contribution, also probably inspired by Aristotle's comments, concerned their structure. Since we are faced once more with propositions whose meaning is not transparent, we will again defer to Epicurus himself: "In a word, the extremity of an atom, being a delicate point which escapes the senses, must be devoid of parts; it is the smallest material body in nature; it has never existed, nor will it ever exist, in isolation, since it is itself part of another body, the first and the last. Joined to other parts of the same type, it forms the mass of the atom. If, then, the elements of the atom cannot exist separately, their union must be so intimate that no force could separate them."

A commentator pointedly observed: "One might wonder if the character of simplicity of the atom (unity, indivisibility) is not compromised by its very spatial extent. If the atom is shaped like a triangle or a trident, how could it not be viewed as comprised of at least three separate parts? It seems inevitable indeed. The atom can be thought of only as being formed of minute parts *(partes minimae)*. But the distinction of each part from the other is imposed only by the structure of the atom as a whole; no part could then be considered to exist separately without contradiction, since the reason for its distinction would then disappear. The parts are undissociable because they can be envisioned only in association with others." The same author adds: "Shapes are correlated with sizes. Indeed, for a given size, parts of the atom can be arranged in relation to one another in many different ways, each of which gives rise to a different shape. The number of possible shapes is, therefore, much greater than the number of sizes, although it remains finite."[15]

This brings us to the views of Epicurus and Lucretius concerning the structure and properties of composite substances. We have to recognize that, on this particular issue, neither one dared venture beyond Leucippus and Democritus. To be fair, they could not have done so even if they had wanted to. The fundamental concept of molecule as we understand it today was completely alien to them. The difficulty, insurmountable at their level of knowledge, had to do with the fact that the ancient atomists had no idea of the bonding forces capable of fusing atoms into profoundly restructured entities. Given their belief in the impenetrability of atoms, they were constrained to stay within the notion of simple juxtaposition, more or less organized and durable, of elementary particles. According to Epicurus: "Atoms move continually in all eternity; some move away from each other [by mutual collisions], while others, on the contrary, enter into a vibration as soon as they happen to be linked by interlacing [mingling, entwining] or when they are surrounded by atoms apt to embrace each other." The concept of orderly vibration viewed as a condition, or at least as a characteristic, of the association of atoms is novel. Epicurus dismissed Leucippus's theory of vortexes.

Moreover, Epicurus was aware of the three principal physical states of matter, which he viewed as determined by the degree of intimacy of atoms and their force of accretion: "Hard and compact bodies [in other words, solids] must have more crooked atoms, more intimately linked and intertwined like branches of a tree." Lucretius stated, "A liquid, being a fluid body, can only be made of smooth and spherical parts." Curiously, gases, such as "smoke, clouds, and flame, are not formed of entirely polished and globular atoms, because they tear our organs; but since they also impregnate rocks, their elements cannot be crooked and encumbered: they must have a moderate figure, exhibiting pointed extremities rather than hooks." Unfortunately, Epicurus did not specify what kind of transformations he envisioned, for instance, for the atoms of water when they change from liquid to ice or vapor.

On the other hand, Epicurus did propose an evolutionary and dynamic conception of the state of the universe and its components. Material bodies undergo a continual loss of atoms (notably through emanations), but also a gain (by means of an inbound flux of external atoms): "The universe renews itself every day" (Lucretius). All worlds, including the celestial bodies, are the seat of these incessant fluxes and are mortal. As M. Cariou summarized elegantly: "What is a beginning? An encounter. What is an end? A separation. Between the two is the duration, always relative, of a world."[16] Only matter as a whole is eternal. What is more, in this endless ballet of atoms, the addition of new particles demands their adaptability to a preexisting organization, which may introduce in nature a mechanism for selection, hence for specificity and evolution. This situation applies in particular to living organisms, where it can create the illusion of a final cause.

Lastly, an important remark is in order. The atomic theory was not the centerpiece of Epicurus's philosophy, in spite of the emphasis on it in this discussion. While it was indeed an integral part of his worldview, it was actually peripheral to his primary preoccupation, which centered on moral issues. Epicurus was "physicist through moralism."[17] His fundamental goal was to make men happy, and in the context of this effort, the gods and atoms found themselves intertwined. "For the gods do exist . . . but not in the way that man ordinarily conceives." This sentence, taken from Epicurus's letter to Menoceus, is one of the keys of his approach to the problems of moral philosophy and human happiness.

Generally speaking, men either admire "the magnificent order of nature . . . that divine wisdom has constructed for the human species" or, more often, experience a profound awe at the phenomena of the universe, of the heaven and of the earth, which they do not understand, in which they see the manifestation of invisible powers, and, above all, the fear of death and of the fate that awaits them beyond: "The ignorance of causes makes the mind perplexed and hesitating." Epicurus had a deep conviction that if the gods do exist, they nonetheless have nothing to do with human affairs. Lucretius articulated this aloofness and the absurdity of believing in a sympathetic interference of the gods into the course of this defective world in some unforgettable pages (unfortunately too long to reproduce here).

And so, what a marvelous sense of serenity is conveyed by the remarks of Hippolytus of Rome (A.D. 170–235) in *Philosophumena* (referring to a single god, but evidently applicable all the same to a multiplicity of gods): "Epicurus asserts that God is eternal and immortal, and that he preordains nothing; that there exists, in a word, neither providence nor destiny, but that all things happen in and of themselves. He puts God's place of existence in the spaces he calls interworlds . . . where he [God] enjoys, in perfect quietude, the ultimate felicity, free of worries and causing none to anyone. . . . If God had wanted to accede to the wishes of men, they would all have perished long ago, since they constantly ask many things that are injurious to their fellow men." Still, the fright caused by the mysteries of the cosmos and the awesome manifestations of nature, as well as an obsessive fear of death, are the inseparable companions of the human race, and no genuine happiness is possible as long as they project their shadows on our lives. It is therefore crucial to free oneself from these fears, and what better way to achieve this goal than to show that these mysteries and frightening events can all be explained in terms of a physics that is resolutely and strictly mechanistic, devoid of any divine purpose, involving only material principles and their interactions? That is the meaning and mission of the atomic theory, whose explicative power is capable of accounting for all aspects of the existence and evolution of the worlds. Such an understanding of the causes of natural phenomena, death being only one example, must serve as the foundation of a moral philosophy leading to wisdom and well-being. Obviously, the infinite, the void, and nothingness, to the extent that they be understood as part of the natural physics of the world, scare the Greek thinkers less than do arbitrary and unpredictable gods. Sartre was to write that in Epicurus's world, "causes systematically replaced designs. . . . Epicurus did not eliminate ghosts, but he turned them into strictly physical phenomena."[18]

The very basis of this moral philosophy can be summarized, unfortunately, in one word: pleasure (*hédoné*). I say "unfortunately," because this word, which the moralists generally frown upon and which is traditionally associated with Epicurus's doctrine, unjustly brought much misfortune both to the doctrine itself and to its author. That fate was most unfair, because in its deeper meaning, pleasure, as it was defined by Epicurus, does not at all conform with the usual connotation of the word. Indeed, Epicurus stated in his letter to Menoceus: "Pleasure is the beginning and the end of a happy existence." He hastened to add: "But precisely because pleasure is our principal and innate gift, we seek not every form of pleasure." In fact, Epicurus's brand of pleasure, intended to be the source of well-being, has nothing to do with the perverted sense generally associated with the term. "When we state that pleasure is our ultimate goal, we do not mean the pleasure of debauchery, nor that which comes with material enjoyment, contrary to the claims of people who ignore our doctrine, or who disagree with it, or who interpret it erroneously."[19] The pleasure advocated by Epicurus has actually a rather monastic slant: One must be self-sufficient, learn to be content with little, live simply and cheaply, eat moderately and wholesomely, enjoy the pleasure of friendship, shun the ambitions of public life, avoid the turmoils of the soul—in a word, live like a sage.[20] "Wisdom is the principle and the greatest good. It is even more precious than philosophy" (Diogenes Laertius).

Twenty centuries later, Michel Eyquem de Montaigne (1532-1592) would pay tribute in his *Essais* (1580) to the authenticity of Epicurus's wisdom. He saw in the dignified and courageous attitude of the philosopher in the face of death—"the ultimate jump from ill-being to nonbeing"—the expression of a "tragic wisdom," displayed by Socrates and Cato as well. Julien Offray de La Mettrie (1709-1751), a materialist philosopher, would express hints of a similar opinion in his 1750 work *Système d'Épicure* (Epicurus's system).[21]

 **4**

A Very Particular Atomist

Plato

All of Western philosophy is but a long commentary on the writings of Plato.

—A. N. Whitehead

By any measure, Plato and Aristotle stand out as the two giants of the philosophy of nature in ancient Greece. Their own concepts and theories, as well as their opinions of and commentaries on those of others, had a profound influence on shaping man's perception of the world. Their attitude toward the atomic doctrine as originally devised by Leucippus, Democritus, and Epicurus, their silences and criticisms, and their attempts to substitute alternative corpuscular theories, whatever their flaws, all merit a careful examination. We will outline first Plato's unorthodox position, and then Aristotle's more conventionally antiatomistic views.

Plato's philosophical output is huge, covering a broad spectrum of preoccupations of the human spirit. It is beyond our scope to summarize it in its entirety. We will touch only on those aspects of his work that deal with the atomic theory.

Plato drew quite liberally from the ideas of his predecessors, and it is easy to recognize in his writings the influences of Heraclitus, Parmenides, Empedocles, and most of all, as we will see, Pythagoras. He melded all those disparate doctrines into a whole cemented together by a number of general concepts that formed the basis of his own philosophy of nature and guided him in formulating an original theory of the inner structure of matter. Briefly, his philosophical views include the following propositions.

1. The changing world of becoming, that which we can observe, is but a reflection and an imperfect copy of a higher world, eternal

and immutable: the world of forms or Ideas, which, alone, represent reality.[1]

2. The world was fashioned and is now governed by an intelligent organizing power, a craftsman or demiurge—God—who operates according to a distinctly final design, and whose purpose is to maximize good. This permanent teleology suffuses Plato's views of the making and workings of the universe. It is important to note that, according to Plato, the world was *fashioned* by the craftsman. Plato's god is not responsible for creation *ex nihilo*, in the sense that he does not create the matter that makes up the world. That matter is eternal, as is God himself. His act of creation consists in molding the world out of this preexisting and primordial stuff.[2] What is more, he is not free to do with it entirely as he wishes; he must take into account a dual set of restrictions imposed by the nature of the materials at his disposal and by the objectives he sets himself. In other words, he does, with the best possible intentions, what he can with what he has (as we all do), under "the effect of necessity" or that of the "errant cause."[3] Nevertheless, since he is God, he creates, at least in Plato's view, the "best of all possible worlds." This belief, which Plato seems to have been the first to articulate, will perpetuate itself all the way through modern times. In Greek antiquity, it competed with other far less optimistic views. For instance, Heraclitus asserted that "the most beautiful world is a heap of garbage strewn haphazardly on the ground." Likewise, Philolaus professed that "God locked up all things in what can be called a guard shack," as cited by Athenagoras.[4] It should be added that Plato's world is finite, unlike that of the classical atomists (but in agreement with the antiatomists; see chapter 5).

3. If no exact explanation of the world in becoming—the perceptible world—is possible, nothing should prevent us from exploring it; in the best of cases, we might even uncover its laws, which are those of a divine scheme imposed by the craftsman. Of perception and reason, the two sources of knowledge to draw from in this endeavor, Plato clearly favors reason. In fact, the possibility of knowledge and its very mechanism are intimately linked to the fundamental doctrine of ideas, as well as the doctrines of memories and of the immortality of the soul. The soul, which existed long before us and will transmigrate to other bodies after our death, has already experienced and witnessed in another world what it discovers, analyzes, and organizes in ours.[5] Without this prior knowledge, the source of which is in the world of ideas, the soul would be incapable of this selective understanding and systematization. But we should not allow ourselves to be deluded, for in the end, this intellectual pursuit can do no more than "save appearances"—in other words, account in an intelligible man-

ner for a set of observed phenomena and their apparent connections. As a result, the study of the perceptible world can, at best, be only a recreation or a pastime.

4. In applying these general ideas to the physical structure of the world and of matter, the teleological and ethical purpose of the craftsman must be complemented by an esthetic motive. The best possible world must also be the most beautiful. We will shortly come to appreciate the weight of this criterion (albeit highly arguable) in Plato's "atomic" reflections.

We now focus more specifically on those Platonic concepts that deal with the central topic of our discussion, namely, primordial substances, particularly as they relate to the atomic theory.

To start with, it is important to recognize that Plato based his entire cosmogony on Empedocles's theory of four primordial substances—fire, earth, air and water—while borrowing at the same time from the proponents of more restrictive doctrines: "A fable, that is evidently what they all tell us, as if we were children. This one speaks of three elements, which sometimes go to war against one another, sometimes make peace, at which time we witness their nuptials and the birth and nurturing of their offspring. Another speaks of only two: the moist and the dry, or the warm and the cold, which he shelters under the same roof and places in the same bed. In our opinion, the noble Eleatic branch, which goes back to Xenophanes and even before him, sees only the One in what we call the All, and in this manner concocts its fables" (*Sophist*, 242c–d).

While Plato proclaimed (*Timaeus*, 31b–33a) that "each of the four elements participated wholly in the composition of the world, because its author crafted it out of all the fire, all the water, all the air, and all the earth, leaving aside no portion or no power of any of these elements," he nevertheless assigned a special role to fire and earth: "The Divine Being first took fire and earth to form the universe." From the start, an esthetic motive comes into play: "When God undertook to organize the whole, in the beginning, fire, water, earth, and air all carried traces of their own essence, but they were in a state of utter disarray in which everything finds itself in the absence of God. In this state he took them, and began to give them a distinct configuration by means of ideas and numbers. That he extracted them from their disorder to assemble them in *the most beautiful* and best way possible, that is the principle which must constantly guide us in our entire exposition" (*Timaeus*, 52c–53c).

The next step of his doctrine also rests on a borrowed concept, namely, the proposition—generally attributed to the Pythagoreans, and more specifically to Philolaus (see chapter 2)—that these four material elements correspond to four of the five regular polyhedra. Plato ascribed a particular characteristic to these elements, as well as specific shapes to their corresponding particles. This was a decisive step in the geometrization of the primordial substances, quite a natural one for someone who ordered carved on the pediment of his Academy the inscription "None is to enter who is not a geometer." The correspondence (one might almost say identification) between substances and polyhedra is as follows: fire = tetrahedron (4 faces), earth = cube (6 faces), air = octahedron (8 faces), and water = icosahedron (20 faces). A special and particularly noble fate is reserved to the fifth regular polyhedron, the dodecahedron (12 faces): "God used it to complete the design of the universe." Plato's demiurge is evidently a mathematician concerned with the geometrical harmony of his work. We will return to this point later.

Next comes the crux of the argument, which is Plato's personal contribution. It specifies an additional—indeed essential, in his mind—aspect of this systematic and constructive process of geometrization. To describe this original idea and what inspired it, we again defer to Plato himself (*Timaeus*, 52c–53c): "First, it is obvious to everyone that fire, earth, water, and air are all material bodies. But corporeal entities always have a certain depth, and depth is, of necessity, enclosed by the nature of its surface, and every surface with rectilinear edges is composed of triangles. Moreover, all triangles derive from two particular triangles, of which one angle is right and the other two are acute. One of these triangles has on either side of the right angle edges divided in equal lengths; the other has edges of the right angle divided into unequal lengths. These are what we hold to be the origin of fire and the other bodies, in accordance with the method that combines plausibility with necessity. As for even more distant origins, they are known only to God and a few privileged men whom he favors."

The preceding argument would suggest that Plato's key idea was rooted in logical reasoning. But the next passage brings out an esthetic foundation as well: "Now, it is necessary to explain the formation of *the most beautiful bodies*, of which there are four, dissimilar, but such that some of them can be engendered one from the other by dissolving themselves. If we succeed, we will then hold the truth about the origin of earth and fire, and of the bodies which constitute their intermediate stages. For we will concede to no one that it is possible to discern bodies more *beautiful* than those, each of them being of a unique kind. Let us strive, then, to compose through harmony these four kinds of higher bodies in *beauty*, so that we will be able to say that we truly understand their nature."

Some clarification of the argument might be helpful. First, although Plato's conception of the structure of matter is indeed corpuscular in nature, the essence of elementary particles is embodied, according to him, in their surfaces: The boundary itself becomes the primordial attribute. As a corollary, this view entails some severe restrictions on what shapes are possible.[6] The two triangles Plato selected as fundamental building blocks of the surfaces of primordial substances—in other words, his two "principles"—are the isosceles (2 equal sides) right-angle triangle (α), and a particular scalene (3 unequal sides) right-angle triangle (β) obtained by slicing an equilateral triangle in half, the hypotenuse of which is, therefore, twice as long as the smaller side (see Figure 3).

Plato could have described his primordial triangles by means of the simple process sketched in Figure 3. Perhaps to give them added grandeur, he chose instead a depiction in terms of a more elaborate scheme, illustrated in Figure 4.

Two β triangles can form an equilateral triangle, which in turn can be used as one of the sides of a tetrahedron, an octahedron, or an icosahedron. The α triangles, on the other hand, can only generate squares, with which one can produce a cube.

Regardless of one's opinion about the objective significance of these propositions, their internal logic leads to a number of interesting consequences. For instance, earth, which is made of isosceles triangles, is clearly separate from the other three elements, which are all formed of scalene triangles. Moreover, the common structure of the latter three elements makes it plausible to envision that they might transform into one another (but never into earth). At the same time, this elegant explanation of the mechanism of change salvages the concept of uniqueness of the

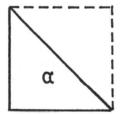
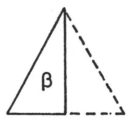

Figure 3. *An elementary representation of Plato's triangular "principles."*

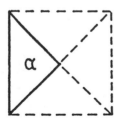
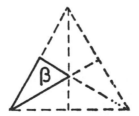

Figure 4. *A more sophisticated depiction of Plato's triangular "principles."*

basic elements.[7] As a matter of fact, Plato constructed an entire arithmetic of triangles. He specifically considered the possibility, for example, that an icosahedron of water (20 faces) could split into two octahedra of air (2×8 = 16 faces) and one tetrahedron of fire (4 faces). An intriguing aspect of this process of transmutation of elements by dissociation and recombination of triangles is that the constituent figures may wear out and fit less readily together. Plato interpreted this mechanical degradation as the cause of aging and death. In so doing, he took away some of the stateliness of his ultimate constituents of matter, in contrast to Democritus and Epicurus, who insisted that atoms were immutable and indestructible.

Moreover, Plato argued that "the two triangles [α and β] constructed in the beginning were not of a unique size: some were large and some were small, in numbers as large as the number of types of each kind. That is why, when these triangles mix together and one with another, an infinite variety of them results, all of which must be studied if one is to engage credibly in a discourse on nature" (*Timaeus*, 57a–58d). The various geometrical properties of the elements thus provide a simple explanation for the multitude of material bodies observed: Congruent triangles can be assembled to produce corpuscles of a common type but of different sizes. For instance, the union of eight isosceles triangles can form a "large" square, composed of the four "small" squares in Figure 3. Lloyd sees there the seed of the concept of isotopes.[8] Furthermore, Plato did not limit himself to a mere exposition of his principles of the structure of elements. He also expounded in detail his views on the nature of numerous chemical substances, to the point that Partington considers the *Timaeus* perhaps the first treatise on chemistry.[9] It would be an interesting point to debate, although that is not within our present scope.

Plato even proposed correlations between the behavior of elements and their shape. For instance, fire would owe its ability to penetrate into and "dissolve" material bodies to the cutting action of the sharp points of its tetrahedral particles.

In Plato's view, particles are surfaces, and surfaces enclose volumes. It then becomes eminently relevant to wonder how the Master of the Academy conceived of the material interior, that is, of the content of these volumes. Aristotle asserted that Plato rejected void, but others have strongly disagreed. C. Mugler summarized these opposing views in his commentary on Aristotle: "Plato's theory of the transformation of elements would imply, according to him, the existence of void both inside and outside the regular polyhedra. Under these conditions, and in contradiction with the belief of Democritus, whose atoms were solid, Plato's assemblages of triangles and squares would merely delineate space. Some assume that the triangles were not just surfaces but consisted of thin layers of matter; others suggest that the polyhedra contained a sort of primordial stuff."[10]

We have already mentioned that Plato admitted only four fundamental substances taking part in the constitution of the universe, while there exist five regular polyhedra. Consequently, one of them remained unused in terms of the equivalence between elements and polyhedra. The polyhedron that was "left over" is the dodecahedron, which, unlike the other polyhedra, is made of pentagons. Plato reserved for it the noble role of being "the image of the organization of the universe," which he regarded as spherical. An esthetic motive is undoubtedly at play in this choice, since of all the regular polyhedra, the dodecahedron is the one most like a sphere. While on the subject of the universe, we should stress that Plato accepted the existence of only a single one, whereas Democritus conceived of an infinite number of them. But, according to Plato, "he [Democritus] was not well versed in matters of knowledge" (*Timaeus*, 55d–56d).

What conclusions emerge from this mixture of reasoning, imaginative speculation, and esthetic drive that seems to form the basis of the Platonic philosophy of nature? Chambry summarizes elegantly the difficulties raised by Plato's works when considered in their totality: "Should a modern reader, unfamiliar with ancient philosophy, come to read the *Timaeus*, he would be struck with complete bewilderment. A world composed of assemblages of triangles, the four elements taken as simple material bodies transforming into each other, a triple soul residing in three different locations of the body, the liver mirroring intelligence and either threatening or calming the appetitive soul, diseases explained by disconcerting fantasies, the metamorphosis of men into women and animals of all kinds, a God who does not create the world but organizes a world as eternal as he, who is helped in his task by subordinate gods, who is modeled after forms or eternal beings existing independently of him, stars that are gods, souls where intelligence turns in circles like stars—all of this would seem quite extravagant, the creation of a delirious author. And yet this system of the world is the product of one of the most profound and brilliant minds to have graced humanity."[11]

Since we want to restrict ourselves to Plato's teachings on the subjects of primordial substances and corpuscular conception of matter, the simplest tack is to try to separate intent from execution, and in the execution, central core from superfluous embellishment.

The intent, quite obviously, is to reach beyond the belief—widespread at the time—in four primordial substances considered the ultimate constituents of matter, inasmuch as such a conception could satisfy neither those who see in nature primarily the immutable nor those who see in it primarily change. Hence Plato's motivation to define a unit of matter at a more elementary level, apt to provide at the same time common building blocks, therefore interchangeable, embodied in his polyhedra. This approach has a definite atomic character. A further motivation, almost certainly inspired by a Pythagorean influence, was to mathematize nature (the best way, according to Plato, to establish a bridge between the world of ideas and the perceptible world). This second approach has a definite mechanistic character.

While such intents are perfectly laudable, what can be said of their execution? It appears to us difficult to escape the conclusion that, although his ubiquitous polyhedra and triangles are a resounding testimony to Plato's brilliant imagination, they do not earn him a place among the great scientists in history. His theory manifestly represents nothing more than mathematical speculation with an esthetic foundation. Bertrand Russell could have cited these fantasies to justify his assertion that in the sciences "preconceived ideas of esthetic origin are just as likely to induce errors as those of moral or theological origin."[12] If one adds to this hodgepodge the teleological argument of a grand design imposed by a creative divine thought from the beginning of the creation of the structured world, one can appreciate the extent to which the role of Plato's polyhedra-atoms differs from that of the corpuscles-atoms of Leucippus-Democritus. As Partington observed, "[Plato's] atoms are not cause but co-cause." The gods of the trio of "classical" atomists remained in the shadow, and the universe of these philosophers was purely "mechanistic." Plato's God does not fit this requirement nearly to the same degree. One senses that the world of Leucippus-Democritus was too "cold," too bereft of transcendent power, to satisfy Plato's more poetic soul. As Crescenzo said: "For both Plato and Aristotle, who were constantly in search of the prime cause and the ultimate purpose, it is as though Democritus had told them the plot of a comedy while skipping the first and last scenes."[13]

Under these conditions, does Plato deserve a place in the history of the atomic movement? I would answer yes in the context of atomic thought, but no with regard to the atomic theory. With his concept of polyhedra-atoms based on two types—and two types only—of triangular surfaces (even though they can come in various sizes), he proffered a system that some view as "geometrical atomism" (a designation he probably would have rejected). On the other hand, his pictorial approach to atomism was so fanciful, his propositions so arbitrary, his esthetic arguments so artificial, that all things considered, it is no wonder that he never became an instrumental figure in the evolution of the atomic theory, at least not in the variant that culminated in its modern formulation.[14] To be sure, the concept we have today of the atomic world differs quite substantially from ancient predictions. Still, it remains rooted in the speculations of those atomists whom Plato eclipsed. In fact, he never mentioned Democritus's name in any of his writings, not even to contradict him. He is said to have wanted to burn all his books. In a recently released work, Jean-François

Revel theorizes that this deliberate omission comes from an intense dislike Plato felt toward Democritus, the kind of loathing "that prevents us from even uttering the name of our most hated enemies, so as not to acknowledge their existence, and because, more important, when it comes to the ultimate hatred—philosophical hatred—a thinker will never find peace by just trying to refute an opponent: He wants him forgotten, he wants to prevent anyone from reading his works, he wants to pretend that he never existed."[15] Be that as it may, Plato directly contributed to the denigration of the atomic theory through countless centuries during which his influence on the philosophical and scientific scenes remained preeminent.

5

The Antiatomists

All of modern philosophy revolves around the difficulty of describing the world in terms of subject and predicate, of substances and qualities, of the particular and the universal.

—A. N. Whitehead

ARISTOTLE

Like Plato's, the philosophical writings of Aristotle (because he hailed from Stagira, he is often called the "Stagirite") are voluminous. His scientific contribution is no less considerable. Even though it is in large measure simply a compilation, it bears the personal seal of the logic and, at times, imagination of its author. As we did in our review of Plato, we will confine our attention to issues of direct relevance to the theory of atoms and related matters. From the outset we want to stress some significant differences between Plato and Aristotle. Although, as we have seen in the last chapter, Plato disagreed with the Leucippus-Democritus-Epicurus conception of atoms, what he proposed amounted to a geometric variation of their theory. Aristotle, on the other hand, adopted on this issue a much more negative attitude, rejecting both the propositions of the atomists and those of his philosophy teacher. The prestige, indeed the virtual aura of infallibility, associated with the body of his teachings made Aristotle the most illustrious of the ancient antiatomists and the head of a long line of followers. In fact, traces of antiatomism in the Aristotelian mold (admittedly modified and somewhat moderated) persisted until the dawn of the twentieth century (they are very much in evidence, for instance, in some of Duhem's views; see chapter 19).

To clarify Aristotle's position, it is helpful to highlight a few of his fundamental philosophical principles. In doing so, the reasons for his conflict with atomism and opposition to Plato's doctrine will become immediately apparent. In fact, Aristotle's philosophy is often described in terms of its contrasts with Plato's. For instance, as we have seen in the previous chapter, Plato put the real world in the realm of Ideas, of which the perceptible world is merely a copy; he explained knowledge as mirroring the prior experience of the soul in a higher world. Aristotle flatly rejected such a separation between sense-observable things and their essence. His entire vision of the world rested on an inextricable link between matter and *form*, which constitute the two principles of being; the term *form* is not restricted here to its common meaning of geometric shape, but refers to the specific properties that define an object or being—in other words, the properties that make it what it is. The link is inextricable in the sense that these two principles could not exist independently of each other. Every object is the combined product of the two, as taught by the doctrine of hylomorphism (from the Greek *hyle,* matter, and *morphé,* form).

The opposition to the atomists is manifest. Aristotle's originality was to elevate form to the level of principle, while in the atomists' view, it was merely a secondary characteristic. Quoting Aristotle himself: "Upon consulting the writings of the ancient philosophers, one could believe that the goal of physics is only to study matter; Democritus and Empedocles barely touched on the questions of form and essence. But if it is true that art imitates nature, it can be said that it is the objective of a single and unique science to study, up to a certain point, both form and matter. . . . The architect is concerned at the same time with the shape of the house and with raw materials, with the stone walls and the wooden parts, in short, with everything; one can conclude from this that physics must study both aspects together."[1] This dualism conforms to Aristotle's precept that opposites are the principles. For Democritus, the principles were, as Aristotle reminded us, the filled and the void, or being and nonbeing. With Aristotle, it may be argued that it is formlessness and form. St. Augustine wrote several centuries later in his *Confessions* (Book 13, 33), referring to God: "Without any time interval, you gave shape to shapeless matter. . . . Matter you made out of nothing, and shape you created from shapeless matter."

The antagonism between Aristotle and the atomists becomes even more pronounced when one considers how each side envisioned the source of our knowledge. For Aristotle, this source resides in our senses, even though perceived impressions may subsequently be exploited by reason for the purpose of describing them in a scientific framework. The world is indeed as we perceive it, see it, touch it, understand it, or feel it. A red object *is* red, a sweet food *is* sweet. This position represents a shift toward a conception of a world of *qualities*, which supposedly constitute its intrinsic reality. Such a conception is, from the start, in conflict not only with Plato's, but also, and quite flagrantly so, with the vision of the atomists. Whether they recognize (as does Epicurus) or deny (as does Democritus) the testimony of the senses, the atomists place reality exclusively in the world of quantities, whereas sense-qualities are defined merely in terms of how we human beings respond to these quantitative data. Koyré will state: "Based as it is on sense perception, Aristotle's physics is decidedly antimathematical."[2]

Intimately related to the matter-form pair of concepts, there exists in Aristotle's doctrine the pair potency-act (or potentiality-actuality) as well, providing a solution to the problem of change. In this vision, matter is a substrate, a pure potentiality, which acquires its explicit and specific expression through an equally specific process of concretization. Thus, actuality realizes potentiality and determines that matter emerging from that potentiality becomes what it is. One can readily appreciate the implications of such a pair of concepts—one permanent and rather static, the other changeable and dynamic—for interpreting the multiplicity of substances and the possibility of their transformation and mutual conversion. It is relevant to add, particularly to shed light on some semantic issues to come up later, that Aristotle made a distinction between *primary matter*, a philosophical concept designating an indeterminate substrate (the form and transformations of which are corporeal), and *matter* (the secondary matter of later scholars), which represents a substantial entity already formed, a substance that is the potential seat of changes of lesser significance referred to as accidental changes. The world of material bodies we normally interact with is essentially "a vast arrangement of substances and accidents."[3] Centuries later, the problem of "substance-accidents" would become the core of the conflict between atomism and the Church.

No presentation of Aristotle's fundamental concepts would be complete without a mention of the four causes that, according to him, preside over the generation of all things and their changes. In addition to the two internal causes identified as the two principles—matter, which becomes material cause, and form, which becomes formal cause—there are also two complementary causes: the effective or motive cause, representing the agent or the initiating and executing instrument of the process; and the final cause, embodying the ultimate purpose, the "why" of things.[4]

This conception of cause is markedly distinct from the common meaning of the term such as we understand it today. The notion of final cause is the one that raises the most questions. This cause (which would act via the effective cause, which it sets in motion) reflects Aristotle's pronounced tendency to see in the world, and more particularly in living beings, the product of a rational intention, which, incidentally, guides the evolution of the world toward good. The similarity of this view with Plato's teleological doctrine and its conflict with the strictly mechanistic concept of the atomists, which rejects any divine design, are evident.

However, the most direct and forceful repudiation of the atomic theory by Aristotle is his affirmation of the continuity of matter, hence his rejection of any limit to its divisibility, and, most important, his denial of void. That will be for him the decisive argument, inasmuch as rejecting void is equivalent, in his mind, to rejecting atoms.

Before examining Aristotle's reasoning on this issue, it is necessary to first review his conception of the inner structure of matter, and, in particular, how he perceived its elementary constituents. If one rejects atoms and, at the same time, proposes to construct the world out of matter and form, what is there to turn to if not the familiar primordial substances?[5] Somewhat disappointingly at first sight, Aristotle appears to have simply embraced Empedocles's teachings, including the traditional doctrine of four primary elements. Upon closer scrutiny, however, a very original twist in Aristotle's approach becomes apparent. Consistent with his own beliefs in the essential role of *qualities* in the way we perceive the world and in the actualization of forms out of the potentialities of matter, what he envisioned as ultimate principles were not the four substances themselves, but four *primordial qualities*, which all other observable qualities can be reduced to. These four primary qualities are the warm and the cold (thought of as active qualities), and the dry and the moist (passive qualities). Disregarding incompatible pairs (warm-cold, dry-moist), the four qualities can be grouped into four pairs: warm-dry, warm-moist, cold-dry, and cold-moist. In Aristotle's view of the transition from potentiality to actuality through the effect of a formal cause acting on a material cause, the four primordial substances thus derive from the action exerted on "primeval matter" by pairs of qualities, according to the following scheme:

warm + dry = fire
warm + moist = air
cold + dry = earth
cold + moist = water

This correspondence will give rise later, particularly in the Middle Ages, to various diagrammatic representations, such as the one illustrated in Figure 5.

A formal advantage of this conception is to suggest almost naturally a mechanism for the mutual transformation of the elements: All that is needed is to interchange the associated qualities. It is even possible to conceive of relatively easy transmutations, involving the exchange of a single quality between adjacent elements in the figure (for instance, the exchange water ⟷ air can be described in terms of the exchange cold ⟷ warm, the moist being common to both elements), as well as

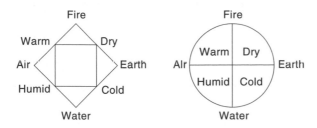

Figure 5. *The relations among elements and qualities according to Aristotle.*

more difficult transmutations involving the exchange of two qualities between elements opposite each other in the figure (for instance, the changing of water into fire). By further assuming that the four elements represent not just four specific substances, but the broader characteristics of certain types of more general compounds, this theory has no difficulty, at least in principle, explaining the great diversity of physical bodies observed in the world.[6]

While diversity is easily accounted for, it is another matter altogether to rationalize the continual and numerous transformations actually taking place in the world. The transmutations alluded to above occur rather seldom. The vast majority of changes in the composition of observable bodies—what we call today chemical reactions—involve combinations and separations of elements, processes that Aristotle sensed might require something more than simple permutations among four partners. Indeed, Aristotle appears to have had a clear intuition of a necessary distinction between a simple mixture of elements, which amounts to a trivial clustering of their particles, and their "intimate" mixing—what today would be referred to as a chemical reaction—in which the constituent elements cease to exist as distinct entities (although they survive in potentiality and can be regenerated by decomposition) and undergo a change giving rise to an entirely new substance, one composed of new uniform particles. The homogeneity of these mixtures, which was thought by the atomists to be merely an illusion (the weakness of our senses preventing us, according to them, from discerning the juxtaposition of the mixed elements), became a reality for Aristotle; he explicitly envisioned mixtures in which the constituent materials no longer remain as they were, but correspond to "a union with alteration of the mixed substances."[7]

Obviously Aristotle had no way to understand the nature of the phenomena involved in this intimate association or how they could be responsible for these transformations. We have to wait until modern times to arrive at the notion of *molecule*—that is evidently what we are talking about—and until the twentieth century and the advent of quantum mechanics for an understanding of the critical role of inner electronic exchanges in such processes. Nor could the atomists account for molecular transformations, *a fortiori*, I would argue, since this decisive step remained impossible as long as atoms were considered impenetrable. Aristotle was probably more aware of this problem than the atomists, and he aimed his most pointed criticisms at them on that particular topic. The issue will resurface repeatedly during the course of the centuries. It was to remain in the eyes of many the Achilles' heel of the atomic theory, to the point that a few die-hard antiatomists of the nineteenth and twentieth centuries would continue to rely on Aristotle's argument in their last-ditch battle against the theory.[8] The modern philosopher and epistemologist Gaston Bachelard (1884–1962) makes this theme the centerpiece of his reflections on the atomic exegesis: "One effectively searches for the atom when one analyzes the phenomenon, but, at the same time, one can justify atomism only through a process of synthesis, by clarifying how one can conceive of a *combination*."[9] In fact, philosophical practice will show that while it is relatively easy for some (and more difficult for others) to decompose matter into atoms, it is invariably much more difficult to perform the reverse operation.

We are now ready to tackle the crucial issue of refuting void and—the two go hand in hand—rejecting atoms. To support his position, Aristotle advanced a series of arguments, the most important of which are related to his theory of motion, an area where he happened to commit his greatest blunders. That was most unfortunate because the consequences of these errors were to prove grievous. Given the abstract nature of the problem of atoms, these errors became a serious impediment to the development of science right up to the dawn of modern times. His fundamental mistake was to have failed to appreciate what was to become known as the law of inertia. Aristotle could simply not conceive why a body in motion should continue to move in a straight line as long as no external force opposes that motion. For Aristotle, any body in motion had to be continually pushed along by some propulsive agent or motive power. We will point out a little later a very peculiar consequence of such an assumption.

A second error, no less weighted with repercussions, was to make a bold distinction between two types of motion: One is "natural" and the other "forced." Aristotle believed that the four elements have a natural and innate predisposition to move spontaneously in the direction of the natural places destined to receive them. This favored direction is upward for fire and air, and downward for earth and water. Another, entirely different natural motion applies to the celestial bodies; that motion is circular. Any other motion was perceived as "forced." In addition to these propositions was the equally false contention that the speed of a body in free fall is in inverse proportion to the density of the medium through which it moves. This series of tenets forms a body of laws at the basis of what can only be described as an erroneous dynamics.

Erroneous or not, this dynamics was in part the basis of Aristotle's objections against the concept of void. While the atomists took the view that void was a prerequisite for motion (since matter would constitute an obstacle in a completely filled medium), Aristotle was convinced not only that it was unnecessary to insist on void, but that motion in void was actually impossible. He claimed that since in empty space there can be no "natural" places or directions, neither up nor down, void could not support any of the natural motions of the four elements. An object placed in void would not know where to go, would have no reason to proceed in one direction rather than another, indeed would have no reason to move at all. Things would be no better should a motion be imposed on that object. Once such an object is set in motion in void, its speed, unopposed by anything, would be infinite, regardless of the weight of the object. There would be no reason to stop at any particular spot, hence no reason to stop at all. Void would therefore be incompatible with both motion and rest. For Aristotle, such consequences made the hypothesis of void absurd. But if there is no void, there can be no atoms, either.[10]

Disagreement with the atomists appears again in connection with another problem, also with important consequences. We have stated that Aristotle adopted Empedocles's four-element theory. The situation is actually rather more complex. While he considered these four elements adequate to account for ordinary reality here on earth, Aristotle felt he had to postulate the existence of a fifth element, which he named *aither* (ether, in modern language), as the constitutive principle of celestial bodies. One of the arguments in favor of this additional element turns out to be related to the problem of the natural motion of elements discussed earlier. Evidently Aristotle realized that the natural motion of celestial objects, which is circular, has nothing in common with any natural motion of the four elements here on earth. Consequently, it became "natural" in this universal perspective to associate it with a new and different element—ether—endowed with its own natural predisposition. The exact nature of this new element was never specified. What was certain, though, is that it could not be transmuted into any of the other four elements.

One of the consequences of such a conception is to invoke different structural and dynamic principles for the heavens and the earth, that is to say, for the celestial and the sublunar worlds. These two domains became fundamentally disparate. This conclusion was very much a regression from the view not only of the atomists, but even of the early natural philosophers, the Milesians. It was to have lasting and terribly negative repercussions on the development of science. One had to wait until Galileo (1564–1642), Kepler (1571–1630), and Newton (1642–1727) for celestial physics and terrestrial physics to merge once again, for the realization that "nature is the same in the heavens and on earth."

In the context of our present focus, Aristotle's positions on three other issues are also significant. They are God, the soul, and the universe.

God. God does have a place in Aristotle's philosophy. The primary argument he advanced in favor of God's existence rests essentially on his theory of motion, in which, as we just saw, anything in motion must receive its impulse from something. By simple extrapolation back to the infinite past, one has to accept that motion must have originated with a primeval motive power, itself immobile, immutable, and eternal.[11] In Aristotle's view, this primeval force has to be God. Not surprisingly, Aristotle's God is pure form and pure potentiality, the only entity to be devoid of material substrate. Much like Plato's God, he is rather like a craftsman, the architect of the world; the matter with which he works is eternal. It is important to add that in Aristotle's conception, God is both primary and final cause. Indeed, God provides the cause of motion by giving it a specific purpose: It enables the world to continually march toward greater good and perfection, a process that is spurred on by the love God evokes, the love men feel for him. This teleological worldview is, of course, diametrically opposed to the strictly mechanistic conception, stripped of any divine design, that was adopted by the atomists.

The soul. While in the view of the atomists the soul, being made of atoms, was degradable, Aristotle considered it the form of the body and believed that it confers on the body its functional unity and purpose. For instance, the purpose of the eye is to see, but an eye is by itself incapable of seeing anything. It is the soul that actually does the seeing. It cannot exist independently of matter, and disappears at the same time as the body, in agreement with the view of the atomists. Aristotle rejected Plato's metempsychosis.

The universe. Whereas the atomists admitted a multitude of worlds, which are born and die according to the movements of atoms, they alone being eternal, Aristotle conceived only of a single universe, ungenerated and indestructible, and therefore eternal.[12] He also believed that our own world is finite.

Finally, having delved into Aristotle's antiatomism, we must say a few words about what some regard as his own corpuscular theory. The contention that he indeed came up with one is based on an incidental remark he made in chapter 4 of book 1 of his *Physics*, where he criticizes Anaxagoras's theory that matter is infinitely divisible but that its qualities are conserved: "Flesh, bones, and the like, are all parts of animals. . . . From there it is evident that neither flesh nor bones, nor anything of a similar nature, could possibly have an infinite dimension, neither in the direction of the large nor in that of the small." The remark applies explicitly to living beings, and there is no evidence that he had a similar opinion about inert matter, although his general conception of chemical compounds (as entities in which the characteristics of the components have fundamentally changed) might encourage some to think so. It should be noted that some scholars will indeed extend this idea to all material bodies, as did notably Albert the Great, as well as Thomas Aquinas, who wrote: "A natural body cannot be divided infinitely, otherwise it would be converted into another."

In any event, it is easy to see how the minimalist particles envisioned by Aristotle differ from Democritus's atoms. Aristotle did not deal with fundamental limits to the division of matter, but with practical ones ensuring that the specific qualities of material bodies are conserved. Beyond these limits, matter simply ceases to be what it was. In this sense, Aristotle's conception could be described as a qualitative pseudo-atomism and is quite different from the quantitative atomism of Democritus.

The notion of minimalist particles will make repeated comebacks at various times in history. While some early commentators, such as Alexander of Aphrodisia (second and third centuries A.D.) and John Philoponus (fifth and sixth centuries A.D.), defended the abstract idea of qualitatively varied minimalist particles (they called them *elachista*, the Greek equivalent of *minima*), later thinkers attempted to give them a more substantial role in the making of physical reality. That was the case, in particular, of some Arabian philosophers (see chapter 11).

Because of Aristotle's great historical importance, it is naturally tempting to make a value judgment about his scientific contribution. But to base such an assessment solely on his position vis-à-vis the atomic theory would do him a disservice. Thus, I prefer to review here three more general opinions, each of which reflects the expert opinion of a very knowledgeable scholar.

According to the historian of science Alexandre Koyré, "Aristotelian physics is wrong. That is quite obvious. It is hopelessly obsolete. But it is a physics nonetheless, that is to say, a theory that is highly, albeit not mathematically, elaborated. It is neither a crude expression of common sense nor a childish fantasy, but a genuine theory, in other words, a doctrine that starts from commonsense data and submits them to an extremely coherent and rigorous systematic analysis."[13]

The eminent philosopher Bertrand Russell wrote: "In reading any important philosopher, but most of all in reading Aristotle, it is necessary to study him in two ways: with reference to his predecessors, and with reference to his successors. In the former aspect, Aristotle's merits are enormous; in the latter, his demerits are equally enormous. For his demerits, however, his successors are more responsible than he is. He came at the end of the creative period in Greek thought, and after his death it was two thousand years before the world produced any philosopher who could be regarded as approximately his equal. Towards the end of this long period, his authority had become almost as unquestioned as that of the Church, and in science, as well as in philosophy, had become a serious obstacle to progress. Ever since the beginning of the seventeenth century, almost every serious intellectual advance has had to begin with an attack on some Aristotelian doctrine; in logic, this is still true at the present day. But it would have been at least as disastrous if any of his predecessors (except perhaps Democritus) had acquired equal authority. To do him justice, we must, to begin with, forget his excessive posthumous fame, and the equally excessive posthumous condemnation to which it led."[14]

Finally, the writer Arthur Koestler expressed the following opinion: "Aristotle's physics is really a pseudo-science, out of which not a single discovery, not a single invention, not a single novel idea emerged in two thousand years; indeed nothing could come out of it, and that was precisely one of its powerful appeals."[15]

To balance somewhat this last verdict, we also reproduce the assessment of J.-M. Aubert, taken from his scholarly analysis of a Christian view of the world: "The myth of Aristotle's responsibility for scientific paralysis is so deeply ingrained in numerous minds that it is imperative to denounce it as inane; Aristotle's reasoning cannot be faulted, for the simple reason that it was virtually eclipsed by many other currents of thought, all the way to St. Thomas. Instead, one might turn the criticism around and wonder whether scientific progress was not hampered in part by the lack of interest displayed by the Greco-Roman world in the doctrine and method of the Stagirite."[16]

THE STOICS

One becomes a Stoic, but one is born Epicurean.
—Diderot

The physical doctrine of the Stoics constitutes probably the clearest expression of opposition to the teachings of the atomists, particularly that of Epicurus. Historians of Stoicism traditionally distinguish three great periods in the evolution of this philosophical school: the period of early Stoicism (fourth and third centuries B.C.), of which the most renowned representatives are Zeno of Cittium, Cleanthes of Assos, and Chrysippus of Soli; the period of middle Stoicism (second and first centuries B.C.), whose main thinkers are Posidonius of Apamea and Panaetius of Rhodes; and the period of imperial Stoicism (first and second centuries A.D.), exemplified by Seneca, Epictetus, and the Roman emperor

Marcus Aurelius. Yet Stoic physics had already reached its quasi-definitive form with Zeno (332-262 B.C.) and Chrysippus (277-204 B.C.).

The universe of the Stoics, like that of the Epicureans, rests on two corporeal principles, unengendered and indestructible. The first of these principles—a passive one—is matter, the structure of which is conceived as the opposite of the corpuscular nature advocated by the atomists. In fact, the Stoics were staunch proponents of a continuist theory of matter, in which matter is viewed as infinitely divisible and having parts that communicate among themselves. It is therefore not surprising that the Stoics rejected the existence of void, at least in "the interior of the world" (which they, like Aristotle and indeed all antiatomists, considered finite); curiously, they were willing to accept void in the exterior. Void, which was regarded as the very prerequisite of motion by the atomists, seemed hardly necessary to the Stoics; they believed motion was possible in filled space, which they thought of as an elastic medium.

The second principle of the Stoics has nothing whatsoever in common with that of the Epicureans. This principle—which is active—is the *pneuma*, representing the vital breath, the universal reason, the soul, destiny, and God, all rolled into one. *Pneuma* and substance are closely intertwined. Several consequences result from this proposition and from its premises. One is a sort of grandiose pantheistic conception of the universe. Another is a predilection for a theory of mixtures that goes beyond a description in terms of simple aggregations or juxtapositions, more or less intermingled, of elementary particles—a description typical, as we have seen, of the ancient atomic theory. Their own concept of mixture takes on the attributes of a genuine process of synthetic reorganization in which the constitutive substances disappear in the formation of a new entity. Such a picture can be construed as a molecular theory in an embryonic stage.

In addition, while the primordial substance is eternal, the world itself is engendered and corruptible. Its genesis is rooted, once again, in Empedocles's four fundamental elements—fire, air, water, and earth. The Stoics effectively ranked them in order of importance. They ascribed a primary role to the first two, fire and air, and between these an even more privileged role to fire, which assumes an almost divine and providential character.[17] Quoting Plutarch: "Designating as prime elements the four bodies, earth and water, air and fire, they turn, I do not fathom how, some into simple and pure entities, and others into composite and mixed entities. For they know that earth and water are capable of sustaining neither themselves nor other beings, but that they preserve their unity while partaking of the power of air and fire. Air and fire maintain themselves through a strength inherent to them; it is by mingling with the other two elements that they confer on them strength, stability, and substantiality."

Diogenes Laertius stated: "The birth of the world takes place when, start-ing from fire, the substance transforms itself, through the mediation of air, into humidity, of which the thick and consistent portion makes the earth, while its subtle parts become air and, sublimating even further, generate fire; later, according to the mixing of the elements, out of them emerge the plants, the animals, and the other kinds of being." This quota-tion proclaims one of the fundamental characteristics of the Stoic vision of the universe, namely, the concept of eternal return and the endless cyclic evolution of the world.

A fundamental difference with the Epicurean view is also apparent in the position of the Stoics with respect to the role of chance in the making, forward march, and evolution of the world. While both doctrines share the same ethical basis and the same basic aspiration to ensure the well-being of mankind through peace of mind and wisdom, the Stoics refuse to accept that the world's destiny is simply the outcome of chance. They distinctly see in it the intervention of an organizing force, the maturation of a seed, with a final purpose that can be qualified as providence. Cicero discussed in several passages of his *Treatises on the Nature of the Gods and on Destiny* the objections the Stoics raised against the Epicurean doctrine, as well as his own views on the subject. Predictably, the celebrated *clinamen* of the Epicureans and its alleged role in matters of volition and free will were favorite targets of his criticism. For instance, he wrote: "Here I am not surprised to see someone convinced that there exist certain solid and indivisible material bodies, moved by their force and weight, and that this world, so rich and so beautiful, results from the fortuitous concert of these bodies. If one believes such a thing possible, I cannot conceive why one would not believe as well that by haphazardly throwing a vast quantity of the twenty-one letters onto the ground, the result could be Ennius's *Annals*, such that they could then be read. I doubt if chance could by itself complete even a single line. How, then, can these people assure us that the world was made of corpuscles devoid of any color and any other qualities, gathering randomly and accidentally? Or that, at any instant in time, innumerable worlds sometimes are born and sometimes perish? And if the coincidence of atoms can create a world, why can it not create a portico, a house, a city, things that are far less laborious and much sim-pler? Certainly, the absurdities they tell about the world are so unreason-able that it does not seem they ever looked at the magnificent beauty of the sky, the subject of my next development." He also wrote in his *Treatise on Destiny*: "The atom deviates, says Epicurus. And to begin with, why does it? Together with Democritus, they believed in another motive force, an impulse which he called collision; you, Epicurus, gave them gravity and weight. What, then, is this new cause in nature that inspires atoms to deviate? Do they draw straws to decide which one will deviate and which one will not? Why do they deviate in the smallest interval and not in the largest? Why in a single interval and not in two or three? That is wishful thinking, not a discussion. For you say that the atom changes place and deviates without external impulsion, that in the void which it traverses there is no cause for it to alter its path, and that there is in the atom itself no modification explaining why it does not retain the natural motion dictated by its weight. Having then proposed no cause capable of producing this deviation, he still believes he says something worthwhile when he proclaims what everyone's reason rejects and despises. And no one, in my opinion, does more to assure not only destiny, but also the overpowering inevitability of things, and no one is more effective in denying the voluntary impulses of the soul, than a man who admits that he could not have resisted destiny had he not had recourse to these imaginary declinations." Cicero continued in the same *Treatise:* "But Epicurus believes he can avoid the inevitability of destiny by means of the declination of the atom; aside from gravity and collisions, a third motion is thus born, when the atom deviates from the vertical in the smallest possible space: this declination, he is forced to admit, if not explicitly, at least in reality, is without cause. . . . More perspicacious was Carneades, who taught the Epicureans that they could buttress their cause without this imaginary declination."[18]

THE NEO-PLATONISTS

Another philosophical movement that exerted a significant influence on human thought, particularly on religious matters in the Middle Ages, also adopted an overt and vigorous antiatomistic position (while also attacking the Stoics): the Neo-Platonic school, founded by Plotinus of Lykopolis (around 205–270 A.D.). It does not seem essential here to describe the modifications or additions introduced by the Neo-Platonists in the teachings of their illustrious master. On the other hand, it is interesting to review a few of the criticisms articulated by Plotinus against specific elements of the atomic doctrine, particularly since his arguments will come up again with other antiatomists at various times in history. Plotinus asserted that it is impossible to rely on atoms to explain the organization and regularity of the world, volition in human actions, or the essence of the soul (which he viewed as incorporeal and indestructible). He also felt compelled to reject the concept of void. A few significant citations deal with several specific issues.[19]

On the topic of chance: "To attribute the formation of the perceptible world to spontaneity and chance is the absurdity of men incapable of understanding and observing. An obvious absurdity, even before any reasoning; an absurdity, too, which can be proven by well established argumentation" (*Enneads* III, 2). "Order is not born out of disorder, nor law out of inequality, contrary to the beliefs of a certain philosopher" (*Enneads* III, 2).

On the topic of free will in human behavior: "Regarding events or those eternal beings that do not always result in the same outcome, it must be said that they are always driven by causes and one must not admit effect without cause; there can be allowance neither for vain 'declinations,' nor for 'motion of a body which takes place without anything preceding it,' nor for an inconstant inclination of the soul which occurs without anything inducing it to do what it was not doing before" (*Enneads* III, 1). "To ascribe everything to material bodies, whether they be atoms or what is called elements, to generate out of the resultant irregular motion things like rule, reason, and the governing soul, is both absurd and impossible; and it is even more impossible, if one may say so, starting with atoms. Many quite correct arguments have been advanced on this point! If one accepts such principles, it does not even necessarily follow that there is for all things a necessity, in other words, a destiny. Assuming first that atoms are these principles, they are prone to a random oblique motion, each in a different direction. None of these movements is regular, since there is no rule, and the result, once produced, would be regular! . . . It must be necessary for material bodies undergoing collisions with atoms to proceed with the motion imparted to them by these atoms; but to what motions of atoms can one attribute the actions and passions of the soul? What is this collision which, by pushing the soul downwards, or by striking it in any other fashion, causes it to reason or to want in a particular manner, gives a necessary existence to reason, will, or thought, and, more generally, is responsible for life itself? And how can one account for the power of the soul to resist the passions of the body? What movements of atoms stir the thoughts of the geometer, the arithmetician, or the astronomer? What movements are the source of wisdom? For what belongs to us in our actions, what makes us living beings, all of this would disappear if we were carried wherever these material bodies take us at the whim of their impulsions, like inanimate things" (*Enneads* III, 1).

Finally, on the topic of the essence of the soul: "It is impossible for the association of material bodies to produce life and for things devoid of intelligence to engender intelligence. Could one argue: yes, but this mixture is not an ordinary one? Then there must exist a regulator and a cause for the mixture; and it is this cause that has the character of soul. For there would be no composite bodies and not even simple bodies in the reality of the world, were it not for the pervasive soul in the universe, since it is the reason behind all the matter comprising every object, and since reason cannot come from anywhere but the soul" (*Enneads* IV, 4). "It is impossible for an entire physical body to wholly penetrate another; but the soul suffuses everything, therefore it is incorporeal. . . . If one claims that atoms or any other indivisible entities create the soul through their concert, the implication that different parts of the soul would have to unite and achieve harmony will refute this thesis, since there is neither penetration nor harmony between impassive objects, incapable of uniting into one; but the soul is in harmony only with itself. Moreover, materiality and size cannot be derived from indivisible entities. . . . Everything that, in order to exist, is predicated on combinations decomposes naturally into the elements which form it; but the nature of the soul is to be one and simple, and to exist in reality as a whole through life itself; as such, it can never perish. Does it decay, as some say, by division and fragmentation? But it is neither mass nor quantity, as we have shown. Is it by alteration that it can come to destruction? Alteration, when it is the cause of destruction, eliminates the form, but preserves matter; and that can only happen to a composite being. If then it can be destroyed by none of these processes, it must necessarily be indestructible. . . . If one claims that every soul is corruptible, everything would have perished long ago" (*Enneads* IV, 7).

This brief account of how the ancient Greek thinkers struggled with the problems raised by philosophy, religion, and atoms highlights the richness of their propositions and the broad range of their intellectual approaches. In fact, there have been many other thinkers who attempted to cross-pollinate elements of the different doctrines we have just reviewed. By way of example, we will mention Strato of Lampascus (340–269 B.C.), the second scholar of the Lyceum, who succeeded Aristotle. Influenced by Epicurean philosophy, he accepted void but rejected atoms (according to Cicero, he called them "Democritus's dreams"). He held that the fundamental principles were the dual qualities of warmth and cold, and, being resolutely "mechanistic," he rejected the Aristotelian teleology. Much to Cicero's approval, he maintained that "there is no need for the work of the gods to construct the world." This prompted Cicero to comment, "He frees God from an enormous task and myself from a great fear. For who could live with the thought that God pays attention even to him and not fear day and night the divine power, not worry that, should adversity befall him (and who is immune from that?), it would not be well deserved?"

Another noteworthy example is Diodorus Cronus. A representative of the school of Megara, founded by Euclid the Socratic, he taught during the latter part of the fourth century B.C. that physical reality is built out of "minimalist bodies without parts," but he rejected the existence of void.

✵ 6

Principles and Primordial Substances

Chapters 3 through 5 guided us across new frontiers of Greek scientific and philosophical thinking. We found ourselves venturing beyond the relatively primitive efforts of the early natural philosophers, whose attention centered rather narrowly on the definition of "primordial substances." As we have seen, many thinkers, from the atomists to Plato and Aristotle, advanced bold propositions that gave their "principles" precedence over such substances. Yet they all still attributed to Empedocles's four primordial substances an intermediate, albeit important, role in the making of the world. The situation is illustrated in Figure 6, which elaborates on Figure 2 by bringing it up-to-date,

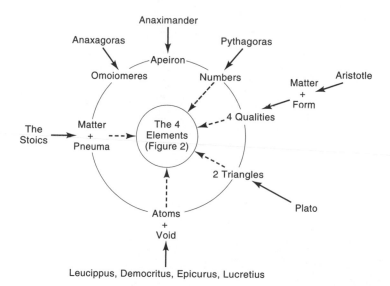

Figure 6. *The principles and primordial substances of the Greek philosophers.*

as it were, at the close of the ancient era. In Figure 6, the inner circle can be viewed as the equivalent of Figure 2. The outer circle identifies the "principles," which superseded "substances" as primordial elements in the structure of matter in the view of individual philosophers. Those among them who continued to regard Empedocles's four substances as essential intermediate agents in the world's organization are indicated by dotted arrows pointing to the inner circle. The near unanimity of opinion on this subject is quite remarkable. Bolstered by the blessing of such luminaries as Plato and Aristotle, who were to take turns dominating the philosophical thinking of the Middle Ages, this picture came to be perceived as virtually indisputable scientific truth throughout that period of human history.

7

Hindu Atomism

Had common experience not exposed us to various phenomena involving dust, we can presume that atomism would not have been so readily accepted by philosophers and that it would not have enjoyed such an abiding destiny.

—Gaston Bachelard

Even though the great adventure of atomism, which was to eventually spread across the globe, originated in Greece, it would be an injustice to fail to acknowledge that the same idea was also burgeoning in the faraway land of India. The atomic theory was as commanding a topic of reflections and debates in Hindu philosophy for two thousand years as in the Western world. Some historians of philosophy and science have even theorized about the possible ties between these two geographically remote variations on the same fundamental theme, particularly in their early stages. More specifically, speculations have persisted about their chronological order and about the possibility that one movement may have influenced the birth of the other. The question has apparently not been definitively resolved. Toward the latter part of the nineteenth century, Mabilleau suggested that Hindu atomism actually came first and that it probably played a role in shaping its Greek counterpart.[1] More recent authors have argued in favor of completely unrelated origins.[2] The present consensus is that Hindu atomism evolved relatively independently and appears to have had no impact of consequence on the development of the atomic theory in the Western world, which is the primary topic of concern to us here. Accordingly, we will give only a brief summary of some key aspects of several atomic doctrines flourishing in Hindustan. These doctrines are generally associated with specific theological systems. Unfortunately, these are of such complexity that discussing them in any detail would spill far beyond the scope of this book (these great systems—Brahmanism, Buddhism, Jainism—are described in detail in a number of excellent treatises).[3] What we will do, instead, is simply

review the fundamental beliefs of the Hindu proponents of the atomic
theory. To the extent possible, we will try to highlight the elements common to the various schools of thought, as well as their differences.[4]

The most ancient sacred texts of Brahmanism, a religion born in
India after the settlement of the Aryans, are the Vedas, composed
between the fourteenth and seventh centuries B.C. They were inspired by
Vedism, an ancient religion in that part of the world, and the Brahmanas
and the Upanishads, which are interpretative outgrowths of Vedism
(eighth to sixth centuries B.C.). The Vedas contain no mention of or allusion to the concept of atom.[5] The Upanishads, on the other hand, tell of
the gradual maturation of the theory of the four elements: earth, water,
fire, and air.[6] The Hindu atomic doctrine is thus not a revealed truth but
the fruit of the reflections and intuitions of many thinkers, who often
combined religious beliefs with secular traditions. Although nowhere in
evidence at the outset, atomism eventually became a common topic in
the teachings of most of the great philosophical systems sweeping the
Indian subcontinent.

As is well known, Brahmanism gave birth to six great dominant philosophical systems, often grouped in pairs according to their doctrinal
similarities: the Mimamsa and the Vedanta, the Nyaya and the Vaisheshika, the Sankhya and the Yoga. Among these, the Nyaya-Vaisheshika movement was the strongest defender of atomism. The fundamental ideas
of this system were expressed in the Vaisheshika Sutra, written in the
first century B.C. by Kanada (which makes him a kind of Indian Democritus in the eyes of some), and in the Nyaya Sutra of Gautama, which
appeared somewhat later. Supporters of this doctrine remained active
until the seventeenth century A.D. While both the Mimamsa and the
Sankhya were receptive to the atomic concept, the Vedanta was steadfastly opposed to it.

Likewise, of the two great systems produced by Buddhism (which
emerged in northern India, near Nepal, in the sixth century B.C.), only
the traditional movement of the Hinayana (Sanskrit word meaning "the
lesser vehicle") was in favor of the atomic theory, while the schismatic
movement of the Mahayana ("the great vehicle") rejected it (especially
the school of Yogacara).

Jainism, the third great philosophical system of India, was founded by
Mahavira at about the same time as Buddhism. Also like Buddhism, it was
a reformist movement aimed against Brahmanism, and it too adopted an
atomistic view of the material world.

The upshot of this rapid overview is that the prevailing philosophical
and theological thought in India, the other great cultural center of antiquity, was substantially and continually infused with a corpuscular view of
nature.

Table 1. Attitude vis-à-vis atomism.

Brahmanism	Mimamsa and Vedanta		±	–
	Nyaya and Vaisheshika		++	++
	Sankhya and Yoga		+	
Buddhism	Hinayana	Sautrantika	+	
		Vaibhasika	+	
	Mahayana	Madhyamaka	–	
		Yogacara	–	
Jainism			+	

Table 1 summarizes schematically where the Hindu philosophical systems just mentioned stood with respect to the atomic theory. The + and – signs indicate positions for or against the theory, double signs imply strong feelings one way or the other, and ± suggests neutrality or lack of consensus.

We now proceed to take a closer look at the fundamental teachings of these different schools.

ATOMISTIC DOCTRINE OF THE NYAYA-VAISHESHIKA

A characteristic feature of this school is the grafting of the atomic theory onto the doctrine of the four elements, which are the same as in Empedocles's theory: earth, water, air, fire. The Nyaya-Vaisheshika philosophy distinguished four types of primordial atom, corresponding to the four material elements. Much like in the Greek doctrine, Hindu atoms were eternal, indestructible, without parts, and innumerable. However, we note the lack of any reference to a homogeneous primitive matter, which prompted Mabilleau to claim: "Kanada did not rise to the level of atomism proper, in which an original uniform matter serves as substratum to all material bodies, and in which specific differences emerge solely out of combinations."[7]

Actually, the traditional elements constitute only four of the nine substances recognized by the adherents of the Nyaya-Vaisheshika. The remaining substances include: ether; space; time (both space and time are all-penetrating and eternal); souls, of which there exist two kinds, namely, God, an omniscient soul, free of pain and pleasure, and individual souls, which are apportioned to each body and are eternal; and the *manas*, an organ of thought that is atomic in nature and exists in unlimited number, since a *manas* is associated with every soul. Its purpose is to act as a link between the soul and external objects.

Besides the four (or five, if one includes the *manas*) types of atom, the Nyaya-Vaisheshika doctrine admits the existence of twenty-four qualities, divided into two groups: "general" qualities, which reside in all substances (e.g., number, extent), and "specific" qualities, associated more directly with some of them (e.g., color, odor, flavor, and touch, each of which can be of several kinds; in particular, there are seven colors, six flavors, etc.). All atoms have a round shape and differ from one another by their specific qualities. For instance, earth atoms are characterized by color (of which there are seven kinds), flavor (six kinds), odor (two kinds), and touch; water atoms are characterized only by color, odor, and touch (one kind of each of these qualities); fire atoms are defined by color and touch; and air atoms involve touch alone. These qualities are latent, or imperceptible, in "simple" atoms taken individually. They affect our senses—and hence manifest themselves to us—only after a minimum number of them have coalesced. Note that, in contrast to air and fire, water and earth are also endowed with weight.

This brings us to the second original characteristic of the Nyaya-Vaisheshika philosophy, namely, its conception of how substances are formed with elementary atoms. This process is gradual, involving intermediate clusters and specific stages of combination. For instance, the first stage is a "dyad" or "binary atom," resulting from the union of two "simple" atoms. The second stage consists of the direct association of three dyads, which forms a "triad" or "ternary atom" and therefore comprises six simple atoms. Neither three atoms nor two dyads are able to produce a substance. The process continues, three dyads producing a "quaternary atom," and so on. In addition, only atoms of similar type can associate, for example two atoms of earth or two of water; an earth atom and a water atom cannot unite to form a dyad, and therefore cannot form a substance.

The triad, as defined above, is the smallest unit of substance that can be perceived by our senses. According to the Vaisheshika Sutra, "the atom is the sixth part of a particle visible in a ray of sunshine." Does this statement imply that the concept of atom was inspired by observing dust floating in a ray of sunshine?[8] If so, it would please Gaston Bachelard, who viewed the atomic theory and its origins essentially as a "metaphysics of dust." He wrote: "Without this special experience [involving dust], atomism could never have evolved into anything more than a clever doctrine, entirely speculative, in which the initial gamble of thought would have been justified by no observation. Instead, by virtue of the existence of dust, atomism was able to receive from the time of its inception an intuitive basis that is both permanent and richly evocative."[9]

Another important aspect of the Nyaya-Vaisheshika philosophy concerns the search—the same one that remained for so long a stumbling block hampering the development of the atomic theory in the West—for what causes atoms to combine into aggregates. The answer to this question is murky. Sometimes the Nyaya-Vaisheshika writings postulate that atoms were set in motion (with an initial fillip, perhaps?) by an intelligent and organizing power, which would have imposed a design on the evolution of the world; sometimes they assert that atoms are endowed with an "invisible and particular virtue," a proclivity to unite according to certain rules. What is clear is that despite the appearance of haphazard and arbitrary properties, atomic combinations actually reveal the teleological purpose of a God who would be essentially the organizer, the demiurge, but not the creator of the universe, "its effectual but not material cause."[10]

Finally, according to the Nyaya-Vaisheshika philosophy, life and conscience appear to involve, as Mabilleau claims, the intervention of "special causes, of which it cannot be said precisely whether or not they belong in the general realm of atomism. . . . Two logical solutions are possible: that of Democritus, who considers thought to be the highly complex and carefully elaborated result of purely mechanical combinations; and that of Leibniz, who ascribes to a conscious monad, that is to say, to an atom of superior essence, the task of moving and stirring lower atoms. Kanada appears to favor the second option," based on a written reference he makes to an "animated atom, infused with conscience," which communicates this faculty to the individual privileged to possess that atom.[11]

In fact, the atomic theory of the Nyaya-Vaisheshika is molded to comply with certain fundamental tenets of the Brahminical doctrine, which include, among other things, a cyclical cosmic process, without beginning or end, a multiplicity of worlds, the retributive consequences of human actions, and the transmigration of souls. As an example, the theory intimates "that in the course of the eternal cosmic process, atoms in turn unite and separate continually. At the conclusion of one cosmic period, atoms isolate themselves from one another. Such a phase of universal rest lasts until atoms are again set in motion and link together into dyads, tryads, etc., allowing the souls that failed to reach salvation in the previous cosmic period to receive the fruits of their actions."[12]

ATOMISTIC DOCTRINE OF THE BUDDHISTS

The atomistic doctrine of the Buddhist school (school of Hinayana) is, by and large, fairly similar to that of the Nyaya-Vaisheshika. Buddhists recognize the four elements—earth, water, air, and fire—and attribute to them an atomic structure; they also divide the properties associated with atoms into "natural" and "derivative" categories. The natural properties of the atoms of the four elements are solidity for earth, viscosity for water, movement for air, and heat for fire. The sensory organs, too, are made of atoms. Generally speaking, atoms are indivisible, invisible, inaudible, flavorless, and intangible. They never exist in isolation but only as aggregates, which are essentially juxtapositions rather than combinations. The smallest perceptible aggregates contain seven or eight atoms, with a central atom surrounded by six or seven others. Moreover, Buddhists oppose any substantial quality ascribed to soul and conscience; these remain outside the realm of atoms. In addition, they deny the existence of a creator or ruler of the world, and affirm that the cosmos obeys an immanent moral law.

ATOMISTIC DOCTRINE OF THE JAINISTS

Jainism, the third philosophical system dealing with atomism, differs from the others on several important points. For instance, while its adherents admit that matter is eternal and composed of atoms, themselves eternal, indestructible, impenetrable, indivisible, indeed punctual, they consider these particles homogeneous in their substance and differentiated solely by qualities, such as color, flavor, taste and touch, that are not associated with them permanently, but that can be exchanged. Unlike the adherents of the Nyaya-Vaisheshika and Buddhism, Jainists do not recognize the original existence of four different types of atoms corresponding to the four elements. In this sense, they bear a kinship to the ancient Greek atomists. Like them, they believe that souls have been endowed in all eternity with matter, actually are "contaminated" by it, according to Glasenapp.[13]

The Jainists have their own distinct conception of what induces atoms to interact and aggregate. They believe that movement is an intrinsic property of these particles, which can cluster in different ways so as to give rise to different spatial arrangements. Their most intriguing proposition, however, concerns the special role of water in the formation of aggregates, where it would effectively act as an interatomic cement. They segregate atoms into humid and dry ones and assume that combinations can occur only between these two types, the likelihood of such combinations being determined by the degree of humidity and dryness of the atoms involved. Thus, they propose an association of opposites, in contrast to the adherents of the Nyaya-Vaisheshika, who favor combinations of like entities. This debate is quite reminiscent of another one taking place in ancient Greece, thousands of kilometers away.

Also of interest is the Jainist belief that the soul is "composed of parts" (which Mabilleau interpreted to mean "particles"), "harmonizing its dimensions with the body." However, to preserve the possibility of its transmigration (a fundamental creed in Hindustan) among various animal species, including man, they also assert that "it can be increased or decreased by addition or separation of the constituent parts, in order to conform with the change of person or animate being." This belief gave the Jainists the advantage of an atomic conception embracing both bodies and souls. That did not spare them, however, from the criticism of Shankara, the eminent spokesman of the Vedanta philosophy, which was opposed to atomism as a whole, as evidenced by several passages of the Brahma Sutra, the fundamental text of that school. Shankara objected that if the soul is allowed to change, in other words, to be transitory, it retains nothing of the eternal principle embodied in its name. We might add that the Jainists propounded an atomic structure of space and time.

The other Brahminical schools aligned with atomism played a much less significant role in developing and disseminating that doctrine. Nevertheless, the adherents of the Sankhya did make a noteworthy contribution. They admitted five elements, but only as particular forms of a primitive matter. They attributed to them an atomic structure, but referred to them as "gross," and postulated that five "subtle" elements of an ethereal nature interpose themselves between the world of material bodies, viewed as emanation of these elements, and our senses.

In summary, the atomic descriptions of the world proffered by the ancient Greek and Hindu philosophies exhibit striking similarities, although they may differ notably in their details. The fact that such nearly identical elemental doctrines existed in two places of the world as distant as Greece and India, at a time when the traffic of travelers must have been minuscule, if it existed at all, is intriguing. One can legitimately wonder whether the parallelism is a case of spontaneous germination or the outcome of unknown contacts. Regardless of the answer, there remain between these two protagonists in the atomic saga some differences concerning the nature of these corpuscles: homogeneous for the Greek atomists, but heterogeneous and quadriform for the adherents of the Hindu schools of Nyaya-Vaisheshika and of Hinayana (although not for the Jainists). While the Greeks were willing to give their atoms different shapes, the Hindus considered all of theirs spherical.

Another significant difference has to do with the nature of sensory qualities (color, odor, etc.) associated with atoms, or, more specifically, with their aggregates. While Hindu philosophers regarded such properties as intrinsic, their Greek cousins saw them merely as "secondary, subjective, corrupted, or conventional."

In addition, the Hindu atomic philosophy sets itself apart by its notion of dyads, triads, and so on, used as fundamental building blocks of all substances, as well as by the rigid and somewhat arbitrary rules of their formation beyond the level of the dyad. Must one see in such complex and peculiar entities the harbinger of the concept of molecule? If so, the

Hindus would have been far more advanced than the Greeks in this respect.[14] On the other hand, the size they attributed to atoms was considerably larger than that envisioned by the Greeks (and quite different from reality).

Their conceptions of the soul are also quite dissimilar. For the Greek atomists, the soul is but a combination of atoms, no less corruptible than any other material, and destined to disintegrate together with the body. Hindu philosophers have a very different view. They see it as a personalized entity, predestined to a long, if not eternal, journey through space and time, with all the adventures that such a peregrination implies.[15] Yet these contradictory conceptions shared a common purpose: to deliver man from the fear of death. This bears witness to the truth that, if one looks beyond any specific doctrinal differences, man seems always to pursue the same goal wherever he happens to be on this earth. This observation remains as valid today as it was centuries ago.

 PART TWO

A Few Scattered Revivals During a Prolonged Suspension (First to Fifteenth Centuries)

History and the Natural Sciences were both necessary against the Middle Ages: they represented knowledge against superstition.

—F. Nietzsche

In the world of theater, a suspension is a period when nothing happens. That was essentially the predicament of the atomic theory during the first fifteen centuries of our era. We could even have symbolized this state of affairs by leaving a few pages of this book blank. We refrained from doing so for three reasons. First, because it is helpful to spell out the fundamental objections that the Church was to raise during this drawn-out era against the propositions of the atomic theory. Second, we wish to review the names of a few, unfortunately rare, thinkers in the Christian world, who in the second half of this period, courageously defended the cause of atomism, at a time when it was almost unanimously rejected. Third, because this suspension was actually interrupted by a period of activity (from the tenth to the twelfth centuries) that saw the atomic idea revived, rejuvenated, and remolded by Muslim philosophers according to their own concepts, to such an extent that we can unhesitatingly speak of an "Arab atomism." This effort constitutes one of the few bright spots against an otherwise depressingly bleak background in the history of medieval atomism. In this context, we will also examine the positions of some medieval Jewish philosophers who were actively involved in this field during that time. Indeed, perhaps the most detailed and accurate account of Arab atomism is to be found in the writings of Moses Maimonides, the most famous of the Jewish philosophers, even though he himself was an antiatomist.

8

Early Medieval Christianity
vis-à-vis the Atoms

*There always comes a time when curiosity becomes
a sin.*

—Anatole France

Before examining how the atomic theory
evolved in the Christian world during the Middle Ages, it is useful to look
in the Holy Scriptures for any mention of the doctrine (in the broad
sense, encompassing both its theological and moral aspects), of its
founders, or of its proponents. Needless to say, the right place to look is
in the New Testament, the Old Testament being far too ancient.

An explicit reference to the Epicureans can be found in the Acts of
the Apostles, in the context of a visit by St. Paul to Athens. It reads (Acts
17:18–25): "A few Epicurean and Stoic philosophers began to speak to
him. And some said: What does this speaker mean? . . . Can we know this
new doctrine you teach? . . . But all Athenians, as well as the foreigners,
who lived in Athens spent their entire time telling or listening to news.
Paul, standing in the middle of the Areopagus, said: Athenians, I find you
in all respects extremely religious. Indeed, while walking through your
city and looking at the objects of your worship, I even discovered an altar
with the following inscription: To an unknown god! What you revere as
unknown, that is what I announce to you." Unfortunately, as we know, this
speech was not very well received. "When they heard him speak of the
resurrection of the dead, some derided him, while others declared: we
will listen to you speak on this topic some other time" (Acts 17:32).

There might even be another similar reference in the Epistle of Paul
to the Colossians (Col. 2:8): "Beware of anyone trying to prey upon you
with his philosophy and with vain deceit, based on the tradition of men,
on the rudiments of the world, instead of on Christ." Some argue that the
remark is aimed at Epicurus's teachings. At least, that is the opinion of
Clement of Alexandria, who wrote in Book II of his *Stromates*: "By this,

Paul does not reject philosophy as a whole, but that of Epicurus, because it denies providence and deifies pleasure." However, this view is obviously only one possible interpretation.

Beyond the biblical texts themselves, we are interested here in the position of the Christian world in general toward atomism during the fifteen centuries stretching from the end of antiquity to the Middle Ages, before the dawn of modern times. The attitude is unambiguous: The Church, whose thinking dominated much of this period, adopted almost unrelentingly a staunchly antiatomistic position. That should come as no surprise, since nothing in its fundamental teachings predisposed it to reduce the entire universe to elements that are simple, indivisible, identical, and eternal, blindly obeying mechanistic laws. On the contrary, many aspects of its doctrine place it in formal opposition to much of the cosmogonical, cosmological, philosophical, and ethical concepts of the atomic theory. This oppositional stance came about quite naturally and was reinforced by the influence of Plato and Aristotle, the two greatest philosophers of antiquity, both of whom also rejected atomism and shared, during the course of the centuries, the favors of Christianity. Plato's influence was stronger until the twelfth century, while Aristotle's predominated thereafter.

The Church's aversion to indivisible entities or any notion of a unique primordial substance as the source of the inanimate, animate, and spiritual worlds is clearly expressed by St. Augustine (354–430), whose opinions will be instrumental in establishing and spreading the Christian doctrine for many centuries: "Let those philosophers disappear, who attributed natural corporeal principles to the intelligence attached to matter, such as Thales, who refers everything to water, Anaximenes to air, the Stoics to fire, Epicurus to atoms, that is to say, to infinitely small objects that can be neither divided nor perceived. Let us reserve the same fate to all the other philosophers, who are too numerous to name and who claimed to have found in simple or combined substances, lifeless or living, indeed in what we call material bodies, the cause and principle of things. Some of them believed that inanimate beings could produce living beings. Such was the case of the Epicureans; these philosophers, and others like them, were incapable of imagining anything but what their heart, corrupted by their senses, allowed them to see."[1]

What is more, the considerable divergence of opinions among the Greek philosophers will in itself be good reason for Christian theologians to reject the entire body of their propositions. For instance, St. Basil (330–379) wrote in the fourth century: "Whatever the allegations of the Greek natural scientists, none of their theses have survived or proved firm, as everything they created was invariably destroyed by their successors; consequently, it is superfluous to contradict them. They destroyed each other. And because they were ignorant of God, they could not agree that an omniscient cause directed the creation of the world. And yet they chose as firm ground this very uncertainty about the origin of things; each constructed his conclusions for himself."[2]

In the next century, Cyril of Alexandria pursued the same line of thought: "Since the sons of the Greeks burst with arrogance at the mention of their masters of thought, since they imagine that they can intimidate us by quoting the names of the likes of Anaximander, Empedocles, Protagoras, and Plato, adding to this list all the other crafters of their impious beliefs and—if I dare say it—of all the other sources of their ignorance, let us proclaim that we can see right through them when they practically brandish against one another their divergent doctrines and invoke about any aspect of the real world irreconcilable justifications."

St. Augustine himself used a similar argument: "In philosophy, you say, nothing can be known with certainty; and, to disseminate your opinion far and wide, you seize upon the quarrels and dissensions of philosophers, in the belief that their disagreements can be used as weapons against them. Indeed, how can we render a judgment in the pending suit between Democritus and the ancient physicists concerning the question of a unique world or a multiplicity of worlds, when harmony failed to reign between Democritus and his own successor, Epicurus? When this philosopher of voluptuousness permits atoms, which are like his devoted servants, when he permits these corpuscles, which he happily imagines to exist in darkness, not to maintain their steady course, but to stray here and there in other directions, he squanders his entire patrimony into quarrels."[3] The assumption of multiple worlds, accepted by the atomists, was a further reason for St. Augustine to reject their doctrine. The Church could not accept the reproduction, in many copies, of the original sin, of reincarnation, and of the Resurrection, all of which a multiplicity of worlds would suggest, indeed demand.

In addition to atoms, the Church also rejected the reality of void conceived as incorporeal three-dimensional space. In the process, it obliterated from its own worldview the two great principles on which the teachings of Democritus and Epicurus were founded. To be sure, in deference to God's omnipotence, one had to consider that he could, if he so desired, create and preserve void, but such a supernatural void would only be some imaginary space devoid of any dimension.

However, the Church did embrace the theory of the four elements and of the pair of qualities associated with each element in Aristotle's system. Acceptance of this view is clearly expressed by the fathers of the Church, notably Basil of Caesarea, as early as the second half of the fourth century.[4] St. Augustine fully aligned himself with this doctrine, which, from that point on, enjoyed quasi-unanimous approval. Centuries

later, some scholarly works, important because of their wide dissemina-
tion and influence (for instance, *De imagine mundi* by Honorius of Autin,
or *Heptateuchon* by Thierry of Chartres, in the twelfth century), further
consolidated this acceptance. Thierry of Chartres even carried out a
deliberate Christianization of the four elements by identifying them with
the heavens (fire + air) and the earth (water + earth) of the Creation
(see, however, chapter 9).

Creation! It is a topic of major discord, unavoidable and insurmountable on principle, between the biblical teachings and the propositions of
the atomists, for whom the world is, as we have seen, eternal and uncreated, on the rationale that it is impossible for something to come from
nothing. As it happens, the eternity of the world has been disputed by
every great theologian and writer of the Church, from Basil the Great
and St. Augustine in the early centuries of our era to Thomas Aquinas in
the thirteenth century. Basil wrote: "As a starting point in discussing the
forms which were given to the world, it is appropriate to first speak of the
order which we perceive in things taken collectively and in each one separately. The heavens and the earth, are, after all, creations, as I must teach
you again. They did not come into being by themselves, contrary to what
many have suggested, but they were created by God. . . . Some of the
Greek scientists have tried to rely on materialistic starting points and have
attributed the origin of the structure of the world to substances which
they called elements. Others plunged into lucubrations, claiming that all
that is visible was formed from indivisible and imperishable atoms, separated by pores, and that life appeared and disappeared as soon as those
atoms united or dispersed. Those who have ascribed the structure of the
firmament and of the earth to such insignificant and perishable causes
have assuredly pursued a chimera."[5] St. Augustine's opinion is expressed
in *The City of God*: "But why did it please the eternal God to make the heavens and the earth then, which he had not made before? If those who ask
this question want to pretend that the world is eternal, without any beginning, and was, therefore, not made by God, they completely turn their
back on truth and are guilty of the mortal folly of impiety."

Another aspect of the battle between eternity and creation of the
world must be pointed out, inasmuch as it further exposes the incompatibility between the atomic doctrine and Christian beliefs. Christians
reject eternity of matter not only because it contradicts the formal teaching of the Bible, but also because affirming that the world has always
existed would impose unacceptable limits to divine freedom. But the
contingence of the world, in other words, the absence of its necessity, is
a key stricture in the Christian dogma of creation. Only God is necessary.
In anticipation of subsequent debates on God's role in the march of the
universe (see, in particular, chapters 11 and 12), it is useful to recall the

teaching of St. Thomas Aquinas, which holds that creation is the communication of what is with what would otherwise not be. As such, it never ceases to operate: "God preserves things in the sense that he never stops infusing existence into them," Aquinas wrote in *Summa theologia.*[6]

Note, however, that against this background of controversy on whether the world is eternal or created, there exists a convergence of views—limited, to be sure—between the atomists and the Church theologians concerning the definition of time. It will suffice to compare Epicurus's opinion, as related by the doxographers, and the words of St. Augustine in *The City of God.* On the one hand, Sextus Empiricus recounted: "According to the interpretation of Demetrius Laconius, Epicurus says that time is the accident of accidents; it accompanies days and nights, the seasons, emotional and impassive stages, movements and states of rest. All these things are accidents of attributes, and time, which accompanies them all, could rightfully be called the accident of accidents." Aëtius concurred with this interpretation: "Epicurus says that time is the accident of accidents, that is to say, the companion of movements." On the other hand, St. Augustine declared: "If, then, the true difference between eternity and time is that time is marked by successive changes, whereas eternity can suffer no change, who could fail to see that time would not have existed had a creature not been made to displace such or such object by a certain movement. For, this change and this movement, in which one element or another, which cannot exist together, take turns and succeed each other in intervals of shorter or longer duration, gave birth to time."

But the consonance of views ends there. Indeed, St. Augustine extended his reasoning further than Epicurus and drew a complementary conclusion: "Therefore, since God, whose eternity precludes the slightest change, is the creator and the organizer of time, how can it be claimed that he created the world after some time interval? . . . It is thus indisputable that the world was made not in time, but encompassing time." In other words, time was created simultaneously with the world. St. Augustine was not the first to express this view, Plato having already proposed a similar argument in the *Timaeus*, nor is he the first among the Christians, as Clement of Alexandria (180–216) enunciated it in the *Stromates.* St. Augustine himself was probably inspired by Philo, another famous Alexandrian philosopher, who was a contemporary of Christ (20 B.C.–A.D. 50) and, interestingly, Jewish. Ten centuries later, Maimonides asserted that the simultaneous birth of time and the universe is implicit in the teachings of the Old Testament. The atomists, for whom both the world and time were eternal and uncreated, obviously did not have to agonize over such questions.[7]

Besides eternity, the Church also rejected the infinitude of the universe and the plurality of worlds contained in it, both of which were championed by the atomists. St. Augustine used these issues to berate Epicurus. For instance, he wrote: "As for those who admit, as we do, that God is the Author of the world, but who criticize us on the question of time in the world, let us now examine what they themselves claim about the breadth of the world. Just as they ask: 'Why was it made at this particular time rather than at another?,' one could turn the question around and ask them: 'Why was it made here rather than elsewhere?' If, indeed, they imagine that before the advent of the world infinite stretches of time existed during which, in their view, God could not have remained idle, let them also imagine infinite reaches of space outside the world. And if they say that the Almighty could not have stayed inactive there either, would they not then be forced to dream, along with Epicurus, of innumerable worlds [the only difference being that, instead of attributing, as he did, their formation and dissolution to haphazard movements of atoms, they would regard them as created by God]? Is this not a logical consequence, if they do not want God to remain inactive in the boundless immensity of these spaces extending on all sides around the world, and if they insist that no cause can destroy these same worlds, as they believed of our own? For we are dealing with those who believe with us that God is an incorporeal Being, creator of all natures distinct from himself. As for the others, it would be revolting to include them in this discussion on religion."

Another tenet of the atomists is in sharp conflict with the Church authorities, namely, that which attributes the formation and evolution of the world to chance alone, thus excluding from them any final design. This is an unacceptable notion for Christians, for whom divine providence alone determines the fate of the universe and of each of its creatures. This belief constitutes one of the underpinnings of their entire faith. As early as the first half of the third century, Lactantius (185–253) had made it a central argument in condemning not only Epicurus's doctrine, whose rejection of providence he qualified as "ridiculous and blasphemous," but Epicurus himself, whom he labels "chief pirate and captain of bandits."[8] He especially lambasted the atomists' refusal to see in the magnificence of the human being the singular solicitude of God. One century later, Basil the Great expressly took to task those who "have not accepted that a rational cause [God] should have presided over the genesis of the universe," but resorted instead either to "worldly elements" or "invisible atoms and substances."[9]

The materiality of the soul, by definition subject to decomposition and death, was equally unacceptable to the Church, as it conflicted head-on with its spiritual doctrine. Schrödinger even expressed the opinion that this disagreement about the nature of the soul was the root cause of the secular divorce between atomism and Christianity.[10]

Yet another tenet of the atomistic doctrine was bound to collide with the Church fathers and, by extension, with Christian theologians of later centuries. It had to do with the nature of the "morality" advocated by the Epicureans, which was deemed profoundly amoral by the Church. On this particular issue, Epicurus was victimized by sometimes incompetent and often malicious interpretations on the part of his adversaries (particularly the Stoics), as well as by the abuses committed by some of his own pupils. For instance, St. Augustine wrote: "Ask Zeno or Chrysippus who has wisdom: they will reply that it is the one of whom they were speaking. By contrast, Epicurus or any other of their adversaries will deny it and pretend that, for them, the one with wisdom is a kind of bird catcher, adept at capturing pleasures. Hence a dispute: Zeno proclaims, and the entire Stoa reverberates in unison, that man is born for no other purpose than virtue, that the latter draws souls to itself because of its splendor, with absolutely no external advantage, without any lure of reward, and that the pleasure advocated by Epicurus is the realm of beasts only, and that it is impious to reject man and wisdom in their society. But now Epicurus, like another Liber, summons from his gardens a throng of his inebriated disciples to his rescue, but only to search frantically what they can tear apart with their dirty fingernails and rotten teeth; he reiterates with great emphasis, using the people as his witnesses, words of pleasure, of suavity, of idleness, he affirms harshly that it is only through pleasure that one can attain happiness."

Even though the argument is not terribly sophisticated, it is hardly surprising that after such comments, picked up and amplified by the successors of St. Augustine, the influential bishop of Hippo, the atomic theory enjoyed so little success in a world suffused with Christian theology and morality.[11] When Gassendi undertook to revalue the philosophical and "scientific" content of the atomic theory during the Renaissance, his first priority was the rehabilitation of Epicurean morality (see chapter 12).

There remains a very specific and quite important disagreement—the most important one in the view of many—dividing Christians and the atomists. It centers on the problem of the Eucharist. The problem is best defined by first quoting the Gospel: "As they were eating, Jesus took some bread; and, after saying grace, he broke it, and gave it to them, saying: 'Take it, this is my body.' Then he took a cup; and after saying grace, he gave it to them, and they all drank from it. And he said to them: 'This is my blood, the blood of the covenant which is spilled for many'" (Mark 14:22–23). "Later, he took some bread, broke it and gave it to them, saying: 'This is my body, which is given for you; do this in my memory.' Likewise, he took a cup after supper, gave it to them, saying: 'This cup is the new covenant in my blood, which is spilled for you'" (Luke 22:19–20).

These words are at the root of one of the thorniest and most enduring aspects of the clash between the teachings of the Church and of the atomists. Indeed, the atomic theory seems to be completely at odds with the philosophical and theological proposition that the Church professed in its interpretation of the mystery of the Eucharist. The delicate problem faced by the Church was to devise an explanation consistent with its own physical picture of the world, while preserving the miraculous character of the Eucharist. The mystery of the transformation of bread and wine into the body and blood of Christ received during the Council of Latran in 1215 the designation of transubstantiation and was proclaimed dogma at the Council of Trent in 1563.[12]

The proper interpretation of the nature of this mystery, debated since the time of St. Augustine, reached the pinnacle of its formulation in the writings of Thomas Aquinas (1225–1274) seven centuries later. It is based on Aristotle's theory of hylomorphism. It constitutes a very important paradigm of an alliance between the teachings of the Church and Aristotelianism. Thomas Aquinas (often called the "Angelic Doctor") cemented the association between these two doctrines and contributed much to its success and durability.

What is the gist of the problem? According to the author of an extensive analysis of this crucial issue, "transubstantiation was the one dogma that flagrantly exposed the contradiction between the testimony of the senses and a doctrinal affirmation of faith. Indeed, the dogma starts with sensible phenomena (color, flavor, odor) and mechanistic and chemical properties identical to those of everyday experience, but beyond this experiential similarity, it postulates a radical change in the substance of the consecrated bread and wine. . . . It was not difficult to anticipate that this position would create numerous victims among philosophers and scientists."[13] And, indeed, the victims were many.

It must be recognized that the theory of hylomorphism seemed to propose an intellectual framework singularly well suited to a physically conceivable interpretation of the Eucharistic miracle. The reader might recall that in Aristotle's physics, every substance is composed of both matter and form, the former being extension and quantity, and the latter encompassing all properties of the substratum. In nature, these two realities are inextricably entwined in the sense that, although conceptually distinct, they could in no way exist independently of each other. The substance presents itself to our senses as a set of accidents (color, consistency, odor, taste, etc.) that are specific to it and enable us to identify it. In this context, the Eucharistic miracle represents a rupture of the link between the substance and its accidents. The essence of the miracle is that the substance of the bread disappears entirely, to be replaced by the substance of the body of Christ, while the accidents of bread continue to exist. There is a decoupling, quite unique in nature, of a normally undissociable whole, the appearance and persistence of "accidents without cause." Our senses have the impression that they are still dealing with bread, although it no longer is bread; they have the impression that they are still dealing with wine, although it no longer is wine.

While this decoupling is perfectly conceivable in Aristotle's physics, even if only in the form of a miracle, it is absolutely inadmissible in the atomic theory, which denies any real existence independent of sense qualities. As we have seen, the only reality in this theory is atoms (and void), and the perception of sense qualities derives solely from the movements of particles, which bring them in contact with our sensory organs and stimulate them. Sense qualities have no independent existence per se. When a substance (bread or wine) disappears, all that is left of these qualities are names. Borrowing the language of Democritus, we might say that they exist only "by convention." Under these conditions, while sensory effects are produced by atoms, the persistence of these effects in the consecrated wafer implies, of necessity, the persistence of the atoms of bread. The substance remains, therefore, bread, squarely in contradiction with Church dogma.[14]

In short, the conflict is overwhelming. After simmering throughout the high Middle Ages, it reached crisis proportions at the dawn of modern times, according to a recent thesis, because of the implications raised by the Galileo case. We shall return to this point in chapter 12. We simply wish to indicate here that the Christian atomists of the Middle Ages, of whom we will speak repeatedly, were recruited precisely from among the opponents to Aristotle's hylomorphism.

Finally, the points of discord between the atomic theory and the teachings of the Church are so numerous and the points of agreement so few that it seems desirable to close this chapter with a rare example of the latter. It concerns the issue of homogeneity, or lack thereof, of the matter forming the universe. More precisely, in the terminology in which the problem was couched at the time, the degree to which celestial matter (that composes the heavenly bodies) and sublunar matter (of which earth is made) might differ. In keeping with the general principles of their philosophy, the atomists believed that they were one and the same. Aristotle, on the contrary, made a distinction between these two types of matter, celestial matter being, according to him, incorruptible and untransmutable, while terrestrial matter is corruptible, subject to transformation and decomposition. On this particular matter, the Church unwittingly took the side of the atomists. In fact, according to St. Augustine (and, centuries later, other Christian thinkers, such as John Duns Scotus [1266–1308] and William of Ockham [1290–1349]), there is nothing in the Book of Genesis to indicate that God might have used different materials to create the heavens and the earth.[15]

9

The Medieval Christian Atomists

Error is the norm; truth is an error gone astray.
—George Duhamel

This chapter contains very few names. All the same, we will divide those who showed an interest in the atomic theory during this drawn-out period into three categories: the chroniclers, the sympathizers, and the proponents.

The chroniclers are essentially historians who, while compiling mankind's philosophical and scientific knowledge, always make it a point to include commentaries on the atomic theory. Their purpose is generally more ambitious than merely cataloging facts: They all confess to a desire to place science at the service of faith, the profane at the service of the sacred. The most prominent individuals in this category are Isidore of Seville (c. 560–636), bishop of that city; Bede the Venerable (672–735), an illustrious representative of monastic science; Rhabanus Maur (784–856), archbishop of Mainz; and, after a gap of several centuries, the Dominican Vincent of Beauvais (1190–1264), author of one of the most voluminous encyclopedias of his time.

The distinction between sympathizers and proponents is not always clear. Among the sympathizers, two philosophers of the twelfth century, one English, the other French, with much in common, stand out. The first, Adelard of Bath (d. c. 1150), was a distinguished Western translator of Arab scientific texts who contributed a great deal to rekindling a general interest in ancient science. In addition, his intellectual distaste for "authorities" and his glorification of reason reflect a strong-willed personality with an interest in all natural phenomena. While he defended the theory of the four elements, as did virtually everyone else during the Middle Ages, he looked beyond that framework and sensed the likelihood of a world based on atoms. Quoting Bréhier: "He puts reason above senses and authority. The senses, which Plato called irrational, are incapable of judgment in the realm of either the very large or the very small (*nec in maximis nec in minimis*): the maxima [the very large] must designate here the universe, and the minima [the very small] those atoms, which, as Lucretius showed so abundantly, can be known only by reason; authority is no less suspect, as only reason can distinguish between truth and falsehood."[1] This is not to say that reason is necessarily in conflict with faith, since God himself is the author of the laws of nature. Adelard proclaimed in his *De eodem et diverso*: "I accept God unconditionally. All that exists comes from him and through him. Yet, physics is not without a certain order and coherence, but it must be constantly reevaluated as human knowledge progresses."

The second sympathizer of the atomic theory is Thierry of Chartres (d. c. 1155), chancellor of an episcopal school that was one of the leading philosophical and scientific beacons in Europe during the eleventh and twelfth centuries. This great teacher of the school contributed much, as did Adelard of Bath, to the revival of the works of antiquity, with a view toward a synthesis between their scientific content and matters of faith. Thierry of Chartres qualifies as a sympathizer to the extent that while he too promoted the theory of the four elements, which he viewed as having been created at a particular instant during the genesis of the world, he seemed to also recognize that these elements could have a corpuscular structure. Jolivet sums up his attitude: "Of these elements, the two crudest ones—earth and water—placed themselves in the center, and the two lightest ones—air and fire—above. Not that these elements are composed of different particles: it is because of their movement that fire and air are light, and it is a similar movement that compresses water and earth and gives them their denseness, which also provides a center of reference and anchor for movement itself."[2]

Among the true proponents of the atomic theory, the first is a physician from Carthage, Constantine the African (twelfth century), who receives dual credit for having translated into Latin the writings of a great many Arab, Jewish, and Greek physicians, and for having explicitly defined atoms as the fundamental constituents of substances.

However, the strongest advocates of atomism during this second part of the Middle Ages appeared later, one in the twelfth century, and two in the fourteenth century. Like Thierry of Chartres, of whom he was a contemporary, William of Conches (1080–1154) came from the same episcopal school, but took a more decisively proatomic stance than Thierry. While he recognized that God's action is the source of the world and of everything it contains and takes place in it, he also professed that this action exerts itself through the laws of nature. According to him, reason alone can uncover these laws; no other source could take precedence, the Bible itself having no competence in scientific matters. As he wrote in his

Dragmaticon: "God could turn a cow into a tree, but did He ever actually do it? And so, it is up to you to discover the reason why an object behaves the way it does." In a rather bold choice for the time, William of Conches repudiated the theory of the four elements. He wrote: "The ordinary image of the four elements is good only for those who, like peasants, ignore the existence of all that cannot be grasped by the senses."[3] To this obsolete picture he deliberately substituted the vision of the world favored by the atomists of antiquity, and he regarded atoms, "simple and extremely small particles," as "first principles." Those are the material elements that, by juxtaposition, form all the things of the world and produce their sensible qualities. The atomists' credo was thus clearly reaffirmed.

Although the school of Chartres produced both a sympathizer and a proponent of the atomic doctrine, it would be a mistake to conclude that it was more or less pledged to that idea. The school actually had strong Platonic roots, and William of Conches encountered much hostility there. For instance, Gilbert de la Porrée (1080–1154), Thierry's successor as chancellor of the school, did not accept the idea of pure elementary particles; he took the view that the smallest particles were already made of a mixture of the four elements.

William of Ockham (1300–1350) deserves credit for having given renewed impetus to the cause of atomism. Ockham is known in the history of philosophy chiefly as the author of a movement known as *nominalism*, according to which only singular objects and concrete individuals are real and can be comprehended. The general terms under which it has become customary to designate objects or entities—the so-called universals—have no reality. For our own purpose, the most significant contribution of this Franciscan monk is his criticism of Aristotle's physics and, in particular, the way he defined substance as being composed of matter and form. For Ockham, substance is reduced to extent, and the qualities through which we perceive it result solely from the various combinations that the "elementary particles" making up matter can assume. One need only replace the phrase "elementary particles" with "atoms" to be faced once again with the fundamental ideas of Democritus and Epicurus. Ockham's attack against Aristotle's physics led him to oppose numerous aspects of an Aristotelian-Thomist description of the world. For instance, Ockham envisioned that the universe can be infinite and eternal. It is no wonder, then, that in 1340 his theses were officially condemned by the Church.[4]

Anti-Aristotelianism also inspired Nicholas of Autrecourt (1300–1350), another defender of the atomic doctrine.[5] While he was convinced that God is the cause of all things, he rejected the authority of teachers and advised the study of nature. More important, he followed the example of William of Conches and William of Ockham in repudiating the Aristotelian concept of substance, which, in his mind, undermined Aristotle's overall view of the world. As a result, he was of the opinion "that in all of Aristotle's physics and metaphysics, no two conclusions are certain, perhaps even not a single one."[6] Nicholas of Autrecourt proposed instead an alternative philosophy and physics. He found them ready-made in the teachings of Democritus and Epicurus and wholeheartedly embraced their fundamental propositions: Matter, which is eternal, is composed of invisible atoms, which themselves are subject to a continual motion that causes them to associate under certain conditions and to dissociate under others. The generation and destruction (i.e., corruption) of substances in no way implies that different forms succeed one another in a given object, contrary to what Aristotle taught; they are simply manifestations of the associations and dissociations of corpuscles. The sensible qualities through which we perceive substances reveal nothing more than the arrangement and movement of their atoms. As a Christian keen on demonstrating the agreement between the atomic hypothesis and the certainties of faith, Nicholas of Autrecourt did take issue with a few points of the classical doctrine of Democritus and Epicurus. For instance, he viewed the soul as consisting of two spirits, which he named intellect and sense. These two spirits are immortal, and they continue to exist even after the atoms of the human body disintegrate. Nicholas of Autrecourt even conceived of rewards and punishments for them, depending on whether they belonged to just or evil beings, in their future associations with either themselves (this possibility was not excluded in his eyes) or with other combinations of atoms.[7] Unfortunately, Nicholas of Autrecourt's efforts on behalf of the soul did not spare him the wrath of the Church: His ideas were condemned, and on November 25, 1347, he was forced to publicly renounce his writings and burn them.

The courageous attitude of these few notable defenders of atomism ensured the continuity of the fundamental message of the theory's founders at a time when independent thinking was clearly not looked upon kindly. Still, in the final analysis, they largely failed to enrich the theory with any original improvements or innovations of their own. Those who succeeded in doing so during this period—without prejudging the importance of their contributions—were the Arab atomists, whom we will discuss in chapter 11.

 **IO**

Medieval Jewish Thought
vis-à-vis the Atoms

*It would be quite a disgraceful detail for philosophy
to expound the pernicious maxims and impious dogmas
of these various sects: the Epicureans denied any
providence, the Academicians doubted the existence
of God, and the Stoics questioned the immortality of
the soul.*

—Jean-Jacques Rousseau

Paradoxically, history has managed to preserve
the ideas of Democritus and Epicurus through the centuries in large part
because of Aristotle's criticism of the atomic theory. In a similar irony, we
owe the most complete and lucid description of the teaching of the Arab
atomists to a philosopher with Aristotelian leanings, an antiatomist who
also happened to be Jewish. His name was Moses Maimonides. Born in
Cordoba, Andalusia, in 1135, he died in Cairo in 1204. Maimonides traveled widely and visited a number of lands then under Arab domination.
He was the most brilliant and renowned Jewish thinker of this remarkable
period (ninth to thirteenth centuries), a time marked by a vigorous intellectual activity fostering a fertile ground for philosophy—Jewish as much
as Arab, the two being often linked in a common dynamic—across a vast
stretch of the world, from Mesopotamia through North Africa all the way
to Spain.

All Jewish philosophers from that period were opposed to atomism,
either explicitly or implicitly.[1] Maimonides was no exception. Judaism
will, as it turns out, be the only great religion to produce not a single
defender of atomism, at least until modern times. Atomism will undergo
attempts at Islamization, followed by Christianization during the Renaissance, but never Judaization.

"Never" is actually not entirely accurate. We must mention here as a
historical footnote the schismatic Jewish sect of the Karaites, which was

founded in the eighth century by Anan ben David (740–800) of the school of Babylonian Judaism. The Karaites did accept the atomic theory, all the more curiously since they ostensibly accepted only the Bible (of which they consistently advocated the free interpretation by each individual) and rejected the oral teachings of rabbis, including the Talmud and the Midrashim. Yet they adopted an atomic theory borrowed directly from the Kalam, the teachings of Moslem philosophers and theologians, and the Mutakallimun (of which we will speak in more detail in the next chapter). Even in its Karaitic religious cloak, the theory was at odds on numerous points with biblical doctrine, and it elicited fierce opposition on the part of the Rabbinites, the then-official religious authority.

One of the fiercest adversaries of the theory was Saadia ben Joseph (882–942), the famous head of the Babylonian Jewish academy of Sura. He was also known by the name Saadia Gaon, a prestigious title designating both spiritual teacher and political leader in the oriental Jewish diaspora. Saadia was not hostile to secular efforts at shedding light on biblical exegesis. Quite the contrary: As documented in his *Book of Beliefs and Opinions*, he advocated the need for a convergence of religious revelation and rational reflection, and he looked in the Greek philosophy of nature for proof of the great dogmas of the Bible. However, his favorite Greek philosopher was Aristotle, in whose footsteps he followed in rejecting an atomic conception of the world. A defender of the creation of the world *ex nihilo*, he faulted, for reasons that will become clear in the next chapter, Arab atomism for affirming the eternity of matter and the recurrence of creation, and for denying the existence of the laws of nature. This did not prevent Saadia from endorsing a number of other tenets of the Arab Kalam, particularly in the version of the Mutazilites, one of the branches of the Mutakallimun—for instance, tenets dealing with God's unity and justice. As such, he is often considered the founder of a Jewish Kalam, differing from the Arab Kalam primarily in the expurgation of atomism.

Saadia Gaon's crusade hindered the development of Karaism in Mesopotamia; its decline in that land coincided, incidentally, with the disappearance of the great academies of Babylonian Judaism, in Sura in 940 and in Pumbadita in 1040. Karaism managed to regain some measure of prosperity in Jerusalem, particularly through the school founded by Daniel ben Moses al-Qumisi. It is in fact in the writings *(The Muhtawi,* or *The Comprehensive Book)* of one member of that school, Yusuf al-Basir (Joseph ben Abraham ha-Cohen ha-Roeh al-Basir), who lived in the eleventh century and is considered one of the most important Karaitic thinkers, that one can find the most explicit affirmation of the existence of atoms.[2] One of his disciples, Jeshua ben Judah, continued his work. Unfortunately, the Jerusalem center would soon decline—not a surprising occurrence given the political events happening in the land of Palestine from the twelfth century on (the rule of the Crusaders, of the Mamelukes, etc.).[3] Another important center existed in Egypt. But in the second half of the twelfth century it became the target of ruthless assaults on the part of Maimonides, who perceived in the Karaites' abandonment of orthodox belief the danger that they would eventually be assimilated into the peoples among whom they had settled. His condemnation was irrevocable: "They abandoned Judaism, neglected its doctrine, and founded a new sect. They pretended to believe in the Torah, but they contested oral tradition. They began to raise objections against its teachings and to advance arbitrary interpretations of biblical verses without the approval of any authority, and, in so doing, they acted against the word of God."[4] In a book published in 1346, titled *The Tree of Life*, a Karaitic thinker of the fourteenth century, Aaron ben Elijah of Nicomedia (1300–1369), attempted to rescue the main theories of the Kalam, including atomism, which had been battered by the criticisms of the great master.[5] Hasdai Crescas (1340–1410) also defended atomism in Spain. Such isolated efforts were, however, largely ineffective against the sweeping condemnation by Maimonides, who dealt a severe blow to the spread of Karaitism. The sect continued nonetheless to enjoy some support in the Jewish diaspora throughout the Middle Ages and beyond.[6] There are today a few thousand Karaites in Israel and a beautiful Karaitic synagogue in the Old City of Jerusalem.[7]

At this point of our discussion, it is incumbent on us to say a few words about Moses Maimonides's personality and convictions. He was at once physician, theologian, philosopher, and often spiritual and political leader of the Jewish community in which he lived. He spent the greatest part of his life in Egypt, where he settled in 1168. While there, he served as physician to the court of Saladin, became a friend of the great vizier Al Fadel, and was elevated to the honorific rank of *nagid*, or head of the community.[8]

The enormous written output of this "eagle of the synagogue," as he was called by Christian scholars, is aimed at demonstrating the concordance between philosophical thought, which he viewed as having attained perfection with Aristotle, and the Bible. The most original aspect of his effort, particularly when viewed through the literalistic prism of contemporary Judaism, was to make a distinction between the explicit teachings of philosophy and those of the Bible, which are often expressed in the guise of parables. Thus, according to him, it is necessary to interpret these words appropriately, as would befit simple writings intended "for children or women." In a visually descriptive sentence, he compared the hidden meaning of the Bible to "a pearl someone has lost somewhere in his house, which is dark and cluttered with furniture. The pearl exists, but he cannot see it, nor does he know where it is."[9]

What to do if the congruence between philosophical thought and the biblical word appears impossible to establish? Maimonides's unconventional position is best illustrated with the example of a major disagreement, one which he considered to be of the utmost gravity, namely, that between the biblical doctrine of the creation of the world by God *ex nihilo* and Aristotle's thesis of the eternity of the world. Maimonides asserted that no evidence could resolve the dilemma one way or the other. He rejected the self-styled "proofs" of creation advanced by certain philosophers (including, as we will see, the Arab atomists and Saadia Gaon). He refused to make creation's verity—indisputable and quintessential in his own mind—contingent upon complicated "demonstrations," forever open to question, as he was acutely aware of the limitations of human intelligence.[10] Where, then, did his conviction about creation come from? A. Neher offers an interesting interpretation of Maimonides's source of faith: "It is only through a criterion outside the spheres of biblical faith or Greek philosophy that the debate can be resolved: that criterion is God's sovereignty and his transcendence over nature. Divine dynamism operates on a different level than natural dynamism, and it is this transnatural dynamism that makes the creation *ex nihilo* conceivable, plausible, and necessary. . . . Revelation prevails, in the final analysis, over the comprehension of revelation."[11]

The truth of creation is for Maimonides the ultimate foundation of the entire religious edifice, for it alone guarantees the contingency of the world, while the eternity of the world would make of it a necessary manifestation of God, limiting his sovereign power to do according to his free will and his pleasure. It is also the only way to conceive of a grand "design," a teleological vision of the future. This example, certainly of great importance, delineates the limits of the open attitude Maimonides was prepared to assume in the face of these perpetual conflicts between philosophy and theology.

Of more immediate relevance to the theme of this book, Maimonides rejected atoms, as we have already pointed out. His reasons come through in his criticisms of the Arab atomists. Moreover, we note that he almost never referred to the ancient Greek atomists. It is not that he was unaware of their work, since his erudition was immense. The reason, surprising for a scholar of his caliber, is clearly spelled out in one of the rare passages of the *Guide of the Perplexed*, his most famous book, in which he explicitly mentions Epicurus: "As for those who did not recognize the existence of God, but who believed that things are born and perish through aggregation and separation, according to chance, and that there is no being who rules and organizes the universe—I refer here to Epicurus, his sect and the likes of him, as told by Alexander—it serves no purpose for us to speak about these sects; since God's existence has been established, and it would be useless to mention the opinions of individuals who constructed their system on a basis that has already been overthrown by proofs."[12] The fallacy of an atheistic attitude being thus taken as an axiom, there is no need to expend any energy at all demonstrating how mistaken its advocates are.

Being antiatomist, Maimonides also championed the theory of the four elements (actually five, as he also accepted ether, while rejecting void), to which he tried to attach a biblical seal of approval by playing on the double meaning of the word *earth* (*eretz* in Hebrew) in the account of Genesis. It is stated in chapter 1, verse 1: "In the beginning God created the heavens and the earth." In the same chapter, verse 9 (the third day of the creation) reads: "God said: Let the waters below the heavens assemble in a single place and let the dry appear. And it was so. To the dry, God gave the name earth." Quite obviously, the first quote is an allusion to the terrestrial globe as a whole, while the second refers to that part of it that exists as firm ground. Now, Maimonides explicitly identified the terrestrial globe with the four elements: "It must be known that *êretç* [earth] is a homonym, which is used both in a general and in a specific way. In a general sense, it applies to everything that is below the lunar sphere, that is to say, to the four elements; and it also refers to the last of those, which is earth. . . . That is why I translated the [first] verse . . . in such a way that the word *êretç* [earth] means the 'lower world,' in other words, the four elements, while the statement 'And God named the dry part *êretç*,' speaks of earth proper. That is now clear."[13]

 I I

Arab Atomism

Philosophy was only a footnote in the history of Arab thought. The true Islamic philosophical stirring is to be found in its theological sects . . . and primarily in the Kalam.

—G. W. F. Hegel

One of the goals of Greek atomism was to free man from the fear inspired in him by the mysteries of a cosmos governed by invisible powers. Its method was to show that these mysteries could be explained in terms of a mechanistic physics involving solely material principles and their interactions, without any interference from capricious gods. In contrast, the goal of Arab atomism is to affirm the omnipotence and absolute freedom of God, whose sovereign will creates and controls all events. This control is permanent and everlasting, but discontinuous, as though, matter and time being discontinuous, God himself could only act in spurts.

From that perspective, Arab atomism is decidedly religious in nature. The concept of the indivisible is used by orthodox theology to reinforce the doctrine of Islam, to affirm its purity in the face of the offensive of Greek philosophy, particularly Neo-Platonism and Aristotelianism. At the core of this doctrine is the unlimited power of a unique God, the willful creation of the world in time, and the revelation by the Prophet of the divine word in the Koran. Bréhier defines these components of Islamic faith as "a series of discontinuous acts," and speaks of the "discontinuousness" of Islam, which he opposes to "Platonic continuousness" and to the Aristotelian (as well as Democritean and Epicurean) concept of the eternity of the world, which "would make creation a necessary outcome."[1]

The Arab atomic doctrine is expressed in the Kalam, which constitutes a sort of rational dialectic operating on theological concepts.[2] In its modern sense, the term designates a scholastic theology specific to Islam, professing a specific type of atomism that is quite different, as we will see, from the Greek version.

The proponents and defenders of the Kalam are called Mutakall-imun, a designation covering various groups, the most important of which are the Mutazilites, considered the most ancient of the Muta-kallimun, established toward the latter part of the eighth century in Basra and Baghdad, and the Ascarites, founded in the tenth century by Abul Hassan al-Ashari (873–935), a defector from the Mutazilites and probably the most famous representative of the radical branch of Moslem atomism.[3] Among the sympathizers who, without formally joining the ranks of the Mutakallimun, nonetheless defended the same ideas often using the same arguments, we must mention al-Gazhali (1058–1111), who taught in Baghdad, Damascus, Jerusalem, and Alexandria, as well as Muhammad ibn Zakariya al Razi (known also as Rhazes, 865–925), a famous Persian physician and philosopher, whose teachings included an atomic description of the world, even though he considered that every-thing issues from preexisting eternal matter.[4] The influence of the Kalam began to wane in the twelfth century.

In spite of sometimes quite significant differences of opinion between the various schools of the Mutakallimun, their general doctrine, notably on the points that matter to us here, are sufficiently similar for us to fol-low Maimonides's example and discuss them as a homogeneous group. On the few occasions when it will seem significant, we will point out some diverging views. Whatever pure intellectual appeal atomism may have had on the Mutakallimun—who were probably familiar with both the Greek and Hindu versions—we must constantly bear in mind that it is for them (as Maimonides believed) the weapon of choice to defend the orthodoxy of the Islamic doctrine and its fundamental tenets: the transcendence, unity, and sovereign independence of God, as well as the affirmation of the Creation.[5] The result is a very distinctive atomic concept. Mabilleau captured the situation masterfully when he wrote: "The Kalam was, above all, a school of religious reaction, and it was only to better combat philosophers that it enlisted the support of a philosophy."[6]

The time has come to introduce the atomism of the Kalam in its explicit form. Maimonides described it as a set of twelve propositions, of which we will summarize here the essential content.[7]

The first proposition affirms that the entire universe, that is, every-thing it enfolds, is composed of atoms, minuscule and indivisible parti-cles. They are devoid of any quantitative properties (they are dimension-less entities, without shape or extent); they acquire such properties only upon uniting (a combination of two is sufficient), at which time they give rise to material bodies and substances.

The second proposition affirms the existence of void.

The third, and very important, proposition proclaims the discontinu-ous structure of time, "which is composed of instants."

The fourth proposition, evidently of Aristotelian inspiration and heralding the heteroclitic character of Kalamic atomism, introduces the notion of "accidents," "which are new ideas added to the idea of substance." Not a single material body can exist without them; of two opposite accidents, one must necessarily prevail. For instance, if there is no accident of motion, then there must be the accident of rest; opposed to the accident of life is the accident of death. Several accidents can exist simultaneously. The accident of life, for example, must be accompanied by other kinds of accidents, such as knowledge or ignorance, power or helplessness, and so on.

The fifth proposition, which hits at the heart of the special character of Arab atomism, states that "it is within each atom that accidents inherently reside, and no atom can be separated from them." This assertion applies not only to accidents of inanimate substances, such as color, odor, and so forth, but also to those specific to animate bodies; thus, life is an accident associated with each and every atom in a living being. The case of the soul is less straightforward; while some Mutakallimun considered that it is composed of subtle atoms that mix with other (ordinary) atoms, most took the view that it is an accident existing in only one of the many atoms forming man. The same is true of the intelligence of beings that are fortunate to have this quality. On the issue of knowledge, Maimonides noted that there is "among the Mutakallimun some disagreement about whether it exists as an accident in each of the atoms of the body endowed with knowledge, or in a single atom." He added that "both views lead to absurd consequences."

The sixth proposition is perhaps the most important, as it is one of the most original of the atomic tenets of the Kalam, and articulates a very unusual worldview. It proclaims that "an accident does not last two beats of time," that is, two instants. "As soon as it is created, the accident vanishes and does not remain." Now, since no substance can exist without accident, if God wants to keep it in its present state, he must re-create at each instant a new accident of the same kind, and he must do so for as long as he wants to preserve the status quo. God can also, at any instant, create a different kind of accident in this substance, the nature and properties of which he would then modify. Should he at any moment interrupt this repetitive production of accidents, the affected substance would simply cease to exist. The entire universe and every material body in it persist only through this continuous action of God, synchronized with the flow of instants. If one recalls that accidents reside within individual atoms (snow is white because every atom of snow is itself white) and that at each instant the associated accidents must be re-created in every atom, one understands how justified the Mutakallimun were in describing God as "efficient." "Overworked" might be more accurate.[8]

This vision of a recurring creation, during which the continued existence of atoms without inherent continuity—these "phantasmagorical
atoms," as Mabilleau called them—requires that "God renew at each
instant the gifts which he bestowed upon them."[9] The state of the world
thus becomes utterly dependent on God's sovereign will. That is wholly
contrary to the more conventional view that laws, either decreed by God
or reflecting the pure randomness of atomic encounters, translate into
and guarantee the immutability of natural phenomena. It refutes the
notion of cause and effect. As Maimonides commented, "There is absolutely no material body capable of exerting an action; the ultimate efficient cause is God himself."[10] E. Gilson described this peculiar and original conception of the universe thus: "Everything was disconnected in
time and space, so as to permit God's omnipotence to roam with ease. A
matter composed of disjointed atoms, lasting for disjointed instants,
doing what they do in a flicker of time independent of the one immediately preceding it and with no effect on the one following it, the whole
existing, holding together, and functioning only through the will of God
who holds it above nothingness and infuses it with his efficacy, such was
roughly the world of the Ach'arites."[11]

How can a universe subject to such perpetual re-creation, free from
any constraint of necessity, in which atoms and accidents intermingle at
the whim of an external sovereign will, project the appearance of permanence in our eyes? That is where the Arab thinkers introduce their inven-
tive concept of the "habitual" character of God's actions as the source of
stability and order in the universe: "The principle of permanence resides
not in things themselves, but in the Creator." What we perceive as a law of
nature, a cause-and-effect relation, only reflects a "habit" developed by
God to connect atoms and accidents, and, through them, events, in a specific manner. Two examples provided by Maimonides himself illustrate
the idea. "They [the Mutakallimun] hold that, when a man moves his
pen, it is actually not he who moves it; for this movement affecting the
pen is an accident created by God. Likewise, the motion of the hand,
which appears to us to move the pen, is an accident created by God in the
hand that moves; God has merely developed the habit that the motion of
the hand accompanies the motion of the pen, without the hand exerting
any influence whatsoever or any causality on the motion of the pen, for,
they claim, the accident does not spread beyond its substratum."[12] The
second illustrative argument reinforces the same idea: "It is God who
established as a habitual fact that the color black, for instance, emerges
only the instant the fabric unites with indigo; however, this black color,
which God created as the cloth to be blackened came in contact with the
black dye, does not remain, but vanishes at that very instant, and God
creates another black. Likewise, after the disappearance of this black,

God has taken to creating not a red or yellow color, but a similar black."

This theory of "habit," which safeguards the permanent order of the universe by means of repeated and exact replications of instantaneous creations, is thus the ultimate expression of the right of an omnipotent God to lay absolute claim to all reality, its creation, its organization, its preservation, and its evolution. Our perception of permanence merely reflects the constancy of divine habit, a "custom" instituted by God, which he could discontinue at any moment should it suddenly please him.[13] This lack of causality, hence of necessity, in the flow of events precludes any certainty about the future of the established order and places the universe as well as man at the mercy of God's will. Moreover, he is not bound by any promise to maintain the organization he imposed at a particular moment on the universe, an organization that he subsequently preserved only out of "habit." This is a stunning outcome for an atomic theory whose original expression by the Greeks explained the perceived order of the world on the basis of chance, acting in all eternity, without any intervention on the part of a higher power.[14]

The next proposition, the seventh, again illustrates the originality of the Mutakallimun: "The deprivations of capacities are accidents which also have a [real] existence and which, consequently, must be perpetually created or re-created." For instance, death is not a deprivation of life; just as there is the accident of life, of fleeting duration and subject to instant renewal, there is also the accident of death. When God wants a living being to die, he creates in him the accident of death. But that accident disappears immediately, as all accidents must do. Thus, "if God wants to maintain death, he creates at once another accident of death, for death could not otherwise last." This point of view, which Maimonides did not endorse, leads to a few consoling thoughts: For example, ignorance, just like knowledge, is a "positive accident."

The eighth proposition specifies that "physical forms are accidents too," and that bodies, whatever they may be—men, angels, even the heavenly throne—are made of identical atoms and differ only by their accidents.

The ninth proposition affirms that "accidents do not support each other," but are supported directly by the substance itself. As a consequence, the occurrence of one accident does not depend on the prior existence of another.

We will simply give a brief synopsis of the remaining three propositions: "What can be imagined can also be admitted by reason" (tenth proposition); the possibility of infiniteness in the universe is rejected "in any manner whatsoever" (eleventh proposition); "the senses do not always provide certainty" (twelfth proposition).

This list of twelve propositions of the Mutakallimun constitutes the foundation of their vision of the world, from which they derive all the arguments in support of their theological and cosmogonical theses. The most important of these concern the creation of the world, and God's existence, unity and incorporeality. They are all buttressed by arguments, of which only those involving the atomic doctrine of the Kalam are of interest to us here. Thus, to illustrate the dialectic method of the Muta-kallimun, we will limit ourselves to discussing the "atomic" arguments they invoke in support of their fundamental propositions.

According to Maimonides, the reasoning of the Mutakallimun on the issue of the creation of the universe goes as follows: "The atoms of the universe, they proclaim, must necessarily be either united or separated, and there are a number of them that now unite, now separate. But it is clear and evident that it is neither union nor separation alone that defines their essence; for if their essence and nature were to demand that they only be united, they could not ever separate. Thus, separation is no more agreeable to them than union, nor is union more desirable than separation; and, therefore, if they are partly united and partly separated, and if some number of them change their condition, being now united and now separated, it is proof that these atoms need someone to unite that which is to be united, and to separate that which is to be separated. Consequently, they assert, that is proof that the world was created."[15] Thus, arguments centered on the process of creation itself inevitably imply for the Mutakallimun the existence of a creator.

Likewise, atoms serve to demonstrate God's unity. The pertinent argument, known as "mutual exclusion," is epitomized in the following excerpt: "If, it is said, the universe had two gods, the atom—which on principle cannot be immune from either one of two opposite accidents—would have to be exempt from both at once, which is inadmissible, or the two opposites would have to be united at the same time into one substratum, which is equally inadmissible. If, for instance, there were an atom or atoms that one [of the two gods] wanted to keep warm and the other wanted cold, it would follow that they would be neither warm nor cold, because the two actions are mutually exclusive—which is inadmissible because each body is assigned one of two opposite accidents—or that the body in question would be both cold and warm [which is impossible]. In like manner, if one of the two wanted to set a body in motion, the other might want to keep it at rest; and it would follow that it would be neither in motion nor at rest, or that it would be at once moving and standing still."

The proof of the incorporeality of God rests on the same type of reasoning: "If, they say, God were a material body, it would be necessary either that the very essence of the divinity reside in all [simple] substances of that body, meaning in each of its atoms, or that it reside in a single atom of this body. But if it resided in a single atom, what purpose would all the other atoms serve? The existence of this body would be senseless. If [on the contrary] it resided in each atom of this body, there would have to be numerous gods, rather than a single one; but we have already proven that God is unique."

This line of circular reasoning is pervasive, no objection being apt to shake the complete faith of the Mutakallimun in the demonstrative power of their dialectic. A few entertaining pages on this topic can be found in *The Guide of the Perplexed*.[16]

The *Guide* also contains the objections and criticisms of its author against the Kalam in general, and its atomic perspective in particular. Maimonides rebuffed any attempt to prove God's existence by demonstrating the creation of the world, with or without the involvement of atoms. In the process, he contemptuously dismissed the central and most original tenet of the Kalam, namely, the unending recurrence of creation. On this subject, he wrote: "That, they claim, is what is rightfully called believing that God is efficient; and he who does not believe that God acts in this way denies, according to them, that God is efficient. But in my own view and in that of any intelligent man, it is about such beliefs that one can retort: 'Would you mock Him [God] as you would mock a mortal?' (Job 13:9). For this is indeed pure mockery."

He also rejected the real existence of negative accidents, which, according to the Mutakallimun, govern the deprivation of capabilities. On the topic of the accident of death, he wrote: "However, I would like to know how long God will continue to create the accident of death in a corpse! Is it for as long as it keeps its external shape, or as long as one of its atoms survives? For God creates the accident of death, according to their belief, in each one of these atoms. Yet we find molar teeth from the dead which are thousands of years old, which proves that God has not reduced that substance to nothingness, and that, consequently, He creates the accident of death for thousands of years, in such a way that, one death disappearing, he creates another."

In short, Maimonides dismissed the intellectual reasoning of the Mutakallimun. He was not the only one among the thinkers of the time to take such an adverse position. Averroës (1126–1198), a prestigious Arab philosopher who was an Aristotelian like Maimonides and also a native of Cordoba, was no less opposed to the Kalam.[17] However, he defended the Aristotelian concept of minimal particles, which he called *minima naturalia*, qualitatively and quantitatively specific to each substance and present in them as constituent blocks. He deemed these *minima* indispensable for substances to exert their specific actions.[18] These concepts were developed further and clarified by a number of eminent Averroists, such as Siger of Brabant (c. 1240–1284), Agostino

Nifo (1473–1546), Francis Toletus (1532–1596), and Giacomo Zarabella (1532–1589). They were also adopted and refined by the philologist and physician Julius Caesar Scaliger (1484–1558). All these successors to the "Commentator"—as Averroës was called in the Middle Ages—saw the *minima* as agents directly implicated in chemical reactions. In their view, interactions between *minima* are more than simple juxtapositions; they give rise to new unified entities, which, in turn, interact with other *minima*. Such a picture presupposes a modification of the constituent *minima*. Their model thus bears some resemblance to the Aristotelian conception of the nature of compounds; it puts them right in the middle of a particularly delicate aspect of the atomic question.[19]

How should one assess the role of the Kalam in the historical evolution of the atomic theory? Among the three great monotheistic religions to flourish in the West, Islam was the first to proclaim that faith in a unique God, master of the universe, is entirely compatible with a corpuscular conception of the structure of matter. Of course, the Christian atomic vision turned out to differ in several respects from its Arab counterpart. This is hardly surprising, since, as everybody knows, believing in a unique God is not enough to agree on how he rules the physical world, let alone how he orchestrates the world of thought and feelings. Nevertheless, an attempt at reconciling these two worldviews was sure to take place sooner or later. The historical merit of the Kalam is to have provided the first hint that such a goal was realistically attainable. If, furthermore, the fundamentally nonreligious atomism of the Greek philosophers is included in this assessment, the complementary legacy of the Mutakallimun, as independent and unintentional as it was, was to prove that one can accept an atomic vision of the world regardless of one's position vis-à-vis God; it is the structure built on this vision that differs, the foundations remaining the same. That was far from obvious at the time.

PART THREE

From the Renaissance
to the Age
of Enlightenment

12

The Resurgence of the Atomic Theory

Christian Atomism

God plays a much more prominent role in modern philosophy than he did in times past.
—G. W. F. Hegel

The theory of atoms never ceased to arouse strong feelings among both friends and foes. Yet, despite all the attention it commanded, it played virtually no active part in philosophical and scientific developments during the fifteen centuries from the beginning of our era to the dawn of modern times. Opposed by religious authorities, whose influence dominated these centuries, the theory, like so many other human yearnings, had to wait until the Renaissance and the ensuing climate of tolerance to recover from a thousand-year-old neglect. Not until then did it begin to reclaim its place within the body of ideas that were to shape the modern world. Remarkably, this revival was stoked by the efforts of a few believers intent on reconciling the atomic doctrine with their Christian view of the world. At the same time, this period was marked by a vigorous antiatomism that was no less Christian in the sense that its advocates, too, used arguments of a religious nature to buttress their position. They continued the tradition of medieval antiatomism, albeit with a modernized flavor.

The distinction about to be made between Catholic and Protestant revivalists has less to do with their beliefs than with the intellectual environment inhabited by each. The religious authorities in France and Italy, the then-dominant Catholic powers (Spain was another, but it could not boast a scientist or thinker of any stature at the time), exercised strict control over the development of new ideas. This situation is to be contrasted with the relative indifference of the Protestant hierarchy toward atomism, especially in England, which was the leading scientific center of the period.

THE CATHOLICS

Gassendi

The main driving force behind the revival of the atomic concept was Pierre Gassendi (1592–1655), who, it must be emphasized, was a man of the Church, serving as canon in the French city of Digne. Working during the first half of the seventeenth century, a time that saw the beginnings of an intellectual renewal throughout the world, he was a contemporary of Galileo, Descartes, Mersenne, and Pascal (see Figure 7). The key to understanding Gassendi's personality and contributions is provided in the preface to a book by Barry Brundell: "During his life, Pierre Gassendi came in contact with three cultures: the medieval clergy, the humanism of the Renaissance, and modern science. He never renounced any of the three."[1] Indeed, Gassendi set himself the goal of devising a philosophy

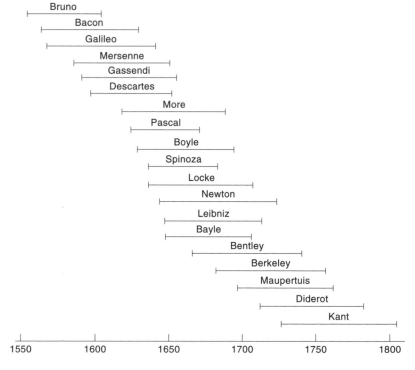

Figure 7. *Scientists and philosophers concerned with the atomic theory during the Renaissance and Age of Enlightenment.*

of synthesis, hence of compromise, between these three currents of thought.

The thread of his intellectual endeavor is twofold, combining a detailed critique of Aristotelianism with a passionate defense of Epicurean philosophy. His objective was to elevate Epicurus's teachings to the level of a doctrine no less acceptable to the Church than Aristotle's was in those days. The goal was ambitious. It required simultaneously the rehabilitation of Epicurean ethics, which Gassendi considered distorted and vilified by its detractors, most notably by the Stoics, and an adaptation of Epicurus's scientific propositions to a Christian view of the world, particularly on the issue of the creation according to biblical revelation.

Gassendi put himself directly at odds with Aristotle by unequivocally accepting the fundamental postulate of the atomists, which holds that atoms and void—the completely filled and the completely empty—are the only principles on which the structure of the cosmos rests. Since rejecting atomism was for Aristotle a natural and inescapable consequence of rejecting void, the major bone of contention between Gassendi and his illustrious predecessor centered specifically on this one issue. This is significant inasmuch as some of Gassendi's contemporaries, such as Francis Bacon, Sébastien Basso, Jean Magnien, and Claude Bérigard, to name a few, although not hostile to the atomic theory and sometimes even openly in favor of atoms, continued to follow Aristotle's lead in rejecting void. Unlike them, Gassendi confidently proclaimed his certainty about the existence of void, a certainty reinforced by the experiments of Evangelista Torricelli (1608–1647) and Blaise Pascal (1623–1662) on the pressure of air.[2] The stakes were important not only for a mathematical description of the world, but in terms of how it should be viewed from a philosophical perspective. Accepting void clearly struck a blow to the very foundation of Aristotle's doctrine, which demands that everything in nature be either substance or accident, and that everything that is not substance or accident be exactly nothing. Quite obviously, void is neither substance nor accident. By recognizing its existence, Gassendi impugned the very foundation of a philosophical system that enjoyed quasi-universal acceptance at the time. Alexandre Koyré, who happened to harbor some strong reservations about Gassendi's work, still conceded: "No one has presented the atomic concept more forcefully and no one has defended the existence of void in all its forms—inside as well as outside the world—with more dedication and persistence than Gassendi; as a result, no one has contributed more to the demise of the classical ontology grounded in the notions of substance and attribute, of potentiality and actuality."[3]

Gassendi also broke with Aristotelianism on the question of "sensible qualities." He fully embraced Epicurus's conception, in which atoms have

three intrinsic properties (dimension, shape, and weight), and are furthermore impenetrable, indivisible, and indestructible.[4] He accepted that
all perceptual qualities (color, warmth, odor, etc.) are associated with
assemblages of atoms. This amounted to a repudiation of the concept of
four primary qualities (warmth, cold, dryness, and humidity), which Aristotle considered the ultimate principles, and whose action on original
matter produced the four primordial substances: fire, air, earth, and
water. In Gassendi's view, however, the atom was definitely not a mathematical point, but a real corpuscle, albeit extremely small.[5] Thus the
shape of atoms explains observable differences: Pointed atoms correspond to spicy or bitter substances, while round atoms give rise to fluid
and malleable bodies. All of this prompted Koyré to speak of Gassendi's
"atom-based qualitative physics." Moreover, Gassendi postulated the existence of atoms specifically associated with certain sensory qualities; there
were luminous atoms, sonorous atoms, atoms of warmth, and atoms of
cold.[6] Although the four primordial substances so dear to every natural
philosopher for two thousand years retained an intermediate role
between atoms and observable matter, only atoms were in Gassendi's view
the true "principles" on which the world is constructed.[7]

Another particularly interesting point is Gassendi's belief that extended bodies were formed by a process involving an intermediate association
of atoms, which he explicitly called "molecules." Given the confusion
about the difference between atoms and molecules, a confusion that was
to prevail in the minds of countless chemists until the middle of the nineteenth century, this distinction constitutes an important intellectual mile-
stone. Gassendi wrote in his *Syntagma Philosophicum*: "Starting with atoms,
certain molecules are first formed, different from one another, which are
the seeds of different things; and each of these things is woven and constituted in such a way that they are not, indeed cannot be, constituted of
any other [seeds]."

In addition to criticizing Aristotle, Gassendi made it a point to defend
the fundamental mechanistic ideas of the ancient atomists, particularly
Epicurus. This did not prevent him, however, from disagreeing with Epicurus on a few specific issues. For instance, while accepting that weight is
directly responsible for the motion of atoms, indeed is one of their intrinsic and inherent properties, he asserted that such motion takes place
naturally in all directions, in agreement with Democritus's views.[8] Thus
Gassendi abandoned the hypothesis of the *clinamen* (the swerve), so central to Epicurus. Similarly, while the ancient atomists believed in an
infinite number of atoms, Gassendi took the view that their number is
finite and in turn dismissed Epicurus's innumerable worlds in favor of
a single one.

Such conceptual differences can be explained by what may be Gassendi's most personal contribution: his attempt to Christianize atomism, to blend the ancient corpuscular doctrine, which seemed to him the most viable basis for a scientific description of the world, with the biblical message transmitted and taught by the Church of Rome. His purpose was to substitute for a purely mechanistic view of the world one that is permeated with divine presence and involvement. Such an undertaking inevitably required the sacrifice of certain basic premises of the materialistic philosophy espoused by Democritus and Epicurus. A cleansing was necessary before proceeding to a synthesis.

One of the thorniest points of contention between these two views of the world—atomistic and Christian—revolved around the dogma of the Creation. The atomic theory postulates that atoms are eternal and uncreated. This position is in flagrant contradiction with the biblical teaching of the Creation, which for the majority of theologians took place *ex nihilo*. As a believer, Gassendi rejected the eternity of atoms and reaffirmed that they were created by God and, consequently, that the material world does have a beginning.[9] Although he did not exclude the possibility that atoms might originally have existed in a state of chaos, he believed nonetheless that they were endowed at the outset of their existence—by God, of course—with all the qualities necessary to their future actions. In particular, it is God who at the time of their creation imbued them with the capacity for motion. How the world evolved out of these atoms is preordained by God and continually guided by his providence. Indeed, Gassendi could not possibly have accepted the materialistic view of the atomists that the fate of the world is governed by mere chance, determined solely by collisions and interactions between atoms according to the random occurrence of their encounters.

For the same reasons, Gassendi refuted the theory of the soul propounded by the atomists, for whom it, like everything else, is composed of atoms subject to a mechanistic fate governing their associations and dissociations, and is therefore perishable. While he admitted that vegetal and animal souls may well be purely corporeal (composed nonetheless of the smallest and most mobile atoms), the human soul was, in his view, immaterial, or at least composed of both matter and spirit, the spiritual part having been created separately by God and infused in man.

Lastly, since the biblical account of creation ostensibly involves only our own world, Gassendi rejected the atomists' claim of an infinite number of worlds. Ours was, as far as he was concerned, unique, isolated, finite, and surrounded by an infinite void.

Although it incorporated a solid atomistic basis, Gassendi's vision of the world appeared shaped by his requirements as a Christian. The intellectual process guiding his thinking was simple. The atomistic tenets he discarded were the very same ones that caused the Church to reject the theory of atoms in its entirety. Still, Gassendi considered this theory the best foundation for a scientific study of the world. He thus strove to embrace it even at the cost of expunging some aspects that were unacceptable to the Church. In the end, atoms and God had to be able to coexist. From a historical perspective, it could be said that Gassendi added God to atoms, although he probably would have preferred to be seen as having returned atoms to God. Under these conditions, Epicurus's theory lost some of its purity and mechanistic rigor. Some, perhaps many, might be of the opinion that it gained in spiritual depth.

It is also worth citing the contribution to this debate by the English philosopher Ralph Cudworth (1617–1688), a contemporary of Gassendi. An atomist like Gassendi, he was prepared to attribute to elementary corpuscles only properties of spatial extent, solidity, shape, and motion. Moreover, in his conception, motion was not an intrinsic property but had to be imparted from the outside. On the other hand, he denied atoms any quality related to life, feelings, or thought (for an opposing point of view, see the section on Maupertuis and Diderot in this chapter). He propounded that God, creator of atoms, was also the creator of immaterial substances responsible for these qualities. These substances, which acted as intermediaries between God and atoms, determined and organized the motion of atoms so as to form all material bodies and living beings. Cudworth named them "plastic natures." He claimed that "it is absurd to assume that everything happening in the world be the result of chance or of a blind and purely mechanistic motion. It is no more reasonable to believe that God intervenes directly in each phenomenon of nature. . . . We are thus forced to accept a certain lower force which . . . imparts to each body its inherent motion, which gives to each organized being its shape, which presides over all phenomena of generation and life." According to Mabilleau's interpretation, this process involves "a plane of the world floating between God who created it and the world that is subject to Him, not a purely abstract plane, but an active order, which links all the separate substances, gives them the unity that atomic matter is incapable of producing by itself, and thereby determines teleological systems which give birth to harmony and life."[10]

Galileo: Christian or Heretical Atomist?

Just as there are men of the world, there are also men of the universe.

—Paul Valéry

The fame enjoyed by Galileo Galilei (1564–1642) among the general public is due to his remarkable contributions to our knowledge and understanding of the macroscopic world. His courageous and tenacious defense of heliocentrism has become the stuff of legend. His views were to bring him before the tribunal of the Holy See and provoke his sentencing in Rome to "ordinary prison" for life. The sentence was later commuted by Pope Urban VIII to house arrest in his home in Arcetri, near Florence. Although forced to renounce his Copernican beliefs, Galileo was to become, and remains to this day, a symbol of the resistance of the human spirit to the oppression of dogmatic ideologies and a shining example of the struggle for freedom of opinion and expression. June 22, 1633, the date of the verdict, will remain a painful memory in the history of the Church, whose reputation was to suffer immensely and persistently from this unjust condemnation.[11] The event amounted to more than a simple mistake. It was a grave error. Not until the end of the twentieth century did the high ecclesiastical authorities acknowledge it as such.

Although it deals more with the grand scale than the infinitesimally small, Galileo's position vis-à-vis the Copernican doctrine bears on our purpose inasmuch as it contradicts some of Aristotle's fundamental propositions. Specifically, heliocentrism demotes the earth to only one of the planets in the solar system, in parity with its companions revolving around a central star. Moreover, this vision is but one element of the almost revolutionary unification achieved by Galileo between the celestial and terrestrial worlds, which reasserted the homogeneity of the entire universe. The tool that made this extraordinary advance possible was a discovery made by Dutch opticians in the early part of the seventeenth century: the telescope. Galileo was to use it as an efficient weapon in a veritable conquest of space.

What extraordinary nights Galileo must have experienced in 1609 and 1610 in Venice, Padua, and Florence, when he trained his primitive but constantly improving telescopes toward the mysteries of the heavens! How astonished and thrilled he must have been when he discovered successively the roughness of the lunar surface, "unequal and covered, like the surface of the earth, with high mountains, deep valleys, and anfractuosities";[12] the multitude of stars in the Milky Way; the satellites of Jupiter; the spots on Venus; and, finally, two large stationary protuberances on either side of Saturn (his telescope lacked the power necessary to resolve these features into a ring).[13] It is no exaggeration to claim that never before in the history of mankind had so many astounding and pivotal discoveries been made in such a short span. Their impact was portentous. They destroyed once and for all the traditional Aristotelian distinction between a celestial world, perfect and incorruptible, and a sublunar world, imperfect and corruptible; in other words, of two worlds qualitatively and hierarchically differentiable. In the process, they paved the way for the proclamation of the universality of the laws of nature. Credit for this achievement will later go to Newton (see the section on him later in this chapter), whose law of gravitation represented the first compelling proof of this universality.

At the same time, the use of the telescope, an instrument built by human hands, demonstrated the crudeness, indeed the imperfect and sometimes misleading perception, of our natural senses. This instrument was able to wrest from the heavens the secrets that the limited power of our natural vision had kept hidden from our knowledge. The world was quite obviously only at the threshold of exploring these novel tools; who could have predicted the stunning discoveries that were to result from increasing the power of this promising instrument? Let us imagine for a moment that Galileo had developed an equal interest in the microscope. He could not, of course, have seen atoms, but while studying the microscopic world, he would undoubtedly have uncovered evidence supporting his vision of the structural unity of nature.

He did not see atoms, but he believed in their existence and said as much, which put him on a collision course with Aristotle and with the Church. The conflict was particularly acute, according to some claims. This is a relatively little-known aspect of the life and work of this most illustrious Tuscan scholar,[14] but of direct relevance to our work, especially in light of a recent assertion made by the historian Pietro Redondi concerning the possible role that Galileo's atomic convictions may have played in his condemnation, an assertion that Redondi bases on a secret document he discovered in 1982 in the Vatican archives of the Congregation for the Doctrine of Faith.[15]

What is it about? Galileo, as we have already noted, openly endorsed the atomism of Democritus and Epicurus. Statements to that effect appear in *Il Saggiatore*, published in 1623,[16] and are reaffirmed in his *Dialogue*, published in 1632.[17] From this, he drew a number of consequences, including some particularly restrictive ones, about the nature of "sensible qualities," a topic that, as pointed out earlier, was a prime source of friction between medieval atomists and the Church. As did his predecessors, he rejected Aristotle's theory on the subject and refused to equate sensory perception with objective reality. Galileo clarified his own position in the following words: "I am simply saying that I feel compelled by necessity, as soon as I imagine an object or a corporeal substance, to imagine at the same time that it is defined and outlined by such and such shape, that it is in this location or in another, at this instant or at another, that it moves or is at rest, that it touches or does not touch another object, that it is unique or multiple, and in small or large number, and no amount of mental effort enables me to separate the object from these circumstances. But that the object must be white or red, bitter or sweet, loud or silent, of a pleasant or offensive odor, I feel no conceptual need to accept that object as being necessarily associated with these qualities; on the contrary, had the senses not detected them, perhaps reason or imagination alone might never do so. I am therefore led to believe that flavor, odor, color, etc., inasmuch as they appear in an object, are nothing more than pure names, and derive their reality solely in the bodies capable of detecting them, so that, in the absence of life, all these qualities are removed and annihilated; but, since we have given them particular names, different from the names of the previous real accidents, we are lulled into believing that they are truly and really distinct from them." He goes on later: "A small piece of paper or a feather . . . brushing between the eyes and the nose, or under the nostrils, provokes an almost unbearable tickle. . . . Yet this tickling sensation is entirely within us, and not in the feather. . . . From such a simple observation I conclude that there can be many qualities attributed to natural bodies, such as flavors, odors, colors, and many others of this type . . . which outside of a living being are, I believe, nothing more than pure names."

Contrary to the Aristotelian (as well as Thomistic) doctrine, which views sensible qualities as real and, as such, considers them linked with specific substances, Galileo insisted that they are merely the result of the association and motion of atoms, which was tantamount to denying their real existence.

But, as we have already stressed, such a position was in irremediable contradiction with the Eucharistic dogma of transubstantiation, which prompted some theologians, particularly the Jesuits, to consider atomism a heretical doctrine and to forbid its teaching in their schools, even though atomism was never officially condemned by the Church.[18] It is not surprising, then, that Galileo's conception of perceptible form was strongly criticized by one of the most skillful Jesuit spokesmen, Father Orazio Grassi, professor at the prestigious Colegio Romano (later to become the Gregorian Pontifical University), in his work *Ratio Ponderum Librae et Simbellae*, published in 1626. In it, the author explicitly condemned Galileo's position as incompatible with the dogma of transubstantiation sanctioned by the Church: "I cannot avoid expressing certain reservations which cause me grave concern. They focus on what we view as incontrovertible on the basis of the precepts of the Fathers, of the Councils, and of the entire Church. I speak of the qualities by virtue of which, although the substance of the bread and wine disappear, through the grace of the omnipotent words, their sensible essence still persist, that is to say, their color, flavor, and their warmth or their coldness. These essences are preserved through divine will and in a miraculous manner, as the authorities have taught us. That is what they affirm. Galileo, on the contrary, expressly asserts that warmth, color, flavor, and other qualities of the same kind, outside of what perceives them, and, hence, within the bread and the wine, are but pure names. Therefore, when the substance of the bread and of the wine disappears, only the name of these qualities remains. But would a perpetual miracle then be required for the preservation of pure names? It is thus clear for all to see how far he strays from those who have expended much effort and care to stipulate the truth of such essences so as to invoke the divine power in such an effect."[19]

This was a public attack, giving Galileo the option to defend himself. He did so by asserting that his propositions concerning sensible qualities simply did not apply to the Eucharist, a reply that evaded the issue, as his detractors must have immediately recognized. But it was the only possible response for Galileo, a Christian scientist and deep believer.

The situation is different with respect to the document discovered, as mentioned above, in 1982 by Pietro Redondi in the secret archives of the Vatican. The document actually appears to be an anonymous denunciation forwarded to the tribunal of the Holy See, calling its attention to the heretical nature of Galileo's views, which contradicted the resolutions of the Council of Trent: "He [Galileo] begins by explaining them [accidents, qualities] with the help of the atoms of Anaxagoras or Democritus, which he calls minima *(minimi)* or minimal particles. It is into those, he claims, that bodies constantly decompose; these particles, however, when interacting with our senses, penetrate our substance, and depending on the diversity of contacts and on the various configurations of these minima, smooth or rough, hard or soft, and depending on whether they are few or many, prickle and pierce us in various ways, dividing more or less, facilitating our breathing, thereby eliciting our discomfort or pleasure. The sense of touch, more material and more corporeal, is properly associated, he claims, with minima of earth. The sense of taste is associated with minima of water, which he calls fluids; the sense of smell with minima of fire, which he calls ignicoles; the sense of hearing with minima of air; he further attributes sight to light, of which he claims he can say little. And he concludes that to stimulate in us flavors, odors, etc., nothing more is required in those bodies that are ordinarily flavorful, odoriferous, etc., than size, shape, and number; and that odors, flavors, colors, etc., exist only in the eyes, tongue, nose, etc., in such a way that, if these organs were removed, said accidents would become indistinguishable from atoms, except by their name. But if one accepts this philosophy of accidents as true, it seems to me that it raises great difficulties for the existence of the bread and wine which are separated from their actual substance in the Holy Sacrament; since one finds in them the predicates and objects of touch, sight, taste, etc., it is necessary, according to this theory, also to say that they contain the minimal particles through which the substance of bread previously stimulated our senses: and if they are substantial, as Anaxagoras claimed, and as this author also seems to accept, it follows that the Holy Sacrament includes the tangible parts of bread or of wine, which is an error condemned by the Sacred Council of Trent."[20]

The case is eloquently made, but it hardly qualifies as a bombshell.

Nevertheless, Redondi uses this document—integrating it in the body of facts, positions, and pronouncements of the numerous individuals involved in Galileo's trial—to advance the hypothesis of a "hidden facet of the condemnation."[21] He claims that the real, or at least primary, indictment that his adversaries, chief among them the Jesuits, wanted to return against Galileo was less his defense of heliocentrism than his atomistic beliefs. The accusation of heliocentrism would then have been a substitute, won by Galileo's sympathizers within the Church, to minimize the danger threatening the illustrious Florentine scientist. According to Redondi, Galileo risked only a prison sentence if convicted of defending heliocentrism, whereas he would have faced the death penalty if found guilty of anti-Eucharistic heresy.

This thesis has prompted much controversy from the moment it was proposed. Even today, it still has supporters and detractors. I am personally inclined to dismiss it. Indeed, if there had been a hidden agenda in this matter, it managed to remain truly hidden; *buried* might be a better word. It is difficult to believe that in a trial with such high visibility, including debates conducted in a carefully staged environment, involving all manner of actors with quite varied opinions, nothing of this delicate game allegedly taking place behind the scenes ever transpired. The discovery of an anonymous letter of denunciation, which incidentally involves only well-known facts, can do little to change this. The now substantial material related to Galileo's trial recently released by the Pontifical Academy of Sciences contains no evidence to substantiate the suspicion of an implicit charge of atomic heresy.[22] Heliocentrism would actually seem a more plausible reason for indictment, since it was a burning topic of intellectual inquiry at the time and monopolized the attention and energy of a great many scholars and philosophers. It certainly was of more interest than the subtle significance of sensible qualities in the Eucharist, which appears to have preoccupied mainly the Jesuits of that era. In addition, it seems rather improbable that, regardless of the charge against him, Galileo would really have faced the death penalty. He had always emphasized that his views on sensible qualities did not apply to the Eucharist. Besides, he was too prominent a scientist, too universally renowned, for such a sentence. Under the circumstances, it would have been difficult to condemn him to death, and virtually impossible to carry out the sentence; it is unlikely that the Church would have dared take such a risk.[23]

That said, one cannot exclude the possibility that in the background of Galileo's indictment, especially prior to the formal trial, there may indeed have been some criticism of his unorthodox position with respect to the philosophical and theological doctrines prevailing at the time. His atomistic convictions may well have raised a few eyebrows. That opinion has been expressed by Egidio Festa: "While the trial of 1633 was brought about at the instigation of the Jesuits, as the testimony of numerous contemporaries tends to prove, the issue of atomism was probably raised during its preliminary phase. But from what we know, it does not appear that Urban VIII attached much importance to it."[24]

Arthur Koestler proposes a thesis that is at once similar and more original, somewhat surprising on the part of someone who can hardly be considered favorably disposed toward the Church: "I am convinced that the conflict between the Church and Galileo (or Copernicus, for that matter) was not inevitable; it was not a collision waiting to happen between two incompatible philosophies, or a war fated to erupt sooner or later. Rather, I believe it was a clash of personalities, of individuals, aggravated by unfortunate coincidences. In other words, I consider it naive and wrong to see in Galileo's trial a kind of Greek tragedy, a one-on-one battle between 'blind faith' and 'enlightened reason.' There was a powerful group whose hostility toward Galileo was relentless: I speak of the university Aristotelians. The inertia of the human mind, its resistance to novel ideas, do not express themselves, contrary to what many believe, in the ignorant masses—easily persuaded as soon as one stimulates their imagination—but among professionals who thrive on tradition and on the monopoly of teaching. Any novelty is doubly threatening to the mediocrity of academics: It challenges their authority as oracles, and it evokes fears of seeing a laboriously erected intellectual edifice collapse. Academic nitwits have been the scourge of greatness from Pythagoras to Darwin to Freud; their wicked and pedantic phalanxes succeed one another century after century."[25]

Be that as it may, it is fair to say that the apparently insurmountable incompatibility between the atomistic concepts and the interpretation of the Eucharist in terms of the Aristotelian-Thomistic doctrine constituted a source of tension destined to endure. As a matter of fact, the animosity, which reached an acrimonious pitch during the seventeenth century, subsequently subsided, without ever completely dying out.

Where do we stand today? Obviously, the Church no longer opposes atoms. Nonetheless, as the encyclical *Mysterium fidei* (1963) proclaims, it continues to cling to its interpretation of the Eucharist.[26] It looks as though all one can do on this particular issue is record an act of divorce between physics and faith. Such a resolution by "amicable separation" is indeed formalized in an analysis by Father P. N. Mayaud of the Society of Jesus, who recently wrote: "Thus the mystery of faith embodied in the Eucharist is a reality which, by its nature, escapes the apprehension of 'physics,' and the term transubstantiation, officialized, if not canonized, at Trent simply contains an affirmation in ordinary words of a mysterious conversion, the reality of which can only be attested by faith."[27] Not all believers are satisfied with this settlement, and some occasionally express a more critical and cynical view. For instance, G. Minois writes: "Thus, until the end of times, the Eucharistic dogma will be expressed in a scholastic Aristotelian-Thomistic vernacular which today is unintelligible to nine-tenths of the European faithful, to say nothing of the Churches of black Africa or Asia. . . . Substance, species, accidents: those are the words which the Church will always be forced to rely on when speaking of the Eucharist, as though concepts in physics, chemistry, and biology had not changed since the thirteenth century."[28]

Let us move on from this controversy and focus now on the innovating stimulus of Galileo's teachings. His atomism is only one aspect of a broader undertaking, namely, the mathematization of the world. For Galileo, nature is mathematical. The science of mathematics is more than a methodological tool to bring order to facts; it is the very foundation of an understanding of nature. His famous words are often quoted: "Philosophy is written in a great book that is always kept open before our eyes; it is the universe. But it can only be understood if one first strives to master its language and to learn the characters with which it is written. It is written in the language of mathematics, and the characters are triangles, circles, and other geometric figures." Apparently ignoring the contributions, admittedly much more primitive, of some of the ancients, Koyré offers this commentary: "Galileo is perhaps the first mind to believe that mathematical forms were actually realized in the world."[29] He further expands on its significance: "Everything in the world is defined by its geometric shape; all motions comply with mathematical laws, not only regular motions and regular shapes, which perhaps cannot be found at all in nature, but irregular shapes as well. An irregular shape is just as geometrical as a regular shape; it is just as precise, only more complicated."[30]

Equally pertinent are the remarks of Crombie: "The fundamental change introduced by Galileo and other mathematicians, such as Kepler, in scientific ontology was to identify the substance of the real world with mathematical entities contained in the theories used to describe appearances."[31] It can thus be claimed that, after Gassendi added God to atoms, Galileo went a step further and imposed mathematics on them.[32]

Considering the problems to be raised later, particularly in the nineteenth century, by the purely theoretical aspect of atomism—the only one accessible until the beginning of the twentieth century, in the absence of any indisputable experimental proof—it is important to point out Galileo's positive perception of the role of theory in scientific research. On this point, Koyré comments: "It is clear that Galileo's conception of a correct scientific method implies the predominance of reason over ordinary experience, the substitution of ideal (mathematical) models for an empirically known reality, the primacy of theory over facts. Only then could the limitations of Aristotelian empiricism be overcome and a genuine experimental method be devised: a method in which a mathematical theory determines the very structure of experimental research, or, in the words of Galileo himself, a method that uses a mathematical (geometric) language to formulate its questions to nature and to interpret the answers."[33]

Finally, having mentioned Galileo's trial and condemnation, one could not conclude this section without an account of more recent events related to this affair, which, after many vicissitudes, finally culminated in a virtual rehabilitation of the Florentine scientist by the highest authorities of the Church. This development played itself out in the broader context of the Church's new and more evenhanded attitude toward science in general.

Galileo's trial was followed by a lengthy "trial of the trial." The absurdity of the condemnation, the magnitude of the injustice committed, and Galileo's growing stature in the history of science combined to make his trial the prime example of a judicial mistake and an abuse of power against which the conscience of a progressive humanity could only revolt. Many voices began to rise and demand a "revision of the trial" and the "rehabilitation" of the condemned. Reparation eventually did happen, three and a half centuries after the verdict. The credit goes to Pope John Paul II and to the positive momentum he imparted to the attitude of the Church toward the fundamental sciences.

The rehabilitation occurred in two stages. The first, a solemn, although largely unpublicized, event, took place on November 10, 1979, marking the observance of the one hundredth anniversary of Albert Einstein's birth. The celebration, organized by the Pontifical Academy of Sciences, included a tribute to Galileo at the conclusion of the proceedings.[34] The second stage was no less solemn and, this time, quite official. It resulted from an initiative taken by John Paul II to see the Galileo case reexamined in a spirit of equity by a body of qualified experts in a variety of fields and with unimpeachable credentials. To that end, the pope instituted on June 3, 1981, a Pontifical Commission for the Study of the Ptolemeo-Copernican Controversy in the Sixteenth and Seventeenth Centuries, a clever euphemism to put the Galileo case in a broader context. The conclusions of the commission were made public by Cardinal Paul Poupard, president of the Pontifical Council for Culture, during a formal ceremony held October 31, 1992, in the Sala Regia. It took place in the presence of members of the Pontifical Academy of Sciences, the College of Cardinals, and representatives of the diplomatic corps, an audience that underscores the universal impact of the event. The purpose of the commission, declared Cardinal Poupard, was "to rethink the entire question, in full adherence to established historical facts and in conformance with the doctrines and the culture of the period, and to acknowledge truthfully, in the spirit of the Vatican II Ecumenical Council, both rights and wrongs, wherever they may have originated. The purpose was not to revise a trial, but to undertake a thoughtful and objective examination, taking into account the historical and cultural conjuncture."[35] The primary conclusion of this multidisciplinary inquiry was summarized in the following words: "In conclusion, a reexamination of the archival documents reveals it once again: All participants in a trial, without exception, should be assumed to act in good faith, in the absence of unofficial documents to the contrary. The philosophical and theological restrictions mistakenly imposed on what was then a new theory of the central position of the sun and the motion of the earth were the consequence of a process of transition in the area of knowledge in astronomy, and of exegetic confusion concerning cosmology.

Heirs to a unitary conception of the world, which prevailed universally until the dawn of the seventeenth century, certain theologians contemporary with Galileo failed to interpret the profound, although not literal, significance of the Scriptures, as they describe the physical structure of the created universe, which led them to unjustly transpose a matter of factual observation into the realm of faith. It is in this historico-cultural conjuncture, far removed from our own time, that Galileo's judges, unable to separate faith from a thousand-year-old cosmology, believed, quite erroneously, that the adoption of the Copernican revolution, not yet definitively proven, was apt to do injury to the Catholic tradition, and that it was their duty to prohibit its teaching. This subjective error of judgment, so clear to us today, led them to take a disciplinary action, causing Galileo 'grave suffering.' This injustice must be acknowledged, as you asked us to do, Holy Father." In his reply to Cardinal Poupard's address, John Paul II accepted its conclusions and clarified the current position of the Church toward scientific investigation: "While contemporary culture is marked by a trend toward scientism, the cultural horizon in Galileo's time was unitary and bore the imprint of a particular philosophical tradition. This unitary character of the culture, which in itself is positive and desirable even today, was one of the root causes of Galileo's condemnation. The majority of theologians did not perceive the formal distinction between the Holy Scripture and its interpretation, which led them to unduly transpose into the sphere of the doctrine of faith a question which constituted in fact a matter of scientific investigation. . . . A tragic lack of mutual understanding has been interpreted as evidence of a conflict between science and faith.

Findings brought about by recent historical studies permit us to affirm that this painful misunderstanding now belongs in the past. . . . There exist two spheres of knowledge: that which is anchored in the Revelation, and that which reason can discover with its own means. To the latter sphere belong notably the experimental sciences and philosophy. The distinction between these two spheres of knowledge must not be perceived as a conflict. These two spheres are not strictly mutually exclusive; they do partially overlap. The methodologies specific to each enable us to uncover different aspects of reality."

And so, as far as the Church is concerned, the Galileo case is closed—this time, it seems, once and for all. Will the lay public concur? Only time will tell.

Giordano Bruno

Time, space, infinity: these are all uncomfortable words.

—Paul Valéry

Giordano Bruno (1548-1600), whose name came up in the previous section, deserves to be mentioned here because he was an avowed atomist, aligning himself explicitly with Democritus and Epicurus. Given his tragic fate—he was burned alive for heresy on February 17, 1600, on the Campo dei Fiori in Rome—it is tempting to compare his case with Galileo's, who was admittedly less severely punished, but condemned nonetheless. This will give us a chance to underline both their commonalities and the nuances, sometimes outright intellectual oppositions, that distinguish these two great humanists. It will also enable us to highlight how Bruno contributed to the revival of the worldview devised by the atomists of antiquity.

Among the convictions common to Galileo and Bruno are a belief in atoms (and in void, although Bruno assumed it to be filled with "ether"), the acceptance of the Copernican doctrine of heliocentrism, and the rejection of certain fundamental Aristotelian propositions, particularly in the domain of cosmology. More specifically, they both rejected the hierarchical distinction advocated by Aristotle between the celestial and the sublunar worlds, which made the earth, although placed at the center of the world, "a suburb of questionable repute." For Bruno, as for Galileo, nature is of a common essence, made of the same matter everywhere.

On the other hand, even in areas where they generally agree, some underlying differences of opinion emerge. Their view on atoms is a case in point. Bruno adopted the atomistic doctrine of antiquity. But in no way does this imply that he endorsed a strictly mechanistic and materialistic conception of the world. He did proclaim, ahead of Galileo, that the "Book of Nature" is written with "characters" or "geometric letters" (triangles, squares, circles, etc.), which represent the microscopic corpuscles characteristic of the different elements, themselves made of atoms. He also assumed logically "that it is not necessary to require many kinds and shapes of elements, just as it does not take many letters to form innumerable entities."[36] Yet Bruno had a spiritualistic conception of the atoms themselves; he attributed to them a soul, which is actually a manifestation of the soul of the universe. Atoms are, in his view, "monads, each of which is an emanation from God and, as such, must contain the infinite from which it emanates." He also rejected the traditional proposition of their chaotic motion and random collisions: "Atoms, stimulated from the inside, do not combine haphazardly or in a disorderly fashion, but according to an organizing will, evolving toward increasingly complex and perfect structures." This is quite different from the purely mechanistic picture of the ancient atomists and from Galileo's view as well. Bruno professed, in fact, a kind of personal pantheism. And that was a major sin in the eyes of the Church.

Likewise, he was not content—unfortunately for him, as this was to be the principal accusation leveled against him by the Church—just accepting the doctrine of a heliocentric but finite world advanced by Copernicus. Instead, he went much beyond this vision—too restrictive, in his mind—and propounded the existence of an eternal and infinite universe, containing an infinitude of worlds similar to ours, inhabited by an infinitude of forms of living beings. Even though all these beings were supposed to continually celebrate the glory of God, the idea of such an infinite universe, of which the earth could not remain the center, was anathema to the Church. Minois has observed that, as far as Bruno was concerned, "Copernicus studied the tree but failed to see the forest."[37]

This infinitization of the universe, undoubtedly the most consequential of Bruno's contributions to the development of modern cosmology, marks a return to one of the pivotal aspects of the ancient atomic doctrine. In this domain, as in many others, the intellectual renewal occurs by a leap across twenty centuries, from antiquity to the Renaissance, over the moat of the Middle Ages. The key role played by Bruno in this renewal is indisputable. Having acknowledged that the seed of the notion of an infinite universe is to be found with the Greeks, A. Koyré, a highly respected expert on the transition "from a closed world to an infinite universe," emphasized that "Giordano Bruno was the first to take Lucretius's cosmology seriously." Alas, what to us would be a credit was perceived as a crime by the ecclesiastical authorities in Bruno's day.[38]

Yet Bruno's argument in favor of the infiniteness of the universe, and of inhabited worlds, rested on a reasoning that was in no way disrespectful of God. On the contrary, he was absolutely convinced that it is impossible to assign limits to God's omnipotence. Indeed, his faith was grounded in the *principle of plenitude*, according to which God could not have failed to create the greatest possible number of diverse worlds, in the largest space possible—hence infinite—and on the *principle of sufficient reason*, which holds that no possibility can be dismissed without a plausible reason. Bruno was expressing these convictions when he wrote in his work, provocatively titled *On the Infinite Universe and Worlds*: "That is how the excellence of God is magnified and the greatness of his empire manifests itself. He does not attain glory in a single sun, but in innumerable suns, not in a single earth and world, but in a thousand thousand, better yet, in an infinitude [of worlds]. . . . I affirm what I cannot deny, namely, that in the infinity of space there could exist an infinitude of worlds similar to this one, or even that this universe could extend its reach and contain a multitude of material bodies, such as those we call stars: moreover, whether these worlds be similar or dissimilar [to ours], it is no less reasonable that the existence of one [of these worlds] be as good as that of another. For there is no less reason for the existence of one than for the existence of the other, and no less reason for the existence of several than [that] of one or the other, and [no less reason] for the existence of an infinitude than for that of a finite multiplicity. Therefore, just as the destruction and the non-existence of our own world would be evil, so would the non-existence of innumerable others be." At one point, he approaches the problem from the opposite angle, exclaiming: "Why would we want to think, or why should we think, that divine efficacy be idle?"[39]

The analogy between Bruno's cosmological conception and that of the ancient atomists is also in evidence in their vision of a dynamic universe in constant reorganization. Bruno himself acknowledges this connection by referring explicitly to Democritus and Epicurus: "There are no ends, bounds, limits, or walls to impede or arrest the infinite abundance of things. . . . For infinity eternally engenders a new abundance of matter. Democritus and Epicurus, who wanted everything to be renewed and rebuilt by the infinite, understood these things better than those who insist on salvaging the substance of the universe as eternal, on the grounds that new [particles] always succeed old ones in the same numbers and that parts of matter are always converted into like parts."[40]

Another parallel between Bruno and Leucippus and Democritus on the respective roles of senses and understanding in our perception of the world is revealed in the dialogues of Bruno:

Philotheo: No corporeal sense can perceive the infinite. None of our senses could be expected to furnish this conclusion; for the infinite cannot be the object of sense-perception; therefore he who demands to obtain this knowledge through the senses is like unto one who would desire to see with his own eyes both substance and essence. And he who would deny the existence of a thing merely because it cannot be apprehended by the senses, nor is visible, would presently be led to the denial of his own substance and being. Wherefore there must be some measure in the demand for evidence from our sense-perception, for this we can accept only in regard to sensible objects, and even there it is not above all suspicion unless it comes before the court aided by good judgment. It is the part of the intellect to judge, yielding due weight to factors absent and separate by distance of time and by space intervals. And in this matter our sense-perception does suffice us and does yield us adequate testimony, since it is unable to gainsay us; moreover it advertises and confesses our own feebleness and inadequacy by the impression it gives us of a finite horizon, an impression moreover which is ever changing. Since then we have experience that sense-perception deceives us concerning the surface of this globe on which we live, much more should we hold suspect the impression it gives us of a limit to the starry sphere.

Elpino: Of what uses then are the sense to us? Tell me that.

Philotheo: Solely to stimulate our reason, to accuse, to indicate, to testify in part; not to testify completely, still less to judge or to condemn. For our senses, however perfect, are never without some perturbation. Wherefore truth is in but very small degree derived from the sense as from a frail origin, and does by no means reside in the sense.

Elpino: Where then?

Philotheo: In the sensible object as in a mirror. In reason, by process of argument and discussion. In the intellect, either through origin or by conclusion. In the mind, in its proper and vital form.[41]

Thus, consciously or not, Bruno owed a great deal to the ancient atomists, and the debt was largely honored by the impetus he imparted to the vindication of their ideas, particularly regarding certain conceptual aspects of their cosmology. However, it does not appear that the atomic root of his theses about the infiniteness of the universe and the infinite multitude of worlds, commonly considered one of the main reasons for his condemnation and execution, was foremost in the mind of his judges. They had many other reasons—or pretexts, if one prefers—to condemn the distinguished Nolan thinker.[42] Indeed, unlike Galileo, Bruno expressed quite openly and repeatedly doubts about the validity of many tenets central to the Christian doctrine, oftentimes to the point of dismissing them. The list includes the thesis of transubstantiation, the concept of original sin, the global deluge, the dogma of the Trinity, the cult of the Virgin Mary, and the divinity of Jesus Christ.[43] In the process, he placed himself in direct conflict with the ecclesiastical authorities, who found his attitude all the more intolerable since he was a former Dominican priest who had left the order. Given the intellectual climate of the time, the tragic outcome of a conflict of this magnitude was inevitable.

This brings us to another fundamental difference between Galileo and Bruno: While Galileo was a genuine and rigorous scientist, though not entirely free of certain contemporary prejudices (who really is?), Bruno—in spite of his astounding, powerful, and prophetic intuitions, which qualify him as a pioneer of several developments of modern science—was essentially a visionary (in the eyes of some) or a seer (in that of others), whose mind failed to shed numerous archaisms and medieval occult beliefs. As articulated in an insightful summary by P. Thuillier: "Bruno entered the future walking backwards."[44]

THE PROTESTANTS

Newton

If I say electricity comes from God, I am both right and wrong. I am in search of a cause in the basic realm of the conditioned. If my answer is God, I have said too much. He is indeed the cause of everything, but I want to discover a causal connection, a specific link with this phenomenon; the answer "God" is too general.

—G. W. F. Hegel

"It seems probable to me that God in the beginning formed matter in solid, massy, hard, impenetrable, moveable particles, of such sizes and figures, and with such other properties, and in such proportion to space, as most conduced to the end for which he formed them; and that these primitive particles being solids, are incomparably harder than any porous bodies compounded of them; even so hard, as never to wear or break in pieces; no ordinary power being able to divide what God himself made one in the first creation. While the particles continue entire, they may compose bodies of one and the same nature and texture in all ages: but should they wear away, or break in pieces, the nature of things depending on them would be changed. Water and earth, composed of old worn particles and fragments of particles, would not be the same nature and texture now, with water and earth composed of entire particles in the beginning. And therefore, that nature may be lasting, the changes of corporeal things are to be placed only in the various separations and new associations and motions of these permanent particles; compound bodies being apt to break, not in the midst of solid particles, but where those particles are laid together, and only touch in a few points."

These often-quoted words of Isaac Newton (1642–1727, born the year of Galileo's death), are taken from Query 31, Book 3, of his work *Opticks*. They capture their author's position as Christian atomist.[45] The essential properties he attributed specifically to fundamental particles are the very same ones assigned to them by the ancient atomists. However, he added to them a new and inherent property called inertia—in the Newtonian sense of the term, of course, meaning "persistence of motion or rest."

Newton also subscribed to the second key aspect of the ancient theory of atoms, namely, the existence of void. He clearly admitted that interstellar spaces might be filled with a highly rarefied ether, which itself could also have a granular structure, with void between its very fine particles. It is interesting to note that while Newton came out in favor of void on the basis of physical arguments—in particular Democritus's argument that motion is impossible in a completely filled medium—his discussions with opponents of void, chief among them Descartes and Leibniz, also involved theological arguments revolving around their conception of the "perfection of the work of God" (see chapter 13). For Newton, "it is quite presumptuous to imagine that a being as small and insignificant as man could still confidently know what God could do better." In his view, to regard the existence of void as impossible was to limit God's power, to impose an obstacle to his freedom of action, to subject him to a demand. This argument takes on its full significance when contrasted to that of the antiatomists cited above.

Although Newton adopted the two central principles of the atomists of antiquity—the corpuscular structure of matter and the existence of void—he nevertheless rejected a purely mechanistic view of the world and, particularly, the determining role that Democritus and Epicurus attributed to chance in its formation and evolution. According to the Newtonian concept, God did not just create the universe, with its remarkable regularity and its intelligible order; he also continually governs its workings. Newton wrote in his most famous book, *Principia Mathematica*: "They are thus compelled to fall back into all the impieties of the most despicable of all sects, of those who are stupid enough to believe that everything happens by chance, and not through a supremely intelligent Providence; of these men who imagine that matter has always necessarily existed everywhere, that it is infinite and eternal."[46] Moreover, not only does the Creator ensure the workings of the world, He also presides, by means of a constant corrective intervention, over its propitious march. To imagine that God could have made and started this world as one builds and starts a machine such as a clock, and then retired into pure contemplation of his achievement, is for Newton an idea that "introduces materialism and fatality, and . . . tends to exclude from the world God's providence and governance." As we will see, this conception of God's continual presence and intervention will also be resisted by the antiatomists, who yet were just as Christian.[47]

However, what is perhaps Newton's most significant contribution to the atomic theory is embodied in his major achievement, which established once and for all the unification of the celestial and terrestrial worlds, of the macroscopic and the microscopic—the universal law of gravitation. It states that any two material bodies attract each other with a force proportional to the product of their masses and inversely proportional to the square of their distance. Indeed, it must be understood to apply to the entire scale of masses, from the infinitely small to the infinitely large.

We will focus here on the significance of this law in terms of the problem of interest to us. It is evident that the single area in which one could anticipate the most profound implications of the law concerns the interactions between elementary corpuscles, interactions that are responsible for their accretion into observable bodies. In modern language, we would say that they are responsible for the formation of chemical bonds and therefore for the structure of molecules. In this respect, several characteristics of the law, often ignored even by those who can recite its formula, must be underlined:

1. Newton did not consider gravity—even though it is ubiquitous in the universe—an immanent property, intrinsic to material bodies or particles, on the same level as, say, their extent, shape, or inertia. The ability of an object to exert by itself an action on another at a distance and across void, "without any emanations or exhalations, or without other corporeal media," but simply through its inherent properties, struck Newton as an unacceptable absurdity. This action could and should be attributed only to the effect of one or several extrinsic forces, the existence and action of which is owed solely to divine will and power. He admitted that he did not understand the nature of these forces, uttering his famous adage, "*Hypotheses non fingo*" (I do not imagine hypotheses). Newton, one of the greatest physicists the world ever produced, did not believe that the universal attraction he had just discovered is due to real, "physical" forces. Rather, he envisioned forces that are supernatural, transmaterial, spiritual, at best mathematical. In his view, "gravitation is God's spirit permeating matter."[48]

2. While the law of gravitation describes satisfactorily the behavior of the world at the macroscopic level, Newton was not convinced that it is equally applicable at the level of elementary corpuscles. He assumed that a law of the same type acts between these corpuscles, but he admitted that the forces involved might well not be the same. Besides the gravitational component, he explicitly envisioned electrical and magnetic components, as well as the involvement of repulsive forces. He stated (Query 31, Book 3 of *Opticks*): "Have not the small particles of bodies certain powers, virtues, or forces, by which they act at a distance . . . upon one another for producing a great part of the phe-

nomena of Nature? For it is well known that bodies act one upon another by the attractions of gravity, magnetism, and electricity; and these instances show the tenor and course of Nature, and make it not improbable but that there may be more attractive powers than these. For nature is very consonant and conformable to her self. . . . And thus Nature will be very conformable to her self and very simple, performing all the great motions of the heavenly bodies by the attraction of gravity which intercedes those bodies, and almost all the small ones of their particles by some other attractive and repelling powers which intercede the particles."

In light of modern theories on the nature of chemical bonds, these words by Newton reveal an astounding prescience of reality. Indeed, we know today that the role of gravity is quite negligible at the level of elementary particles, in terms of both their internal structure and their mutual interactions, as they turn out to be governed essentially by electromagnetic forces. Thus, the distinction predicted by Newton is to be entirely confirmed.

3. The introduction of an attractive force, particularly one of a transphysical nature, in the microcorpuscular domain tends to eliminate certain overly materialistic descriptions of the origin of interatomic interactions, of the type advocated by the ancient atomists and even by a few much more recent ones. In the context of these interactions, the role attributed by Democritus and Epicurus to hooks, points, barbs, and other irregularities on the surface of atoms is well known. These concepts were revived by Gassendi, and also, as we will see in the next section, by Robert Boyle, a contemporary of Newton and the principal champion of the theory of atoms in England. Newton rejected this relatively primitive view: "The parts of all homogeneal hard bodies which fully touch one another stick together very strongly. And for explaining how this may be, some have invented hooked atoms, which is begging the question; and others tell us that bodies are glued together by rest, that is, by an occult quality, or rather by nothing; and others, that they stick together by conspiring motions, that is, by relative rest amongst themselves. I had rather infer from their cohesion, that their particles attract one another by some force, which in immediate contact is extremely strong, at small distances performs the chemical operations above-mentioned, and reaches not far from the particles with any sensible effect."

As simplistic as the hypothesis of hooks and barbs may appear, excluding it fails to resolve the problem. Specifically, Newton's law (or a law of the same type, as he surmised) suggested a global and definitive solution to the problem of attraction between corpuscles (or their aggregates), and, as such, constituted a novel advance with far-reaching consequences. It did not, however, explain one of the essential characteristics of these interactions, namely, their selectivity or specificity, responsible for the formation of the great variety of different material bodies making up the world. At least one of the reasons for this drawback is easy to see: Even though Newton had the remarkable insight to identify quantity of matter with mass—a crucial step in the formulation of his law of attraction—he neglected to address the problem of the diverse volumes and shapes of particles and substances, which alone could account for the diversity and variety of their associations. As a result, the practical consequences of the law were limited, especially in chemistry.[49] In this respect, J. Merleau-Ponty voiced the opinion that "gravitation was as much a problem as a solution."[50] The problem was compounded by the fact that the issue of shape at the corpuscular level was beyond the reach of available experimental techniques. Even at the dawn of the nineteenth century, Laplace (1749–1827), an enthusiastic supporter of the Newtonian theory, would state: "Affinities would thus depend on the shape of the associating molecules and on their relative positions, and through the variety of these shapes, one could explain all the varieties of attractive forces, and thereby reduce to a single general law all the phenomena in physics and astronomy. But the inability to know the shapes of molecules and their mutual distances makes these explanations vague and useless to the progress of the sciences."[51] It is surprising to note how many thinkers and scientists, some even quite prominent, have untowardly, indeed erroneously, resorted to the word *impossible*. Oftentimes, as observed by the modern chemist C. Weizmann (1874–1952), "the impossible just takes longer," a statement that takes on its full relevance in the domain of atomic and molecular structures.

Boyle

Robert Boyle (1627–1691) was, together with Newton, one of the founders and leading protagonists of the theory of atoms in England. He learned the theory essentially from the works of Gassendi and, in all fairness, did not add much to it in the way of original contributions. However, he was an eminent chemist—perhaps the first important chemist among all the atomists mentioned up to this point—and his opinions were shaped by his experimental approach to science, and that does constitute an important innovation.

Boyle took a noteworthy although unsuccessful step toward the modern concept of element. In his book *Sceptical Chymist*, published in 1661, he steadfastly refused to apply the term *element* to the four primary substances of Empedocles (water, air, earth, and fire) or, for that matter, to the three principles of Paracelsus (mercury, sulfur, salt), arguing that none of these could be considered the fundamental constitutive elements of existing bodies.[52] At this point, one would naturally have expected him to disclose his own choice of elements, in the modern sense he himself gave to the word, namely, "primitive, simple, or perfectly homogeneous bodies, which, being constituted of no other, nor one of the other, are the ingredients of which all the bodies called mixed are made, and into which these can ultimately be decomposed." Surprisingly, he stopped at his negative verdict and refrained from proposing any concrete alternative; none of the substances he mentions in his books are described as an "element." Many of those he came in contact with in his experiments would have deserved that designation. Evidently he was unsure about it.

Although an earnest atomist, Boyle referred to his own conception of the structure of matter as a "corpuscular philosophy," which tells much about the nature of his own thinking. In his view, the fundamental particles created by God out of a primary matter, which have an innumerable multitude of different shapes, aggregate to form stable assemblages—"primary concretions"—with specific properties. They constitute the particles of chemical elements, and are simple material bodies whose association produces composite bodies. These "primary concretions" represent the building blocks effectively involved in the formation of chemical combinations. Although divisible in principle, they would not divide in practice during ordinary chemical and physical processes.

Boyle's corpuscular philosophy reflects his opposition to the Aristotelian doctrine, particularly to the concept of material forms. The sensible qualities of material bodies (their "secondary characteristics") are related in his mind, as in that of all atomists, to the effects of the "primary properties," inherent to their constituent particles: dimension, shape, position, and motion.[53] On the question of motion, however, he rejected his ancient predecessors' concepts of innate, disordered motion and random collisions. His Christian convictions prevented him from accepting that the world might be governed by chance. The movements of atoms could not be spontaneous and chaotic; they must, instead, be imposed and directed by God in order to achieve the will of providence, the plan that he conceived for the existence and harmonious development of the universe.

Because Boyle was an experimentalist interested in the behavior of substances, it was in the postulated *shapes* of atoms or "corpuscles" that he looked for an explanation of their chemical and physical properties. In his mind, the mutual affinities of atoms and physical bodies resulted from their spatial configurations, which enable them to adhere one to another. Thus he borrowed from the ancient theory of accretions due to the irregularities of corpuscular surfaces (hooks, cavities, bumps of various kinds, etc.) and rejected the Newtonian idea of a force of mutual attraction. He also resurrected Plato's old notion about the possible wearing out of the primary concretions. For instance, he attributed the sublimation of substances to the shedding of the outer protuberances of corpuscles—the appropriate interactions of which ensure the stability of the solid phase—through an increase of friction between them and a grinding effect promoted by heat, which would facilitate their separation. In liquids, the corpuscles are supposedly mobile and separated by pores.

By and large, Boyle's vision of the world is rather traditionally mechanistic, but somewhat tinged by the effect of a divine purpose. It also contains an intriguing social and political component. Indeed, some scholars consider this the most important aspect of Boyle's corpuscular philosophy.[54] In their view, Boyle felt that teaching and defending a knowledge grounded in an authentic philosophy fulfilled a specific mission, which was to provide experimental confirmation of a divine providence and, in the process, to prevent the spread of atheism (and other rebellious movements brewing in England at the time). We are a long way from the objective of Democritus and Epicurus, for whom a mechanistic vision of a world made of atoms was intended to liberate man from subservience to the gods. That said, their conviction that the progress of science should ensure the improvement of the human condition brings the English Christian atomists one step closer to their Greek pagan counterparts.[55]

Bentley

Richard Bentley (1661–1742), chaplain to the bishop of Worcester, and master of Trinity College in Cambridge, was at once a fervent atomist, accepting the reduction of the material world to atoms and void, and an enthusiastic supporter of Newton's theory (he exchanged with him a sustained correspondence), believing with him that purely material elements could not suffice to account for the ordered structure of the world, and that a teleological spiritual cause must be involved. He also participated in Robert Boyle's crusade against atheism, as evidenced by the eight sermons he delivered in 1692 at public lectures organized by Boyle.

Bentley's original motive for wanting to include a divine factor in the strictly mechanistic vision of the world of the atomists of antiquity is of particular interest to us. In essence, he was struck by the immensity of void. He realized that it is far more prevalent than matter throughout the universe. Through calculations that are as elaborate as they are fanciful (the details of which we will skip), he arrived at the conclusion that "every single particle would have a sphere of void space around it 8,575 hundred thousand million million times bigger than the dimensions of that particle."[56] While he was aware of the porosity of physical bodies, he, of course, had no concept of what today we call interatomic vacuum. Whatever the significance—or, rather, the lack thereof—of the number he proposed, Bentley was acutely aware of a very fundamental characteristic of the structure of the universe: that it is made mostly of void. He appears to be the first to have articulated this assertion so explicitly.

It is this very awareness of the immensity of void in comparison with the quantity of available matter that inspired in Bentley an absolute confidence that Democritus's purely mechanistic and statistical conception was unsatisfactory. The probability of interatomic contacts is so minuscule that it could never account for the formation and harmonious evolution of an organized world such as we perceive it. Certainly, as an admirer of Newton, he believed that atoms, even when dispersed in the infinitude of space, are not strictly independent of one another, since they are subject to universal gravitation. But, following Newton's precepts to the letter, he professed his conviction "that such a mutual gravitation or spontaneous attraction can neither be inherent and essential to matter, nor ever supervene to it, unless impressed and infused into it by a divine power," since any action at a distance "is repugnant to common sense and reason." 'Tis utterly inconceivable, that inanimate brute matter, without the mediation of some immaterial being, should operate upon and affect other matter without mutual contact; that distant bodies should act upon each other through a *vacuum* without the intervention of something else, by and through which the action may be conveyed from one to the other."[57] As Koyré summed it up: "Even if reciprocal attraction were essential to matter, or if it were simply a blind law of action of some immaterial agent, it would not suffice to explain the actual fabric of our world, or even the existence of any world whatever. Indeed, under the unhampered influence of mutual gravitation, would not all matter convene together into the middle of the world?"[58]

Under these conditions, the formation, existence, order, and harmonious course of our world would combine to provide "a new and invincible argument for the being of God, being a direct and positive proof that an immaterial living mind does inform and actuate the dead matter and support the frame of the world." Not only did he thus share Newton's belief in the spiritual nature of gravitation, but he expressed with particularly strong emphasis his faith in a teleological will, which alone, in his judgment, is capable of producing a complex but harmonious world out of the primitive chaos of an extraordinarily diluted matter, scattered through the immensity of an infinite void. His religious convictions even included a genuine "anthropic" flavor.[59] For example, he wrote: "For matter has no life nor perception, is not conscious of its own existence, not capable of happiness, nor gives the sacrifice of praise and worship to the Author of its being. It remains, therefore, that all bodies were formed for the sake of intelligent minds."[60]

JOHN LOCKE, OR AGNOSTIC ATOMISM

Never was there a wiser, more methodical mind,
a more exact logician, than Mr. Locke; however,
he was not a great mathematician.

—Voltaire

John Locke (1632–1704) is one of the founders of empiricism, advocating the primacy of observation and experimentation as the basis for the study and comprehension of nature. As such, he was opposed to the deductive rationalism of Descartes and Leibniz, who believed it possible to infer nature theoretically from the mere knowledge of its "cause" (which they understood to mean God). He is also the author of a "corpuscular theory" of the structure of matter, which, while it articulates with elegance the general ideas of an atomic conception of the world, provides few new elements. Instead, Locke's real originality lies in his professing a lack of confidence in the power of theory to elucidate the fundamental nature of things.

At the basis of Locke's own "corpuscular" variant of the atomic theory, unveiled in his work *Essays on Human Understanding*, published in 1690, are three components of a dualistic vision—or at least description—of the world. They resulted largely from his conviction that our knowledge of physical reality is essentially indirect. Refuting the notion, so dear to Descartes, of innate ideas, he stressed the determining role played by our senses in the way we perceive the world. At the root of our comprehension, Locke postulated two types of ideas: "simple ideas" and "complex ideas." Simple ideas are themselves of two categories: "ideas of sensation," resulting from the impressions produced by external objects on our senses (ideas of heat, color, odor, etc.) and "ideas of reflection," derived from mental processes occurring within ourselves (ideas of reasoning, doubt, etc.), which he ascribed to an "internal sense." "Complex ideas," such as the idea of substance, arise from a combination in our mind of a certain number of simple ideas (the term *combination* implying here various types of mental operations). Likewise, the properties of physical bodies are divided into primary and secondary qualities. The former, in which Locke included shape, size, and motion, constitute the "real," inherent, inseparable characteristics of material bodies; the latter, examples of which are color, odor, heat, and so on, depend entirely on the capabilities and refinement of our sensory perception. This duality of properties, which, in truth, was not invented by Locke but actually goes back to the teachings of the early atomists, was even reinforced by a description of things in terms of their "nominal essence" and their "real essence," and by the interdependence of the two.

It is this dependence that is the linchpin of Locke's "corpuscular theory." Actually, on this subject Locke merely adopted the views of Boyle and Gassendi, accepting as basis of the structure of matter particles endowed with "extent and impenetrability," which are their essential properties, as well as "dimension, shape, motion, or rest," which constitute their "inseparable accidents." The secondary qualities result from the effect of these structural units on our sensory organs, an effect that exerts itself essentially via various types of "impulsions."

Locke's views are original in at least two respects. He placed a stronger emphasis than in any prior theory on the link between our perception of the world and our senses, and, more important, his own reasoning left him deeply skeptical about any prospect of perceiving "true reality."[61] Indeed, even though he endorsed the principle of a corpuscular structure of matter, he regarded it as no more than a hypothesis and believed that, in any event, we will never be in a position to know how the particles making up this structure produce the properties exhibited by matter, of which each of us can only form his own picture. Thus Locke expressed a profound doubt about our ability to ever demonstrate experimentally the existence of these elementary corpuscles. This pessimism induced him to assert that even with a microscope able to reveal their structure and motion, our understanding of the workings of nature would hardly progress, and that such a knowledge might even be harmful to our daily lives—unwise words, as those of the pessimists in sciences often are.[62] Admittedly, Locke could not have foreseen the advent of the tunnel microscope, nor of nanotechnology, but, nevertheless, his judgment appears singularly timorous (see chapter 20). Evidently the scientists of that period failed to understand that the technological development of new tools of research would open up amazing new horizons and accelerate the progress of science. Locke's skepticism, even outright pessimism, earned his doctrine the label "agnostic atomism," coined by Jean Deprun.[63]

Finally, as did other Christian philosophers, Locke rejected the materialistic notion of a dull game of chance as the motive drive in the evolution of the world, particularly as it applies to the human mind. It appeared to him impossible that "things entirely devoid of knowledge, acting blindly, could produce a being endowed with awareness." These doubts gave him something in common with the advocates of the theory of "animate atoms," of whom we will speak next.

MAUPERTUIS AND DIDEROT, OR ANIMATE, SENSITIVE, AND INTELLIGENT ATOMS

If two men were to like the same things, would they
also necessarily have the same aversions?

—Paul Valéry

From the perspective of the Greek founders of atomism, nothing distinguished atoms of living matter or souls from those of inert matter (other than they might simply be lighter, rounder, or smoother). Such a conception was unlikely to satisfy everyone, as indeed we have already established on several occasions. In point of fact, this perceived weakness of the theory was frequently used as a reason for rejecting it, as did the Christian theologians and philosophers of the Middle Ages. That was also what inspired the Mutakallimun, Arab theologian-philosophers from the ninth to the eleventh century, to try to resolve the dilemma with a very original proposition of their own.

In the eighteenth century, the problem took on a renewed urgency, prompted by the ongoing debate about the validity not of the theory of atoms itself, but of a homogeneous and strictly materialistic brand of atomism, particularly as it might be adapted to a description of the so-called organic world. This term had, at the time, a far broader meaning than it conveys today, encompassing the entire biological world.[64] Cassirer would speak of a search for the foundations of a "new philosophy of the organic."[65]

The duality of organic matter and mineral matter constitutes the backdrop against which Maupertuis and Diderot tried to devise a description of the atomic world that could account at once for the limited requirements of the inanimate world, the more elaborate needs of the animate world, and even the subtle essence of man's spiritual endeavors.

Pierre Louis Moreau de Maupertuis (1698–1759) adhered to the corpuscular theory of the structure of matter. He wrote in his *Réponse aux objections de M. Diderot*, "No material substance is continuous."[66] However, he repudiated the purely materialistic and mechanistic character of the ancient theory of atoms, as well as more recent mechanistic systems. Such theories appeared to him incapable of explaining the formation of organized physical structures, and specifically of living matter. He wrote in his *Système de la nature*: "Should we speak here of this absurd system, if a system it is, imagined by an impious philosopher, adorned by a great poet with all the richness of his art, a system that the libertines of our time would like to revive? This system admits as principles of the universe only eternal atoms, devoid of feelings and intelligence, the fortuitous encounters of which formed all things: An accidental organization generates the soul, which is destroyed as soon as the organization ceases. . . . Never will the formation of any organized body be explained solely by the physical properties of matter; to be convinced of this, one need only read the writings of all the philosophers who have tried this explanation, from Epicurus to Descartes."[67]

Maupertuis included gravitation among these physical properties.[68] Although he was instrumental in introducing Newton's theories in France, he believed that the law of universal attraction, verified in physics and astronomy, was inadequate to interpret the phenomena of chemistry (such as the processes of association and dissociation, which are the basis of combinations, transformations, and the increasing complexity of physical bodies) and *a fortiori* the phenomena of biology (such as the appearance, development, and behavior of organized animal structures): "A uniform and blind attraction, spread in all parts of matter, could in no way explain how these parts arrange themselves to form a body with even the simplest organization. If they are all subject to the same tendency, why would some form an eye and others form an ear? Why this marvelous arrangement? And why do they not unite haphazardly?"[69]

This inability of purely material physical forces to account for the elementary manifestations of life appeared even more obvious to Maupertuis when it came to the emergence of the conscious from the unconscious. Such a transformation was utterly impossible for him.

Under these conditions, the only solution was "to transfer conscience into the atoms themselves, as a genuine primitive phenomenon. Not for the atoms to engender conscience, but for them to nurture it and raise it to ever higher levels of clarity,"[70] or, in Maupertuis's own words, "to have recourse [in each of the smallest parts of matter] to some principle of intelligence, to something akin to what we call desire, aversion, memory."[71]

Needless to say, this conception runs diametrically counter to the ideas of Descartes, who clearly separated body (whose attribute is extent) from soul (whose attribute is conscience and thought). In Maupertuis's perspective, "If thought and extent are mere properties, they can very well belong to the same subject whose intrinsic essence remains unknown to us."[72] This conviction clarifies the reason for his reliance on "conscious" atoms.

In keeping with his faith, Maupertuis had no doubt that God imposed "intelligence" on the fundamental elements themselves. He expressed this certitude by writing, "In creating the world, God endowed each part of matter with this property, through which he wanted the beings that he had formed to reproduce. And since intelligence is necessary for the formation of organized bodies, it seems more noble and worthy of the Almighty that they should form through the properties with which he had once infused the elements, rather than for these bodies to be always the instantaneous creation of his power." He explicitly contrasted this conception of the universe, "where the elements, themselves endowed with intelligence, arrange themselves and unite to fulfill the designs of the Creator," with the view, which he rejected, that "the supreme Being, or some beings subordinate to him, distinct from matter, would have used elements the way an architect uses stones to build an edifice." Once "intelligence" is affixed at the elementary level, it participates in the general evolution of the universe: "The elements appropriate to each body, existing in sufficient quantities and at distances compatible with the exertion of their action, will naturally unite."[73] "It seems that, out of all the perceptions of the assembled atoms, a unique and much stronger perception emerges, far more perfect than any of the elementary perceptions."[74]

Moreover, consistent with his overall cosmological vision, Maupertuis had a "modern" notion of what this evolution must have been: a long succession of failures and successes. Only successes were destined to survive so as to produce this universe, which, all things considered, appeared to him marvelous, yet without necessarily being "the best of all possible worlds."[75]

A similar conception of the animate character of elementary particles was also advocated by Denis Diderot (1713–1784) in several of his works: *Pensées philosophiques* (Philosophical thoughts) (1746), *Lettre sur les aveugles* (Letter on the blind) (1749), *De l'interprétation de la nature* (On the interpretation of nature) (1753), and *Entretiens entre d'Alembert et Diderot, Le Rêve de d'Alembert* (Conversations between d'Alembert and Diderot: D'Alembert's dream) (written in 1769, but not published until 1830). The parallel with Maupertuis may seem arguable, inasmuch as Diderot, a deist in his philosophical youth, declared himself openly and resolutely anti-Christian and materialist in his mature years. The two philosophers actually differed on a number of points, particularly on the issue of "initial and final causes." Nevertheless, their general agreement on the nature of atoms, as well as their chronological overlap, supports the view that they belong in the same school of thought. Emboldened by Maupertuis himself, who expressed the opinion that "the main result of any system is a miracle" (in the sense of being arbitrary), we feel a measure of justification in minimizing their differences in religious opinion and lumping them together on the basis of their similar conceptions of numerous other aspects of the atomic universe.

Like Maupertuis, Diderot was convinced that no combination of inert particles could ever produce an organized being, and that it would be impossible to create anything living from dead matter. To him, the decisive proof was the structural, functional, and evolutionary complexity he saw in living matter; for instance in how an egg changes from embryo to a living, organized, and sensitive being. "To suppose that by placing next to a dead particle one, two, or three other dead particles, one can form the system of a living body amounts, it seems to me, to a flagrant absurdity, or I am grievously mistaken. . . . Are you to claim, together with Descartes, that [an animal] is merely an imitative machine? If so, the little children will laugh at you . . . and you will feel sorry for yourself; you will realize that by not accepting sensitivity as a general property of matter, a simple assumption explaining everything, you are renouncing common sense and are hurtling toward an abyss of mysteries, contradictions, and absurdity."[76]

To account for the existence of life, it is then necessary to instill life into matter from the outset, and to do so at the level of the fundamental constituents, which Diderot referred to as "living points" in *Le Rêve de d'Alembert* and, more prosaically, as "organic molecules" in his *Pensées philosophiques.*[77]

Still, there are different ways to envision the emergence and development of life into its present complexity, including its most elaborate quality, which, as we have seen, Diderot refers to as "sensitivity." The term essentially describes the ability to transmit peripheral stimuli toward the central organ, the brain, and, conversely, to transmit the commands of the brain toward the periphery, and to harmonize the entire process.[78] One could conceive of a heterogeneous atomic universe composed, from its beginning, of animate elements scattered across a mass of inert elements. Are these conditions sufficient for the encounters of such particles to generate our complex world, including life? This possibility seemed acceptable to Diderot at one time, but it did not satisfy him for long.[79] He came to see the unity of matter as an absolute necessity, regardless of the diversity of phenomena it exhibits. This demanded that sensitivity be a universal and eternal property of all matter. "Sensitivity and life are eternal," he was to write in a letter to Sophie Volland (October 15, 1759). On the other hand, it is possible to consider that sensitivity is latent in "dead" matter and reveals itself only in "living" matter. The difference between these two stages would have to be determined by the degree of manifestation of a universal sensitivity. The transition from "inert" sensitivity to "active" sensitivity, a routine event, occurs easily—for instance, through assimilation of inert bodies into a living, animal substance. Who could fail to recall the charming dialogue at the beginning of the *Entretiens entre d'Alembert et Diderot?*

D'Alembert: If [sensitivity] is a general and essential quality of matter, then it is necessary for a rock to be able to feel.

Diderot: Why not?

D'Alembert: That is difficult to believe.

Diderot: Indeed, it is for those who cut it, chisel it, crush it, and fail to hear it scream. . . .

D'Alembert: I find it difficult to see how a body can be made to pass from a state of inert sensitivity to one of active sensitivity.

Diderot: That is because you do not want to see it. It is quite an ordinary phenomenon.

D'Alembert: And this common phenomenon, what is it, if you please?

Diderot: I shall explain it to you, since you want to be put to shame. It occurs every time you eat.

D'Alembert: Every time I eat!

Diderot: Yes; for what do you do when you eat? You remove the obstacles that impeded the active sensitivity of the food. You assimilate it into yourself; you turn it into flesh; you animalize it; you render it sensitive.

This transition from a state of inert sensitivity to one of living sensitivity is but one, admittedly crucial, component a great "perpetual flux" presiding over the organization and increasingly elaborate evolution of matter.[80] This flux is determined exclusively by the outcome of chance and probability, which is sufficient, according to Diderot, to generate, through a kind of natural selection, the world as it is, was, and will be at different stages of its evolution: "Kindly be disposed to agree with me that matter has existed in all eternity and that motion is essential to it. I, in return, shall suppose with you that the world has no bounds, that the multitude of atoms is infinite, and that nowhere is this astonishing order belied. From this mutual agreement, it can only follow that the possibility of fortuitously engendering the universe is exceedingly small, but that the number of streams of particles is infinite, that is to say, that the unlikelihood of such an event is more than offset by the multitude of streams. Hence, matter having been in motion in all eternity, and being presented with the choice of perhaps an infinite number of admirable arrangements commensurate with the infinite number of possible combinations, if there is anything repugnant to reason, it is the supposition that none of these admirable arrangements should have occurred among the infinite multitude of those it successively assumed."[81]

Life is undoubtedly the most marvelous manifestation of the creative effect of this accretion of elementary corpuscles: "Life is a series of actions and reactions. . . . While living, I act and react as a whole. . . . When dead, I act and react as a collection of molecules." His concept of selection was, incidentally, quite similar to Maupertuis's: "I can assure you . . . that all injurious combinations of matter have disappeared, and that the only ones to have remained are those in which the mechanism entailed no important contradiction, those which could subsist by themselves and endure."[82] This evolution unfolds on a cosmic level: "How many crippled and failed worlds have dissipated in the past, re-form and dissipate again even now, perhaps at every instant, in distant places that we can neither touch nor see, but where motion continues and will continue to combine masses of matter, until they produce some arrangement compatible with their survival." The diversity of possible arrangements is all the greater since, for Diderot, no two atoms in nature are "rigorously alike!"[83]

Diderot's atomic world is thus essentially dynamic or, more precisely, self-dynamic.[84] The order we perceive in nature results from the persistence of successful associations, whose cumulative effect continues to create the appearance of deliberate choice. Diderot, however, repudiated any notion of "final cause." And since nothing in the processes of nature or in the continual gamble of atomic (or molecular) associations and dissociations is ever final, the image of a future world remains concealed from us.[85]

Diderot's belief in animate atoms (or molecules) is further articulated in a marvelous and moving wakeful dream: "The remainder of the evening was spent teasing me about my paradox. . . . I was offered fine pears that lived, grapes that thought; and I said: Those who loved each other during their life and have themselves interred side by side are perhaps not as foolish as one might think. Perhaps their ashes come into contact, mingle and unite! Who am I to know? Perhaps they have not lost all feeling, all memory of their past state; perhaps they retain a remnant of warmth and life, which they enjoy in their own way at the bottom of the cold urn that holds them. We tend to judge the life of elements in the image of the life of clumsy masses; perhaps they are very different things. . . . O dear Sophie, I thus cling to the hope that I may touch you, feel you, love you, seek you, unite with you, and meld into you when we no longer are, if there were a law of affinity in our principles, if it were our prerogative to form a single being, if it were my fate to be one with you through the course of the centuries, if the molecules of your erstwhile lover were destined to become inspired, aroused, and to seek yours scattered in nature! Allow me this reverie, so sweet to me; it would assure me eternity in you and with you."[86]

After animate, sensitive, and intelligent atoms, here now are atoms in love. And why not, indeed?[87]

HOLBACH, OR RIGGED ATOMS, AND
DE LA METTRIE, OR MATERIALISTIC ATOMS

Paul Henri Dietrich (or Thiry), baron of Holbach (1723–1789), was a profoundly anticlerical materialist philosopher. Inspired largely by the atomists of antiquity, he embellished his own worldview with a touch of whimsical imagination that enlivened the staid solemnity of the ancient doctrine. In keeping with his distant predecessors, he considered matter eternal, uncreated, and destined to exist forever, "because it contains within itself the sufficient reason for its existence." It is composed of the traditional complement of four elements or primordial substances—fire, earth, air, and water—but its constitution implies the existence of fundamental entities, which he indiscriminately referred to as "molecules" or "atoms," without ever defining these two terms. The universe results from interactions between these particles. He stated: "Indestructible elements, Epicurus's atoms, the movements, interactions and combinations of which produced all that exists, are undoubtedly more believable causes than the God claimed by theology."

In his search for what drives this creative mechanical process, he made a clear distinction between chance and necessity. He rejected the first— "Nothing happens randomly"—and accepted the second: "There exists something that is necessary: it is nature, or the universe, and nature acts necessarily the way it does. . . . Everything it produces is made according to laws that are certain, uniform, and permanent." In the tradition of ancient atomism, necessity is not to be confused with grand purpose, since "nature is devoid of design and intelligence."

As a rational evolutionist, he joined Maupertuis and Diderot in trying to track the continuous path—bumpy as it may be—followed by matter from its most primitive form all the way to man. Unlike his colleagues, however, he refused to attribute a sensitivity to individual atoms. He believed that this quality could appear only in specific combinations of atoms: "All of nature persists and is preserved only through the circulation, transmigration, exchange, and perpetual displacement of insensitive molecules and atoms or of the sensitive parts of matter. . . . That is what causes plants, animals, and men to come into existence: organized, sensitive and thinking beings, as well as beings devoid of feelings or thought." He added even more explicitly: "Inanimate matter can come to life, which itself is nothing but a collection of movements. . . . The difference between man and beast comes only from the way they are organized . . . [because] far from being distinct from the body, the soul is but the body itself considered in relation to some of its functions. . . . Man is a necessary outcome of the evolution of our globe."

It is on the issue of "evolution" that Holbach advanced a rather unusual proposition. It deals with the source of what we have come to call the *specificity* of interactions between atoms or molecules, a trait that is widely recognized as essential to the generation of organized structures. Holbach attributed this specificity to a property few of us would be likely to think of: "Would we be amazed if, out of a dice box containing one hundred thousand dice, we were to draw one hundred thousand sixes in a single throw? We most certainly would; unless the dice were loaded, of course! Well, molecules of matter can be compared to loaded dice that invariably produce the same predetermined effects: Since these molecules are fundamentally different individually and in combinations, they are rigged in an infinite number of ways. . . . What is man made of, in the end, if not loaded dice or mechanisms that nature has predestined to produce results of a particular type?" (The idea of dice will reappear much later [see chapter 20] in the musings of a great twentieth-century scientist [Einstein] about whether or not God is disposed to play dice.)

In addition to his scientific views, Holbach is also noted for having fought ferociously against the "theological God" and the clergy, accusing them of causing much of the suffering inflicted on humanity throughout history. He argued that "the peaceful Epicurus never disturbed Greece, and Lucretius's poem never incited civil strife in Rome." This explains, if such a word is appropriate, why Holbach's book *Système de la nature* (The System of nature), in which he expounded his philosophy, was condemned by Parliamentary decree to be burned the very same year (1770) it was published.[88] The charge was "lese majesty against mankind and God." The indictment spelled out explicitly the Epicurean and Lucretian inspiration of Holbach's theories.

To some degree, the conceptions of Julien Offray de La Mettrie (1709–1751) mirror those of the thinkers just mentioned. He was a moderately enthusiastic atomist, but a dedicated believer in an eternal and indestructible matter, endowed by nature with movement, which gives it an innate ability to organize itself.[89] Also an evolutionist, he was interested in the thread of the ever-increasing complexity of the world. On this topic, he wrote: "How many infinite combinations did matter have to go through before it arrived at the only one capable of creating a perfect animal?" He expressly considered that this originative dynamic process can lead to "animate bodies that possess all that is required to move, feel, think, and behave . . . in the physical world as well as in the moral world, which depends on the physical." He observed: "We must accept . . . that we do not know whether matter has an inherent ability to feel, or only the power to acquire that ability through transformations or the various forms it can adopt."[90] Opposed to the concept of final cause, he explicitly argued that organs create their own function, and not the other way around: "Nature did not set out to create the eye in order to see, any more than it created water to serve as a mirror for the humble shepherdess."[91] He rejected both chance and the intervention of an external will to explain the organization of the world. Instead, he simply saw the world as the outcome of natural laws yet to be discovered: "To destroy chance does not prove the existence of a supreme being, since there can be something else that is neither chance nor God: I am talking about Nature. . . . Causes hidden in its midst might have produced everything."[92]

MAXWELL

James Clerk Maxwell (1831–1879) lived a full two centuries after Newton and Boyle, at a time when the foundations of the scientific theory of the atom were beginning to be formulated. He was an outstanding scholar who benefited science with extremely important, if not crucial, contributions in the fields of electromagnetism (Maxwell's equations) and the kinetic theory of gases. Thus it might seem odd to speak of him in this particular chapter. There is, however, a justification. Though he belongs in a period when the controversy about the atomic theory involved for the most part scientific and philosophical arguments only, he did inject into the debate a religious dimension via a curious argument: He believed he could demonstrate that atoms owe their existence to the action of a Creator. For that reason alone, we deem it appropriate to include Maxwell at this stage of our discussion of Christian atomism in England, all the more so since a very similar argument, although in a less elaborate form, would be advanced in the eighteenth century by the famous astronomer Sir William Herschel (1738–1822).[93]

Maxwell openly came out in favor of the theory of atoms as it was formulated by the ancients. During a lecture delivered in 1873 before the British Association, he declared: "According to Democritus and the atomic school, . . . after a certain number of subdivisions, the drop [of water] would be divided into a number of parts each of which is incapable of further subdivision. We should thus, in imagination, arrive at the atom, which, as its name literally signifies, cannot be cut in two. This is the atomic doctrine of Democritus, Epicurus, and Lucretius, and, I may add, of your lecturer."[94]

This places Maxwell unambiguously in the camp of the atomists. In order to follow the remainder of this discussion, however, it is useful to bear in mind that at the time of Maxwell's lecture, the difference between atoms and molecules was not yet understood as clearly and universally as it is today. Maxwell defined his own position in the following statement: "A drop of water . . . may be divided into a certain number, and no more, of portions similar to each other. Each of these the modern chemist calls a molecule of water. But it is by no means an atom, for it contains two different substances, oxygen and hydrogen, and by a certain process the molecule may be actually divided into two parts, one consisting of oxygen and the other of hydrogen. According to the received doctrine, in each molecule of water there are two molecules of hydrogen and one of oxygen. Whether these are or are not ultimate atoms I shall not attempt to decide. . . . Every substance, simple or compound, has its own molecule. If this molecule be divided, its parts are molecules of a different substance or substances from that of which the whole is a molecule. An atom, if there is such a thing, must be a molecule of an elementary substance. Since, therefore, every molecule is not an atom, but every atom is a molecule, I shall use the word molecule as the more general term."

Maxwell's own definition of "molecule" must be kept in mind in what follows. He proceeded to make a number of eminently pertinent observations, which even today we can only subscribe to: "The molecules are conformed to a constant type with a precision which is not to be found in the sensible properties of the bodies which they constitute. In the first place, the mass of each individual molecule, and all its other properties, are absolutely unalterable. In the second place, the properties of all molecules of the same kind are absolutely identical." He stressed that this identicalness is observed whatever the source of the molecular specimens used. "The history of these specimens has been very different, and if, during thousands of years, difference of circumstances could produce difference of properties, these specimens of oxygen would show it." He broadened the significance of this observation by extending it beyond the confines of our own world, giving it a universal impact: "The molecule, though indestructible, is not a hard rigid body, but is capable of internal movements, and when these are excited, it emits rays, the wavelength of which is a measure of the time of vibration of the molecule. By means of the spectroscope, the wave-lengths of different kinds of light may be compared to within one ten-thousandth part. In this way it has been ascertained, not only that molecules taken from every specimen of hydrogen in our laboratories have the same set of periods of vibration, but that light, having the same set of periods of vibration, is emitted from the sun and from the fixed stars. We are thus assured that molecules of the same nature as those of our hydrogen exist in those distant regions, or at least did exist when the light by which we see them was emitted."

The next point is the crucial argument concerning the reason for the invariance of molecular properties and, beyond that, the origin of molecules themselves: "No theory of evolution can be formed to account for the similarity of molecules, for evolution necessarily implies continuous change, and the molecule is incapable of growth or decay, of generation or destruction. None of the processes of Nature, since the time when Nature began, have produced the slightest difference in the properties of any molecule. We are therefore unable to ascribe either the existence of the molecules or the identity of their properties to the operation of any of the causes which we call natural. On the other hand, the exact equality of each molecule to all others of the same kind gives it, as Sir John Herschel has well said, the essential character of a manufactured article, and precludes the idea of its being eternal and self-existent.

"Thus we have been led, along a strictly scientific path, very near to the point at which Science must stop. . . . But in tracing back the history of matter, Science is arrested when she assures herself, on the one hand, that the molecule has been made, and on the other, that it has not been made by any of the processes we call natural. Science is incompetent to reason upon the creation of matter itself out of nothing. We have reached the utmost limit of our thinking faculties when we have admitted that, because matter cannot be eternal and self-existent, it must have been created."

He concluded: "But though in the course of ages catastrophes have occurred and may yet occur in the heavens, though ancient systems may be dissolved and new systems evolved out of their ruins, the molecules out of which these systems are built—the foundation stones of the material universe—remain unbroken and unworn. They continue to this day as they were created—perfect in number and measure and weight—and from the ineffaceable character impressed on them we may learn that those aspirations after accuracy in measurements, truth in statement, and justice in action, which we reckon among our noblest attributes as men, are ours because they are essential constituents of the image of Him who in the beginning created, not only the heaven and the earth, but the materials of which heaven and earth consist."

In summary, it is in essence the absolute identicalness of all "molecules" of like chemical species, wherever they may be found in space and in measurable times, that led Maxwell to refute the idea of their eternal existence or of their natural generation. In his view, this identicalness is the guarantee of a "manufactured" origin. They could only have been fashioned by God's hand. This is a crucial difference from the doctrine of his predecessors from antiquity.

From a purely scientific point of view, which was presumably Maxwell's intent, we might note that, given our present ability to reproduce almost at will any existing molecule, and even to create new ones, all copies of which will necessarily be identical, a cosmic mechanical machinery appears in fact to be a credible alternative to a supreme artisan.

13

The Christian Antiatomists

Whereas the term "Christian atomism" causes no confusion in anyone's mind, the same cannot be said of the label "Christian antiatomism." We are to understand this designation in a more restrictive sense than simply signifying a Christian who rejects the atomic theory. Instead, we will apply it specifically to philosophers or scientists who proclaim their Christian faith and *use arguments of a religious nature* (among others) to refute the theory. At the dawn of modern times, the most noteworthy in this class of thinkers are Descartes and Leibniz.

DESCARTES (1596–1650)

Descartes, born to uncover the errors of antiquity, and to substitute his own.

—Voltaire

If I err, I exist.

—St. Augustine

You might well be one of the most renowned philosophers, perhaps even be revered by some as the founder of modern philosophy. You might also be a distinguished mathematician intent on constructing an entire physics in accordance with the strict criteria of language and mathematical methods. You may have a very clear conception of uniform matter (it was only beginning to take hold after the long reign of Aristotle's dualistic theory, which made a distinction between supralunar and sublunar worlds). You may give of the material world, including the vegetal and animal worlds, an essentially mechanistic description. In short, you may be boldly marching on the global path to rationalism and modernism, and yet all the while agree with the views of Aristotle and the medieval scholastics on certain fundamental scientific problems, even though that agreement might be based on entirely different premises. Nor should that stop you from opposing their beliefs, either! Such was the wholly paradoxical situation Descartes found himself in with respect to the two principal and complementary propositions of the theory of atoms: the existence of void and the finite divisibility of matter. We will confine our attention to this particular aspect of Descartes's vision of the world, which is but a small part of the prodigious scope of his work.[1]

The fundamental premises on which Descartes built his entire natural philosophy led him to reject the atomic theory and, foremost, to follow Aristotle's footsteps in repudiating the concept of void.

The centerpiece of his intellectual construct is the following excerpt from Article 7, Part 1, of his *Principes de la philosophie*: "While we reject everything that we can doubt, or that we even suspect of being false, we can easily assume that there is no God, nor heavens, nor earth, and that we have no body; but we could never likewise assume that we do not exist, while we doubt the truth of all these things: for we so abhor the notion that he who thinks does not truly exist, even as he thinks, that, all the most extravagant suppositions notwithstanding, we cannot help but believe that the conclusion 'I think, therefore I am' has to be true, and thus must be the first and most certain conclusion of anyone who conducts his thoughts logically."[2] The next step in his reasoning is summarized in a few lines taken from the "Letter from the Author" to a translator of his book: "I adopted the being or the existence of thought as first principle, from which I concluded quite logically the following: namely, that there is a God, who is the author of all that exists in the world, and who, being the source of all truth, did not create our intellect in such a way that it could err in its judgment of the things it perceives so very clearly and distinctly. Those are the only principles to which I resort concerning immaterial or metaphysical things, from which I infer quite logically those of corporeal or physical things, namely, that there are material bodies with extent in length, width, and depth, that have various shapes and move in various ways. Those are, in short, all the principles from which I infer the truth about other things."

The last sentences strike at the very core of the debate of relevance to us. They are amplified in the next affirmation: "But, although each attribute is sufficient to define a substance, there is, however, in each substance one [attribute] that constitutes its nature and essence, on which all the others depend. Specifically, the extent in length, width, and depth constitutes the nature of a corporeal substance, while thought constitutes the nature of the thinking substance. For everything else that can be attributed to a material body presupposes extent, and merely depends on that which has extent. Likewise, all the properties found in the entity that thinks are merely different ways of thinking."

Two fundamental propositions stand out in these citations. First, the crucial distinction made by Descartes between matter, on the one hand, and thought or soul (the two words being synonymous for him), on the other: "The soul is of a substance entirely distinct from the body."[3] Spiritual entities have nothing in common with material objects. This duality precludes the application of the laws of extent to the sphere of the spirit. It also reserves for man, the repository of thought, a special and privileged place in the universe. The difference with the unitary vision of the world espoused by the ancient atomists is profound. Second, in keeping with the language of the philosophers of ancient Greece, Descartes constructed the material world on two principles: extent (with shape), and movement. All other sensible qualities, such as warmth, cold, dryness, humidity, which Aristotle considered primary, could not qualify as such principles, since they do not represent the intrinsic properties of things, of which we have an innate knowledge, but depend both on the nature of objects and of the perception we have of them. Extent and motion thus become the two pillars on which the entire architecture of the universe rests. All physical phenomena must be reducible to the combined effects of these two fundamental concepts. Moreover, matter is identified with space or extent: Matter is space, and space is matter. In this respect, Descartes broke with the traditional and intuitive idea of space as a receptacle, necessarily predating the formation and existence of the material world. In his view, space cannot exist separately from matter. Instead, the existence of material objects implies that of space.

In this context, rejecting void is merely a consequence of identifying matter with space. In the words of Descartes himself: "On the question of void, in the sense that philosophers attach to this word, namely, space entirely depleted of substance, it is evident that there is no such place in the universe, because the extent of space or of any location in it is no different from the extent of a material body. And since, by virtue that a material body has extent in length, width, and depth, we are justified in concluding that it is substance, because we consider it impossible for what is nothing to have extent, we must reach the same conclusion about space assumed to be empty: namely, that, since it has extent, it must necessarily encompass substance."[4]

Evidently, Descartes rejected void not just as a physical impossibility, but as a more fundamental absurdity amounting to "existing nothingness," a self-contradiction. Descartes argued the senselessness of such a concept with the following proof, which will later prompt objections from a prominent rebutter (see chapter 19): "We have almost all been preoccupied by this error . . . to believe that, since there is no necessary link between the vase and the object it contains, it appears to us that God could remove the entire physical object from the vase and still conserve the vase in its initial state, without it being necessary for another object to replace the one he had just removed. But, in order to now correct such a faulty opinion, we will note that there indeed exists no necessary link between the vase and the object that fills it, but that such a link is absolutely required between the concave shape of the vase and the extent implied within this concavity, that it is no more repugnant to conceive of a mountain without valley than to conceive of such a concavity without its extent, or this extent without something with extension, because nothingness, as we have already remarked several times, cannot have any extent. That is why, if someone were to ask us what would happen should God remove the entire object that is in the vase, without replacing it with another, our reply would be that the sides of the vase would be in such close proximity that they would immediately touch. For two objects must touch when nothing stands in between, because it would be a contradiction to claim that these two objects be apart, that is, that there be a distance between one and the other, and yet that this distance be nothing: for distance is a property of extent, which could not subsist without something with extent."

The second component of Descartes's antiatomism, also the result of his identifying matter with extent, was his denial that atoms, those "small indivisible bodies," can exist. This denial is a logical corollary of his rejection of void, but Descartes added to it a religious argument related to the highly controversial problem of God's omnipotence: "It is also quite easy to realize that there can be no atoms, or parts of an object that are indivisible, contrary to what some philosophers have imagined. No matter how small these parts are assumed to be, because they must nonetheless have some extent, we hold that not a single one among them cannot be further divided into two or more smaller parts, from which it follows that it is divisible. From our clear and distinct insight that an object can be divided, we must conclude that it is divisible, because any other conclusion would render our judgment of this object contrary to the insight we have of it. And even if we were to assume that God had reduced some part of matter to such extreme smallness that it could not be further divided into smaller parts, we could still not conclude that it is indivisible because, had God made this part so tiny as to be beyond the power of any creature to divide it, he could not have deprived himself of his own power to do so, since it is not possible that he limit his omnipotence, as we have already remarked. That is why we hold that the smallest extended part that can be found in the world can always be divided, because such is its nature."[5]

Descartes's antiatomistic stance, which assumes a completely filled space, is thus clearly enunciated. It then may come as a complete surprise to realize that he proceeded nevertheless, without so much as batting an eye, to develop a corpuscular theory of the structure of matter! According to him, it would be composed of small parts, infinitely divisible in theory, but actually of finite dimension, the only important characteristics of which are, of course, shape, position, and motion. Conscious of the risks that his theory might be likened to ancient atomism, Descartes was anxious to launch a preemptive strike and nip any such attempt in the bud. He did so in one of the few passages of his *Principes de la philosophie* in which he articulated explicitly his objections to Democritus's teachings: "Some will perhaps say that I consider various parts in every object that are so small that they cannot be perceived; and I am aware that it will be frowned upon by those who use their senses as the measure of things that can be known. But it would, it seems to me, do great injustice to human reasoning not to allow it to explore what is beyond the reach of one's eyes; and no one can doubt that there can be objects so small as to be inaccessible to any of our senses . . . . Perhaps some will also say that Democritus had already proposed small objects with various sizes, shapes, and motions, various mixtures of which composed all sensible bodies, and that, yet, his philosophy is commonly rejected. To this I retort that it was never rejected by anyone on the grounds that he [Democritus] considered objects smaller than those perceived by our senses, and that he attributed to them various sizes, shapes, and movements, since no one could tell whether such objects truly exist or not, as we have already proven. And yet it was rejected, first because it postulated that these small objects were indivisible, which I too categorically reject. Next, because he imagined void between them, and I have proven that its existence is impossible; and also because he attributed gravity to them, while I deny that such a thing exists in any object, inasmuch as it is considered by itself, because it is a quality that depends on the mutual interaction that several objects exert on one another; last, it was rejected because it failed to explain specifically how all things were formed solely by the encounters of these small objects, or rather, if he explained it in some cases, the reasons he gave were not very consistent with one another, and that made everyone realize that nature could be explained in a more coherent fashion. . . . And because shapes, sizes, and movements were considered by Aristotle and all the others, including Democritus, and because I reject everything the latter postulated on this and other subjects, just as I generally reject everything that was postulated by the others, it is evident that this brand of philosophy has no more likeness to that of Democritus than to that of any other particular sect."

Moreover, Descartes put forth theories about the properties of his corpuscles that can best be characterized as sheer fantasy. For instance, he believed that while at first these small parts were similar and angular, the wearing effect of their movement led to the formation of particles with various dimensions and shapes. Probably inspired by the ideas of Beeckman (see chapter 17), Descartes went so far as to propose individualizing the particles of the four classical elements: air, fire, water, earth.

He even resurrected the ancient idea of protrusions and fissures in the structure of earth particles, which supposedly enabled them to associate through a purely mechanical effect.[6] In a similar vein, he considered that elongated particles, in the shape of left-handed and right-handed screws, are responsible for magnetism. Descartes also asserted that, in a completely filled world, any displacement of one part of matter implies a corresponding displacement of another—which produces his famous "vortexes" creating and undoing the worlds—and that the intervals or interstices between the parts of matter (which cannot be void) are filled with a "subtle matter" with unspecified characteristics. Taken together, Descartes's postulates constitute a vision of the world that is as fanciful as it is flawed.[7] This justifies Bertrand Russell's aphorism that "a philosophy [of nature] that is not self-consistent cannot be entirely correct, but one that is self-consistent may well be completely false." Modern philosophy, it seems, started its coexistence with the theory of atoms on the wrong foot. Voltaire, an ardent proponent of Newtonian physics, captured the differences between Newton's and Descartes's physics in a few sarcastic remarks in his fourteenth "Lettre philosophique" (Philosophical letter): "A Frenchman arriving in London finds things changed quite radically in philosophy as in everything else; he left a world filled; he comes back to an empty one. In Paris, one sees a universe composed of vortexes of subtle matter; no such vision in London." Still, this critique does not prevent Voltaire from recognizing that Descartes "was honorable even in his aberrations. He was wrong, but at least he erred with a method." Which vindicates Diderot's opinion, expressed in his *Traité de botanique* (Treatise on botany), that "in certain circumstances, nothing is more burdensome and prejudicial than a method." D'Alembert will state more kindly in his *Encyclopédie*, under the heading "experimental," that "Descartes . . . explored new methods in experimental physics, but he was more inclined to recommend them to others than to practice them himself, which perhaps caused him to commit several blunders." Jean-Jacques Rousseau would put it more bluntly: "The less one knows, the more one thinks one knows. Did the Peripatetics doubt anything? Did Descartes not construct the universe with his cubes and vortexes?"[8]

This is an opportune place to say a few words about Marin Mersenne (1588–1648), a devoted friend of Descartes's and one of his most important correspondents—"tenant of Mr. Descartes in Paris," as some called him. His position is particularly interesting in that he was also on very good terms with Gassendi. In fact, it would be more appropriate to speak of his lack of position, as Mersenne adopted an essentially pragmatic stance toward science, seeing his role as limited to "saving appearances" and being "useful." In his opinion, science does not make us understand nature, but only gives us the means to use it. Decidedly mechanistic, but resolutely defiant toward any dogmatic proposition, he embraced neither Cartesianism nor atomism, the two competing theories grounded in mechanism. Nor did he ever take sides between the subtle matter of his friend Descartes and the void in which the atoms of his other friend Gassendi move about. His personal scientific contribution is almost insignificant. He was a member of the Minim order, which gave Voltaire the irresistible opportunity to refer to him, unjustly, as the very minimal Father Mersenne.[9]

Finally, let us examine Descartes's attitude vis-à-vis two related problems central to the debate about the theory of atoms. The first centers on the dilemma about whether the world is eternal or was created. Descartes adopted a conciliatory stance by asserting that, in his view, the two conceptions need not be contradictory: "As for me, I do not see why a creature could not have been created by God in all eternity; God having had his power in all eternity, nothing seems to preclude that he might have exercised it in all eternity."[10] He borrowed the argument of Godefroy of Fontaines, who had already written in the thirteenth century: "In all eternity, God could make the world or not make it; God's power to create and his power to not create, the power to resolve to create or to resolve not to create, or, in other words, to choose whether he wanted to create or he wanted not to create, have coexisted in all eternity. Even though the world could have existed in all eternity, God, however, always had the prerogative to make it or not to make it."[11]

The second problem is whether the world is finite or infinite. On this issue, too, Descartes took a position of compromise. Obviously leaning in favor of infiniteness, he nonetheless declined to apply the term to the material world, preferring to substitute the designation of "indefinite," so as to "reserve the word 'infinite' for God only." In doing so, he took his cue from Nicholas of Cusa (1401–1464), a philosopher and cardinal of the Church, who had talked in his work *De docta ignorantia* of a "reduced infinite universe," in which "centers are everywhere and the circumference nowhere." Fénelon (François de Salignac de La Mothe, 1651–1715), a convinced atomist who regarded Epicurus's theories as "plausible physics," was not too enamored of such terms. He wrote in the fourth of his *Lettres sur la religion* (Letters on religion): "I must confess that there are some things in Descartes that do not seem worthy of him; for instance, his indefinite world, the meaning of which is ridiculous, if he does not mean a real infiniteness."

A CARTESIAN ATOMIST: HENRY MORE

The insights of philosophers have occasionally
benefited physicists, but generally in a negative
fashion—by protecting them from the preconceptions
of other philosophers.

—Steven Weinberg, *Dreams of a Final Theory*

Henry More (1614–1687), a contemporary of Descartes and the founder of the Neo-Platonic school of Cambridge, was one of the first champions of Cartesianism in England. But his vision of the world was also influenced by the ideas of Democritus, by the Stoics, by hermetism, and even by the cabala.[12] His extra-Cartesian ideas predominate on certain issues, notably the theory of atoms, where More clearly disagreed with Descartes. Even though he rejected the strict materialism of the ancient doctrine, he was an avowed atomist, and declared his conviction that both atoms and void are real.[13] He even couched his belief in a poetic style in his work *Democritus Platonissans*, published in 1646, in which he undertook a synthesis of the propositions of Plato with those of the atomists. In his mind, particles were created and are subject to divine intervention.

More had specific objections against Descartes's antiatomism. He scoffed at the idea of identifying space with matter, which had led Descartes to reject void. In More's view, if all matter has extent—and this applies not only to material substance, but, contrary to Descartes's belief, to spiritual substance as well, including the soul and God—space devoid of matter was for him just as real as filled space.[14] Matter occupies space; it moves through space, which is itself motionless. More was unconvinced by Descartes's assertion that the walls of a vase emptied of all the matter that fills it would have to touch, since there can be no distance "which is nothing." He countered this idea with a simple argument: "For, if God annihilated this universe and then, after a certain time, created from nothing another one, this *inter-mundium* or this absence of the world would have its duration which would be measured by a cetain number of days, years, or centuries. There is thus a duration of something that does not exist, which duration is a kind of extension. Consequently, the amplitude of nothing, that is of void, can be measured by ells or leagues, just as the duration of what does not exist can be measured in its inexistence by hours, days and months."[15]

In fact, God intervenes even more explicitly and directly in More's own version of void: "For if after the removal of *corporeal Matter* out of the world, there will still be *Space* and *distance*, in which this very matter, while it was there, was also conceived to lie, and this *distant Space* cannot be something, and yet not corporeal, because neither penetrable nor tangible, it must of necessity be a substance Incorporeal, necessarily and eternally existent of it self: which the clearer *Idea* of a *Being absolutely perfect* will more fully and punctually inform us to be the *Self-subsisting* God."[16] In other words, void is filled with God. It is a spiritual and extended substance. It *is* God. It is no surprise, then, that More should have regarded rejecting void as worse than a mistaken idea. Indeed, he displayed little patience for those who let themselves be impressed by Descartes and his system: "They prefer to rave and rage with Descartes, than to yield to most solid arguments if the *Principles of Philosophy* are opposed to them. Among the most important [tenets] that he himself mentions is that one I have so diligently combated [elsewhere], namely, that not even by Divine virtue could it happen that there should be in the Universe any interval which, in reality, would not be matter or body. Which opinion I have always considered false; now however I impugn it also as impious." Paraphrasing Voltaire, one might wonder about this troublesome habit of men (and not only theologians) to believe that one offends God when one is not in agreement with their opinions.

Void having thus been elevated to the status of reality of the divine nature, More proceeded to proclaim the existence of atoms by turning Descartes's own argument about the relation between the divisibility of matter and God's omnipotence around against him. Descartes rejected atoms, on the grounds that limiting how far matter can be divided would entail a lessening of God's omnipotence, a deprivation that God would have had to impose on himself. More shrewdly remarked that the same objection would hold if matter were infinitely divisible: It would mean that God could never put a stop to the ability to divide, and that too would restrict his omnipotence. In the end, the only way for God to preserve his unconditional omnipotence would have been for him never to have created matter at all. But would he then not be vulnerable to charges of laziness?[17]

More was not the only thinker of that period to be torn between opposing views. The French philosopher Géraud de Cordemoy (1626–1684) was also a Cartesian atomist. While, overall, he followed the line of Descartes's philosophy, he challenged the infinite divisibility of matter and opted in favor of atoms and void. "I recognized that one could conceive of material bodies only as indivisible substances, and of matter only as an aggregation of these same substances," he wrote in 1666 in his "Lettre au père Cossart sur la conformité du système de M. Descartes" (Letter to Father Cossart on the propriety of Mr. Descartes's system). His originality was to view atoms as completely passive. Their movement is not innate. Only God, the ultimate motive cause, can transfer motion from atom to atom during their encounters.

Others, such as the French Huguenot Pierre Bayle (1647–1706), although a Cartesian by formation and of the belief that "the invisible atom is a chimera," admitted at the same time that an infinitely divisible continuum is "plagued with insurmountable difficulties for the human creature."

Thomas Hobbes (1588–1679), another prominent English philosopher of the same period, is also sometimes mentioned in connection with the concept of atoms. He initially based his philosophical thinking on the principle that everything in the universe (including man and thought) is essentially matter in motion, in accord with the mechanistic view of the atomists. In fact, he openly supported atomism in his early works. He eventually adopted a more cautious stance toward the atomic theory, rejecting void but remaining relatively uncommitted on the notion of indivisible atoms. He did, however, accept that matter in nature could be divided into physically uncuttable corpuscles of different sizes, and even sizes that vary depending on the circumstances and the surrounding medium.

One century ago, A. Hannequin voiced the following opinion: "Hobbes pronounced the ultimate word on corpuscular physics; to explain all forms of phenomena, it has as many types of subtle matter, as many types of atom, as are necessary, these atoms being to one another as are infinitely small quantities of various consecutive order. As such, matter, like an infinitesimal element, flees endlessly."[18]

More recently, R. Lenoble expressed a more guarded judgment: "We have seen that the principles of mechanism exploded everywhere at once. There is then no reason to doubt Hobbes's words when he assures us that, around 1630, he arrived independently at the theory of the subjectivity of qualitative sensations. He probably arrived at it via *a priori* speculation about Democritus's doctrine: Qualities are subjective manifestations of the arrangement of atoms. While he is not enough of a physicist to derive from atomism applications on the scale of a Galileo, while he populates extent with subtle spirits which are neither atoms nor Descartes's subtle matter, their nature not being terribly clear to us—nor to him either, for that matter—at least he substantially freed himself from qualitative physics."[19]

LEIBNIZ AND METAPHYSICAL ATOMS

God saw everything he had made; and indeed it was
very good.

—Genesis 1:31

Among the thinkers of the beginning of modern times, Gottfried Wilhelm Leibniz (1646–1716), a distinguished philosopher and a true scholar, an eclectic mind of enormous erudition (Fontenelle would call him a "living dictionary," and Diderot "a thinking machine"), is probably the one who relied the most on arguments of a religious nature to combat and refute the atomic theory. This stance will drag him into a public intellectual duel not with Newton himself, but with his spokesman (or, rather, secretary), Samuel Clarke, a faithful student and friend of the renowned English physicist (who declined or did not condescend to step into the fray personally).[20] The tone of the exchange was further inflamed by opposing views on God's role in the creation and march of the universe, and a bitter dispute about who should get credit for inventing differential calculus. The central issue of the polemic was the existence or nonexistence of void. The possibility of existence of atoms, solid corpuscles whose dimension would restrict the ability of matter to divide, was also a topic of contention. On both questions, the views and arguments of Leibniz were quite original.

We start by reviewing the broad outlines of Leibniz's overall philosophy. In the process, we will have an opportunity to appreciate how similar intellectual positions can resurface after hundreds or thousands of years. For instance, the quest for "primordial substances" pursued by the Greek natural philosophers, whether they advocated a small or infinite number of such substances, appears to undergo a rebirth in more or less the same form at the dawn of modern times. We have already seen how forcefully Descartes affirmed that the reality of the world rests essentially on two substances (in the sense of entities with strictly independent existences, requiring nothing else in order to be): matter (extent), and spirit (thought). At about the same time, the great Dutch philosopher Baruch Spinoza (1632–1677), a Jew who ended up excommunicated by the synagogue, admitted only one such principle: God. He regarded extent and thought as but two among the infinitude of God's attributes, an attribute being defined by what "our understanding perceives in a substance as its essence."[21] Leibniz, on the other hand, defended the idea of an infinite number of substances (originally advanced by Anaxagoras of Clazomenae in antiquity). The substances, which in his mind constituted the reality of the world, and which he called "monads," are absolutely simple, nonmaterial (analogous to souls), devoid of extent or parts and therefore indivisible, infinite in number, without natural beginning or end. Created by God, they can be annihilated only by him. They are all different from one another (since for Leibniz, if two things are identical, they are the same thing) and hierarchized according to the degree to which we perceive them, which is a measure of their ability to mirror the universe (in his words, "the soul is a small world reduced to a point"). They are immune to any external influence, since "they lack windows through which something could enter or exit." Therefore, they exert no action on one another, but each monad contains within itself an image of all the others. All their potentialities are specified from the time of their creation, which implies their harmonious coexistence. Indeed—and this is a fundamental idea of Leibnizian philosophy—while God can conceive of and could have created an infinite number of worlds, the one he did actually create is the best of all possible worlds (not perfect, which would bring another God into existence, but the best that could be realized). Its evolution and fate are predetermined by God during its creation.

In brief, Leibniz was a fierce antiatomist, that is, an avowed opponent of the ancient atomic theory of Leucippus, Democritus, and Epicurus, which postulated the existence of indivisible and undeformable fundamental corpuscles. Yet he was not necessarily adverse to the concept of an atomic structure of the world. In fact, in his view, "monads are the true atoms of nature and, in a word, the elements of things."[22] However, these are "points of energy,"[23] "formal" or "metaphysical atoms."[24] They should not be confused with the material corpuscles of the ancient philosophers, which are never perfectly indivisible because, even though undissociable, they are made up of diverse parts: "Only metaphysical points (as opposed to mathematical points, which are indivisible, but not real) are exact and real." The fierceness with which he was to attack the ideas of the ancient atomists is a testimony to the chasm that separates these two conceptions.

Can this best of all possible worlds, formed of spiritual monads, accommodate the concept of void? Leibniz's response was no, buttressed by arguments rooted in his concept of the "perfection of divine actions." At the basis of his reasoning was his assertion that matter is more important and "more perfect" than void, coupled with the "principle of sufficient reason," to which he resorted a great deal, "by virtue of which I hold that no fact can be deemed true or existing, no enunciation authentic, without a reason why it must be so and not otherwise."[25] In Leibniz's own words: "There is no possible reason, that can limit the quantity of matter; and therefore such limitation can have no place. And supposing an arbitrary limitation of the quantity of matter, something might always be added to it without derogating from the perfection of those things which do already exist; and consequently something must always be added, in order to act according to the principle of the perfection of the divine operations. . . . In like manner, to admit a vacuum in nature, is ascribing to God a very imperfect work: 'tis violating the grand principle of the necessity of a sufficient reason; which many have talked of, without understanding its true meaning. . . . To omit many other arguments against a vacuum and atoms, I shall here mention those which I ground upon the necessity of a sufficient reason. I lay it down as a principle, that every perfection, which God could impart to things without derogating from other perfections, has actually been imparted to them. Now let us fancy a space wholly empty. God could have placed some matter in it, without derogating in any respect from all other things: therefore he has actually placed some matter in that space. Therefore, there is no space wholly empty: therefore all is full. The same argument proves that there is no corpuscule, but what is subdivided."[26] In other words, it is through the existence and organization of matter that God can exert and display his power and wisdom. Void would be of no use to him. It follows logically that the more matter there is in the universe, the greater his possibilities to display these qualities, which argues in favor of a completely filled universe. Therefore, the universe must be filled, QED. Leibniz himself stated: "All those who maintain a vacuum are more influenced by imagination than by reason. When I was a young man, I also gave into the notion of a vacuum and atoms; but reason brought me into the right way."[27]

As is nearly always the case when considerations of God's omnipotence and perfection are involved, the argument of one side is easily turned around by the opposing side. Indeed, Clarke contended that Leibniz's reasoning prevents God from creating a limited amount of matter. That in itself restricts his freedom of action. We will return to this debate at the end of this chapter. But first we must examine Leibniz's arguments against atoms.

These arguments are twofold. First, he voiced a specific objection against any limitation of God's power to subdivide matter infinitely. Based on the principle of sufficient reason, this objection is grounded in the same logic that prompted him to reject void. To the cited quotations illustrating this position one could add a corollary inspired by the principle of perfection of divine actions: "The least corpuscule is actually subdivided in infinitum, and contains a world of other creatures, which would be wanting in the universe, if that corpuscule was an atom, that is, a body of one entire piece without subdivision."[28]

The second argument centers on the impossibility, according to Leibniz, for absolutely identical, hence indistinguishable, corpuscles to exist. This argument is more original than the first, and is in fact a consequence of the principle of sufficient reason. A. Koyré commented that, from Leibniz's perspective, "there is no action without choice, no choice without determining motive, no motive without a difference between conflicting possibilities; and therefore—an affirmation of overwhelming importance—no two identical objects, or equivalent situations are real, or even possible, in the world."[29] Leibniz himself expressed the same thought: "It must be confessed, that though this great principle has been acknowledged, yet it has not been sufficiently made use of. Which is, in great measure, the reason why the *prima philosophia* has not been hitherto so fruitful and demonstrative, as it should have been. I infer from that principle, among other consequences, that there are not in nature two real, absolute beings, indiscernible from each other; because if there were, God and nature would act without reason, in ordering the one otherwise than the other; and that therefore God does not produce two pieces of matter perfectly equal and alike. . . . The vulgar fancy such things, because they content themselves with incomplete notions. And this is one of the faults of the atomists. . . . Simple bodies, and even perfectly similar ones, are a consequence of the false hypothesis of a vacuum and of atoms, or of lazy philosophy, which does not sufficiently carry on the analysis of things, and fancies it can attain to the first material elements of nature, because our imagination would be therewith satisfied."[30]

The debate with Clarke on this subject has all the trappings of a dialogue among the deaf. "This argument, if it was true," replied Clarke, "would prove that God neither has created, nor can possibly create any matter at all. For the perfectly solid parts of all matter, if you take them of equal figure and dimensions (which is always possible in supposition), are exactly alike; and therefore it would be perfectly indifferent if they were transposed in place; and consequently it was impossible (according to this learned author's argument) for God to place them in those places wherein he did actually place them at the creation, because he might as easily have transposed their situation. 'Tis very true, that no two leaves, and perhaps no two drops of water are exactly alike; because they are bodies very much compounded. But the case is very different in the parts of simple solid matter. And even in compounds, there is no impossibility for God to make two drops of water exactly alike."[31] To which Leibniz countered that to hypothesize that some parts of matter could have equal shapes and dimensions "begs the question" and amounts to "a supposition of convenience," which, as far as he was concerned, could not be accepted. As he observed: "To keep multiplying the same object, no matter how noble, would be superfluous and fruitless."

Thus, the presumed impossibility that identical and indistinguishable corpuscles could exist was for Leibniz cause for rejecting the theory of atoms. It is interesting to contrast this point of view with Maxwell's. The English scientist, like his German peer, was troubled by the problem posed by the absolute identicalness of corpuscles or elementary compounds, be they atoms or molecules. One and a half centuries after Leibniz, Maxwell accepted the existence of atoms and molecules that were strictly identical if they belonged to the same chemical species. It was, in fact, this very identicalness that he interpreted as proof that these entities could not have been "manufactured" or fashioned by other than God's hand. So, either refusing to admit the existence of identical particles (as did Leibniz), or accepting the same (as did Newton) can serve equally well to glorify the Lord. While, naturally, the act of glorification itself does not set off any controversy, its contradictory implications can only raise questions about the soundness of the philosophical arguments surrounding this issue.

The sharpest disagreement between Leibniz and Newton concerned God's role in the universe, and Leibniz's affirmation that ours is the best of all possible worlds.

We have seen that in Newton's world, God not only created the universe with its remarkable regularity and intelligible order, but that he also constantly directs its workings. Better yet, he oversees its auspicious march through a series of continual corrective actions.

Leibniz's God is more inclined to keep his distance from the world he fashioned and whose evolution and destiny he preordained once and for all. Since, in any event, it is the best possible world, it is difficult to see why he would want to intervene in its workings. Once launched, this world needs hardly any guidance, and even less repairs. Again quoting Leibniz himself: "Sir Isaac Newton, and his followers, have also a very odd opinion concerning the work of God. According to their doctrine, God Almighty wants to wind up his watch from time to time; otherwise it would cease to move. He had not, it seems, sufficient foresight to make it a perpetual motion. Nay, the machine of God's making, is so imperfect, according to these gentlemen, that he is obliged to clean it now and then by an extraordinary concourse, and even to mend it, as a clockmaker mends his work, who must consequently be so much the more unskilled a workman, as he is oftener obliged to mend his work and to set it right. According to my opinion, the same force and vigor remains always in the world, and only passes from one part of matter to another, agreeably to the laws of nature, and the beautiful pre-established order."[32]

A. Koyré offers an insightful analysis of the fundamental opposition between Newton and Leibniz on this subject: "The God of Leibniz is not the Newtonian Overlord who makes the world as he wants it and continues to act upon it as the Biblical God did in the first six days of Creation. He is ... the Biblical God on the Sabbath Day, the God who has finished his work and who finds it good, nay, the very best of all possible worlds, and who, therefore, has no more to act upon it, or in it, but only to conserve it and to preserve it in being."[33]

Obviously, Newton and his followers failed to accept Leibniz's arguments. Clarke even accused the German philosopher of impiety: "The notion of the world's being a great machine, going on without the interposition of God, as a clock continues to go without the assistance of a clockmaker; is the notion of materialism and fate, and tends (under the pretence of making God a *supra-mundane* intelligence) to exclude providence and God's government in reality out of the world."[34]

One last question must be asked: Why is the best possible world not perfect? Why are suffering, disease, misery so prevalent in it? This eternal question, which man has posed since time immemorial, has prevented many from referring to the world in which we live as the best *possible*. Leibniz's response to this perennial query (which appeared in *Essays on Theodicy*, published in French in 1710) proposes that this world is the best possible because it contains the greatest surplus of good over evil. This affirmation seems to imply that a world in which good were to reign unchallenged would be somehow less desirable than the real world. The explanation for this peculiar paradox is contained in the proposition that good is often more pronounced and more potent when it is in some way opposed to evil. For instance, the more debilitating the illness, the greater the relief when one recovers from it. The surplus of pleasure overcompensates for the prior experience of pain: "Dissonance magnifies the pleasure of consonance."[35] Plotinus had articulated the very same thought thirteen centuries before Leibniz when he wrote: "Let us add that worst things exist precisely because better things exist. How could, in such a multifaceted work, the worst exist without the best, or the best without the worst? We should then not blame the worst for being in the best; rather, we should feel fortunate that the best gave of itself in the face of the worst. To want to destroy the worst in the universe is to destroy providence itself. To what would it then attend? Certainly not to itself, nor to the worst."[36]

I am not convinced that this kind of reasoning is apt to satisfy everyone. As Bertrand Russell noted with his characteristically dry wit: "This argument apparently satisfied the Queen of Prussia. Her serfs continued to suffer the evil, while she continued to enjoy the good, and it was comforting to be assured by a great philosopher that this was just and right."[37]

As for Voltaire's opinion, a rereading of *Candide* is in order. Schopenhauer, who believed that this world is actually the worst possible because "an even worse world, being incapable of surviving, is utterly impossible," wrote this commentary: "I fail to recognize in *Theodicy*, as a large and methodical expression of optimism, any merit other than to have subsequently provided the clever Voltaire with material for his immortal *Candide*."[38]

Nicolas Malebranche, who happened to be an Oratorian priest, had a different opinion on the reasons for the imperfections of the world. In his view, they resulted from God's will to combine the perfection of the design conceived by his wisdom with the requirement that it be implemented with well-defined practical means. He expressed his conviction in his *Entretiens sur la métaphysique* (Conversations on metaphysics), published in 1688: "You would not be mistaken if you believed that . . . God wants to make his work as perfect as possible. . . . But you would have identified only half the principle. . . . Not contented by letting the universe honor him through its excellence and beauty, he also wants its course to glorify him through its simplicity, productiveness, universality, and uniformity. . . . Therefore, do not imagine that God was absolutely determined to produce the most perfect work possible, but only the most perfect relative to the course that is the most worthy of him."

It is also relevant to mention here the attempt of Maupertuis, presented in his *Essai de philosophie morale* (Essay on moral philosophy) to work out a mathematical balance sheet of the amount of good and evil in human existence. To that end, he adopted an appropriate definition of happiness and misery, of which he proceeded to "measure" the intensity and duration. Nothing seemed then more logical than to use the product of the intensity and the duration as a quantitative measure of the magnitude of a state of happiness or unhappiness. On this basis, Maupertuis estimated the worth of various ethical systems (Epicurianism, Stoicism, and others) from the point of view of the equilibrium, or, rather, the disequilibrium, assigned to them by this "mathematics of intensive values." Unfortunately for mankind (and for Leibniz), the calculation turned out to consistently produce a predominance of evil over good.

Finally, d'Alembert rendered a verdict on Leibniz's work in his *Discours préliminaire à l'encyclopédie* (Discourse preliminary to the encyclopedia), worthy of mention: "Not as wise as Locke or Newton, Leibniz did not just form doubts, he tried to dispel them, and in this respect, he was perhaps not as effective as Descartes. His principle of sufficient reason, quite admirable and true in itself, does not appear very useful to beings as poorly enlightened as we are about the prime reasons of things; at best, his monads prove that he realized more than anyone else how elusive a clear idea of matter is, but they appear rather worthless in actually forming one; his preestablished harmony seems only to add another difficulty to Descartes's view of the union of body and soul; lastly, his system of optimism is perhaps dangerous because it purports to explain everything. This great man seems to have brought to metaphysics more instinct than light."

 **14**

Boscovitch, or
Punctual Atomism

Determined to rise above the ongoing controversy between Newton and Leibniz, Roger Joseph Boscovitch (1711–1787), a student of Newton's, attempted in his *Theory of Natural Philosophy*, published in 1758, to blend certain aspects of both philosophies. He described the world as the seat of a multitude of elementary particles viewed as points of force without extent, which make them reminiscent at first blush of Leibniz's "points of energy." Their behavior would be governed by Newton's laws of attraction and repulsion. Whereas Newton was mainly concerned with forces of attraction, Boscovitch emphasized the importance of repulsive forces: This point of view added an original twist to his vision of atomic reality, all the more interesting since, after being forgotten for some time, Boscovitch's name would eventually reappear in certain accounts of the history of atomism.[1]

Starting from the premise that attractive forces dominate when the distance between objects is large and repulsive forces take over when that distance is small (the shorter the distance, the greater the repulsion), Boscovitch concluded that it is impossible for atoms, these presumed ultimate material elements, to touch one another in space. Moreover, he believed that they must be simple, meaning without parts. As such, they must be without spatial extent, in other words, reduced to points. The repulsive force precludes matter from being continuous and makes atoms impenetrable. These properties form the basis of this "punctual atomism," as Bachelard would later call it. The rigid surface of the Democritean atom is replaced by a region of equilibrium between the forces of attraction and repulsion associated with the dynamic field surrounding the atom collapsed into a material point. On the short side of this region, repulsive forces dominate and are responsible for the impenetrability of atoms, while on the far side, attractive forces win out and account for interatomic cohesion. In addition, these dimensionless atoms, or material points, are absolutely identical. The diversity of things observed in nature, even in the organic world, is due to differences in the number, position, and relative distance of their constitutive atoms.

The material character attributed by Boscovitch to these atom-points is of particular interest. Pillon, an enthusiastic supporter of this philosophy, observed: "Stripped of all properties except force, reduced to points, atoms remain genuine material elements, whose existence is distinct from and independent of ours. . . . Hence, we are dealing here with a cosmological atomism, and not at all with a system that could be compared either to Berkeley's immaterialism or to Leibniz's monadism."[2] As such, Boscovitch's effort to meld the views of Newton and Leibniz is only partially successful. The "cosmological" significance ascribed by Pillon to Boscovitch's theory, which he considers its greatest virtue, resides in "the precise and marked distinction between the order of the possible and the order of the existing. To the order of the possible belong the continuous infinitude of space and the geometric requirement that extent be infinitely divisible. . . . The order of the existing includes discontinuous and finite matter, and the arithmetic requirement for a real number—in other words, finite—of elements. . . . Every material body will have a finite number of elements, while infinite divisibility applies to space intervals." As Boscovitch himself stated: "Each interval will certainly be infinitely divisible by interposition of additional material points, followed by still more, and so on, which, however, once compounded, will also be in finite number, and will leave room for many others . . . in such a way that the infinite will subsist only in the realm of the possible, but not of the existing."

Whether, as Pillon insists, this distinction "resolves the insoluble contradiction between the two necessities, the two mathematical evidences . . . and gets rid of a problem that had always confounded philosophers" is another question. Is it even possible to resolve an insoluble contradiction?

The reader might recall that the concept of a punctual atom, without spatial extent, had already been proposed by the Hindu atomists of the Nyaya-Vaisheshika school at the beginning of our era (see chapter 7) and by the Arab Mutakallimun in the Middle Ages (see chapter 11). It turned out to have great appeal among a number of other philosophers contemporary with Boscovitch. Several promptly embraced his dynamic theory assimilating atoms with punctual centers of forces, even if they sometimes introduced nuances traceable to Leibniz's influence. One such example was Christian Wolff (1679–1754). Kant adopted a similar position, indicating in his *Metaphysical Foundations of Natural Science*, published in 1786, his preference for a dynamic philosophy of nature, as opposed to the mechanistic philosophy of the atomists. While, according to him, the latter explains "the specific differences of matter by means of the constitution and composition of the smallest parts, considered to be machines . . .

mere instruments for external motive forces," the former "infers [nature] from motive forces of attraction and repulsion that are inherent and inceptive." To his mind, a dynamic vision of nature made it possible to dispense with void. Boscovitch's theory would be revived in the nineteenth century by Gustav Theodor Fechner (1801–1881), a theorist in psychophysics and author of a book entitled *Physikalische und philosophische Atomlehre* (Physical and philosophical atomic theory), published in 1864, as well as by André-Marie Ampère (1775–1836) and Augustin Cauchy (1789–1857).[3] Almost a century and a half later, the renowned high-energy physicist Leon Lederman (who received the Noble prize in 1988) will underscore the relevance of Boscovitch's concept of the punctual atom, center of a force field, in modern physics.[4]

In addition, Boscovitch shared with both Newton and Leibniz the view that the world could not be the result of chance. According to Pillon's analysis, "It could not have been produced at the outset by mere fortuitous convergence of a staggering number of specific conditions, each of which would have materialized from among an infinite number of possibilities. It could not exist by itself, under its own power. . . . It was created, in other words, culled through a free and intelligent selection process from an infinitude of choices."[5]

Lastly, is this world the best possible? Boscovitch did not think so. In his own words: "It is not contrary to the wisdom and solicitude of God to not choose the best, for the best does not exist. If there were *optima*, their creation would become necessary, divine freedom would be suppressed, and creatures would have a genuine right to existence in their number and in their virtue." That Boscovitch could not accept. The world cannot be the outcome of necessity; it can only be contingent.

 **15**

Berkeley, or
Atoms Dismissed

*Everyone knows that "mind" is what an idealist thinks
there is nothing else but, and "matter" is what a
materialist thinks the same about.*

—Bertrand Russell

The reality of material corpuscles forms the
underlying basis of the theory of atoms. From the time of its inception,
the Greeks had made a distinction between primary and secondary properties of matter, objective and subjective attributes, or those that exist by
essence and those that exist by convention. The primary qualities included dimension, shape, and, eventually, weight; the secondary qualities
were color, odor, flavor, and so on. As we have seen (chapter 12), this distinction received a renewed and more polished expression with the work
of Locke. Yet it was far from universally accepted.

Recall that Aristotle had elevated the secondary qualities to the level
of the primary ones: He saw color, odor, flavor, and so forth, as intrinsic
properties of matter participating in the reality of the world. A red body
simply was red. By virtue of the rule that demands that every philosophical doctrine be eventually refuted by an opposing doctrine, it would be
logical to expect sooner or later the emergence of a theory demoting the
primary qualities to the level of the secondary ones.[1] Such a philosophy
was in fact developed, twenty centuries after Aristotle, by George Berkeley
(1685–1753). He introduced it in several works, the most important of
which are *Principles of Human Knowledge* (1710) and *Three Dialogues Between
Hylas and Philonous* (1783). Berkeley's position was to lead him to a radical, albeit hazardous, conclusion: He rejected matter and, by extension,
the reality of material corpuscles. And since he also refused to accept
void, it is not surprising that he was no great champion of the theory of
atoms.

Four excerpts from his *Principles of Human Knowledge* encapsulate the essential points of Berkeley's thesis of immaterialism (or idealism, as some will call it).[2]

At the basis of his theory is a strong conviction that being is nothing more than perceiving and being perceived *(esse est percipere et percipi)*: "Some truths are so near and obvious to the mind, that a man need only open his eyes to see them. Such I take this important one to be, to wit, that all the choir of heaven and furniture of the earth, in a word all these bodies which compose the mighty frame of the world, have not any subsistence without a mind, that their being is to be perceived or known; that consequently so long as they are not actually perceived by me, or do not exist in my mind or that of any other created spirit, they must either have no existence at all, or else subsist in the mind of some eternal spirit: it being perfectly unintelligible and involving all the absurdity of abstraction, to attribute to any single part of them an existence independent of a spirit."

In such a picture, objects are nothing but a "collection of ideas." In support of this conviction, Berkeley advanced several arguments that speak to the problem of the hierarchization or ranking of ideas: "Some there are who make a distinction betwixt *primary* and *secondary* qualities; by the former, they mean extension, figure, motion, rest, solidity or impenetrability, and number; by the latter they denote all other sensible qualities, as colors, sounds, tastes, and so forth.[3] The ideas we have of these they acknowledge not to be the resemblances of anything existing without the mind or unperceived; but they will have our ideas of the primary qualities to be patterns or images of things which exist without the mind, in an unthinking sustance which they call *matter*. By matter therefore we are to understand an inert, senseless substance, in which extension, figure, and motion, do actually subsist. But it is evident from what we have already shown, that extension, figure, and motion are only ideas existing in the mind, and that an idea can be like nothing but another idea, and that consequently neither they nor their archetypes can exist in an unperceiving substance. Hence it is plain, that the very notion of what is called matter or corporeal substance, involves a contradiction in it."

This argument is central to Berkeley's doctrine: The so-called primary qualities differ in no way from the secondary qualities, since they depend just as much on the objective sensitivity of the subject that perceives them. No discrimination between them is warranted. The distinction claimed by some is based solely on emotional reactions: The secondary qualities evoke in us feelings that are pleasant or unpleasant, while the primary qualities leave us completely indifferent. And in Berkeley's view, that was not a sufficient reason to consider them fundamentally dissimilar.

These arguments entail a number of consequences. "Having done with the objections, which I endeavored to propose in the clearest light, and gave them all the force and weight I could, we proceed in the next place to take a view of our tenets in their consequences. Some of these appear at first sight, as that several difficult and obscure questions, on which abundance of speculation has been thrown away, are entirely banished from philosophy. Whether corporeal substance can think? Whether matter be infinitely divisible? And how it operates on spirit? These and the like inquiries have given infinite amusement to philosophers in all ages. But depending on the existence of *matter*, they have no longer any place on our principles."

Hence, the conclusion: "Matter being once expelled out of Nature, drags with it so many skeptical and impious notions, such an incredible number of disputes and puzzling questions, which have been thorns in the sides of divines, as well as philosophers, and made so much fruitless work for mankind, that if the arguments we have produced against it are not found equal to demonstration (as to me they evidently seem), yet I am sure all friends to knowledge, peace, and religion, have reason to wish they were." So, exit matter and, along with it, atoms. All debates past, present, and future about them are fruitless and irrelevant; a sword slashing air, or, more precisely, the idea of a sword slashing across the idea of air.

Still, no philosopher can be satisfied casually affirming that matter does not exist, "that we eat and drink ideas, that we are clothed with ideas," nor can he simply ignore the laws we have discovered about how nature behaves. Berkeley resolved this dilemma by asserting that things exist because they are perceived by God, who, through pure benevolence, implants in our mind the strong and rational ideas that constitute our own perception of the outside world. Berkeley proceeded to define "reality" with these words: "The ideas imprinted on the senses by the Author of Nature are called real things; and those excited in the imagination being less regular, vivid and constant, are more properly termed ideas, or images of things, which they copy and represent." On the subject of the laws of nature, he declared: "The ideas of sense are more strong, lively, and distinct than those of the imagination; they have likewise a steadiness, order, coherence, and are not excited at random, as those which are the effects of human wills often are, but in a regular train or series, the admirable connexion whereof sufficiently testifies the wisdom and benevolence of its Author. Now the set rules or established methods, wherein the mind we depend on excites in us the ideas of sense, are called the *Laws of Nature*." He further clarified this point of view: "If the word substance be taken in the vulgar sense, for a combination of sensible qualities, such as extension, solidity, weight, and the like, this we cannot be accused of taking away. But if it be taken in a philosophic sense, for the support of accidents or qualities without the mind, then indeed I acknowledge that we take it away, if one may be said to take away that which never had any existence, not even in the imagination."

Berkeley's thesis has met with numerous criticisms. The harshest were expressed by Diderot: "The designation of idealist is given to those philosophers who, being conscious only of their own existence and of sensations occurring within themselves, admit nothing else: an extravagant system which, it seems to me, could owe its existence only to the blind; a system that, to the shame of the human mind and of philosophy, is the most difficult to combat, even though it is the most absurd of all. It is exposed with as much explicitness as clarity in three dialogues by Dr. Berkeley, bishop of Cloyne."[4] The German philosopher Friedrich Wilhelm Joseph von Schelling (1775-1847) will be less diplomatic: "That the ideal as such also be the real, that the affirmative as such be simultaneously the negative, and vice versa, is quite obviously impossible."[5]

16

Kant

An Atomist Turned Antiatomist

Truth at one time, error at another.

—Charles de Montesquieu

We perceive things very differently under various conditions: young or old; hungry or full; by day or night; angry or happy. We can be swayed at any moment because of thousands of circumstances that keep us in a perpetual state of vacillation and instability.

—La Mothe Le Vayer

Experts have traditionally divided the long intellectual activity and abundant literary production of Immanuel Kant (1724–1804) into two great periods: the precritical and the critical. The transition from the first to the second occurred during the decade between 1770 and 1780 and received its concrete, if not definitive, expression with the publication in 1781 (with the second edition following in 1787) of his famous work *The Critique of Pure Reason*. This division is of particular importance to us because it corresponds to two diametrically opposed attitudes toward the atomic theory.

The position of the great philosopher from Königsberg (Kant never left the city in which he was born) during the precritical period is laid out in his book *Universal Natural History and Theory of the Heavens* (1755), in which he compares his own cosmological views with those of the Greek philosophers. Kant was then thirty-one years old and declared himself a believer in the corpuscular structure of matter.

Kant's cosmology is of little importance in the context of the present discussion, and we will not dwell on it, particularly since he summarized it

himself in the course of comparing his own philosophy with the propositions of the ancient atomists. For instance, he wrote: "I will not contest
that Lucretius' theory or that of his predecessors, Epicurus, Leucippus,
and Democritus, bear much resemblance to mine. As these philosophers
did, I postulate the earliest state of nature to be a general dispersion of
the original substance of all celestial bodies, as well as of atoms, as they
are called by them. Epicurus assumed a gravity that caused these elementary particles to fall, and that does not seem very different from Newtonian attraction, which I accept; he also attributed to them a certain deviation from a rectilinear fall, although he had absurd ideas about the
causes of this deviation and about its consequences. To a certain extent,
this deviation conforms with the deflections from a straight fall that we
deduce from repulsive forces between particles; finally, vortexes resulting
from the perturbed motion of atoms were a central feature in the theory
of Leucippus and Democritus, and they will be in ours as well."[1]

Kant's general endorsement of the atomic theory included void, the
second essential component of the ancient corpuscular theory. On this
issue, he affirmed that "matter is distributed at the outset according to
an infinite diversity in the infinite extent of space before the creation,"
but that subsequently "the density of matter decreases as one gets farther
from the center, and the condensation of galactic masses requires a
greater volume as the distance increases. Thus, as the mass of the fashioned universe increases, void becomes more and more prevalent."

However, Kant's favorable, or at least tolerant, stand toward the theory of atoms was rather narrowly limited to its mechanistic aspects. Aside
from recognizing atoms and void, he broke sharply with the philosophy
of the ancient Greeks. His unconditional religious faith prevented him
from accepting their unqualified materialism. In particular, he challenged the decisive role attributed by Democritus and Epicurus to
chance in the formation and evolution of the world, to the exclusion of
any teleological cause: "So much kinship with a conception that was a
genuine theory of the negation of God in antiquity does not, however,
lead mine to share in its errors. . . . The aforementioned theorists of a
mechanical origin of the universe deduced all the order perceived in it
from random chance causing atoms to collect so auspiciously as to form a
perfectly ordered whole. Epicurus was impudent enough to claim that
atoms deviate from their motion without any cause in order that they
come into contact. He and others pushed this absurdity so far that they
believed this blind convergence to be the cause of all living creatures, and
even of reason and folly. In my conception, on the contrary, I find that
matter is subject to certain necessary laws. In all its decomposition and
dispersion, I see a magnificent and ordered whole emerging quite naturally from these laws. This whole is not produced by chance or contin-

gency, but it is clear that natural properties engender it by necessity. Would one then not be inspired to ask: why was matter subject to precisely such laws aimed at order and measure? Was it possible that so many things, each with their own independent nature, had to spontaneously arrange themselves so as to produce a highly ordered whole? And if they did, is it not the undeniable proof of their common origin, which must be a supreme and absolutely self-sufficient comprehension in which the nature of things was planned in accordance with pre-established designs?" Kant further amplified the point: "One cannot look at the universe without recognizing the highly superior order of its organization and the sure sign of the hand of God in the perfection of its relations. After having considered and admired such magnificence, such excellence, reason is rightfully indignant at the reckless folly that outrageously attributes it all to chance and propitious accidents. It is not the fortuitous encounters of Lucretius' atoms that formed the world; forces implanted in its midst and laws issued from the most wise intellect were the immutable source of this order which had to emanate from them not accidentally, but necessarily."

Even though he deplored his "profane philosophy," Kant nonetheless maintained great respect for Epicurus, whose sole error was to draw erroneous conclusions from perfectly valid principles: "The consequences a distorted comprehension infers from unimpeachable principles are quite often suspect, and that was the case of Epicurus' conclusions, although his purpose was worthy of the dignity of a great mind."

The transition to the "critical" period was marked by a complete reversal of Kant's position toward the fundamental postulates of the theory of atoms: He became a firm advocate of the continuous structure of matter and a staunch opponent of void. He made no reference to his prior views, which he dismissed as a "youthful mistake" best left forgotten. In short, Kant embraced all the traditional views of the antiatomists, adding, however, a specific argument that is intriguing and worthwhile to examine.

Kant articulated his new beliefs most clearly and explicitly in his *Metaphysical Foundations of Natural Science*, published in 1786.[2]

In the "Euclidean" style characteristic of his *First Principles*, Kant formally repudiated a corpuscular conception of the world in theorem 4 of chapter 2: "Matter is infinitely divisible in parts, each of which is, in turn, matter." The proof of this theorem, however, requires a definition of what Kant calls "matter," which he gave as follows: "Matter is *mobile* inasmuch as it *fills* space." Kant himself described this definition as dynamic, by which he meant that matter is "associated" with two fundamental "motive forces": a "force of attraction, responsible for one matter to cause another to be drawn to it," and a "force of repulsion through which one matter causes another to move away." Kant's view was that matter is not inherently endowed with such forces, but identifies itself with them; it "is constructed in concert with them and is nothing in its essence save the mutual restrictions of attraction and repulsion."[3] As such, these two forces are its ultimate attributes. Should one of them be missing, space could not contain any matter; absent the attractive force, matter would scatter to infinity, and absent the repulsive force, it would collapse to a point. The degree to which matter fills space is determined by the balance between these two forces.

We are now ready to tackle the proof Kant gave of his theorem on the infinite divisibility of matter: "Matter is impenetrable, because of its original force of expansion, but this force is only the consequence of the repulsive forces of each point of a space filled with matter. But the space that matter fills is mathematically divisible to infinity, that is to say, its parts can be distinguished to infinity (in accordance with geometrical proofs), even though they cannot be moved, and hence not separated. On the other hand, in a space filled with matter, each part of space contains a repulsive force that acts against the other parts in all directions, and thus repels them and is repelled by them, that is to say, is pushed away from them. It follows that each part of a space filled with matter is mobile by itself, and consequently, by virtue of being material substance, is separable from the others by means of a physical division. Hence, the possible physical division of the substance that fills space extends just as far as the mathematical divisibility of space filled with matter. But just as mathematical divisibility extends to infinity, so too does physical division; in other words, all matter is infinitely divisible, and in such a way that each part is itself material substance."[4]

Having so defined the nature of matter, the question of void remained to be dealt with: It was promptly discarded. According to Kant, whereas void is necessary in a mechanistic philosophy of nature, as it explains "the specific differences between various types of matter, which can vary infinitely" through the "combination of the absolutely filled and absolutely empty," there no longer is any need for it in his own dynamic philosophy, which, "on the contrary, explains them solely through different degrees of combinations of the original forces of attraction and repulsion." In this conception, matter fills space in varying degrees and can be diluted indefinitely without ever leading to absolute void. As a matter of fact, Kant adopted a harshly critical attitude toward the notion of void, which he had once supported: "Anything that can spare us from the need to look for an escape route in empty spaces is a genuine gain for science and nature. Empty spaces present the imagination with too much temptation to replace what is missing from our intimate knowledge of nature with fictions. Absolute void and absolute density are to the theory of nature roughly what chance and blind fate are to a metaphysical description of the universe; in short, an obstacle to the victory of reason, giving precedence to fantasy, and lulling sanity asleep onto the pillow of the occult."

And so, from a proponent—admittedly diffident and tentative—of the atomic theory in the early part of his career, Kant became in his philosophical maturity one of its fiercest adversaries. This rather unusual intellectual transformation is worth noting, particularly when contrasted with that of several other philosophers and scientists of the nineteenth century who will follow precisely the reverse path: Antiatomists for a long time, they will feel compelled to concede their error toward the latter part of their lives. This change of heart will allow them to "save face," sometimes at the very last moment. In Kant's case, it is, instead, what he clearly considered a "youthful aberration" that redeems to some degree the errors of judgment in his adulthood.[5]

Having given a fairly comprehensive description of Kant's position regarding the key propositions of the atomic theory proper, it might be interesting, in light of some prior (and subsequent) discussions in this book, to also clarify his views on certain related problems, particularly since some of these views turn out to be central to his philosophy. Aside from *The Critique of Pure Reason*, already mentioned, Kant expressed his ideas with particular clarity in the *Prolegomena to Any Future Metaphysics*, published in 1783.[6]

Perhaps the most important of Kant's contributions is to have clearly discriminated between "phenomena," which are things such as they appear to us and belong in the world of senses, and "noumena," or things-in-themselves, which Kant also called "transcendental objects." Moreover, there no longer was, in his mind, a distinction between primary and secondary qualities in the sense defined by Locke, since they all are an integral part of the phenomenalistic aspect of the thing-in-itself. In addition, Kant eliminated from the thing-in-itself the element furnished to our perception by our cerebral functions. For him, instead of finding the laws of physics in nature, our intellect imposes on nature its own laws, indeed organizes it according to *a priori* categories of human reasoning, which is predisposed to conceive of the external world in specific ways; for instance, by establishing relations of causality.

Kant believed that the reality of objects of knowledge is restricted to that of phenomena. "Things as they are" are beyond the sphere of possible experience. Even though they do exist, we are not now, nor will we ever be, able to know on an absolute level anything definite about their real nature, "because our pure concepts of comprehension and our pure intentions can deal only with objects of possible experience, hence with simple perceptible entities, and because as soon as one ventures beyond that realm these concepts lose all meaning."

Kant consistently refused the label of idealism that some attached to his relentless drive to question the sense in which an intellectual grasp of the world around us is possible. He himself described any philosophy denying the reality of this world as "empirical" or "psychological idealism." He preferred to call his own philosophy "transcendental idealism," to emphasize that while it does make a distinction between phenomena and things-in-themselves, it is not opposed to "the objective reality of external intuition."

⚛ 17

The Rank-and-File Atomists

In parallel with the great philosophical and theological debates revolving around the theory of atoms, a few scientists of that period also made more prosaic, but nonetheless significant, contributions in support of the theory. It would be unfair not to mention them. Five among these individuals are particularly noteworthy: three Frenchmen, Sébastien Basso, Claude Bérigard, and Jean Magnien, and two Germans, Daniel Sennert and Joachim Junge (Jungius). They all carried out their work mostly during the first half of the eighteenth century. Rather than present their individual contributions, which were after all rather limited, we will endeavor to extract their general features.[1]

Leaving aside their religious convictions, which are barely mentioned in their writings, none of these philosophers fully embraced the theory of atoms in its classical form, such as it was defined by Democritus and Epicurus. We have already mentioned (note 7 to chapter 12) the attempts by Basso, Magnien, and Sennert to reconcile the theory of atoms with that of the four elements. Both Basso and Sennert explicitly associated elementary particles with these elements. Starting with ultimate corpuscles, they also envisioned possible associations of a higher order— referred to as *prima mista* by Sennert, and as secondary particles (*particulae secundae*) and tertiary particles by Basso—which can be viewed as the precursors of today's molecules. Sennert expressly considered that composite substances can be dissociated into "atoms," although those are not necessarily atoms in the modern sense.

Furthermore, they all shared a conviction that atoms, while indivisible, are composed of parts whose natural configurations can vary, producing atoms of different and changing types (Epicurus agreed with this view, but Democritus did not). Basso and Magnien specifically supported this conception, although they both rejected the possibility of transmutations, the specific nature of elementary particles being unalterable. Basso visualized the different parts of these corpuscles as surfaces, not unlike the model proposed by Plato (for whom transmutations were perfectly acceptable).

Most atomists of that era tended to shy away from a purely mechanistic conception of atoms and attributed to these particles complementary properties or "qualities," in the Aristotelian sense. Both Sennert and Junge fit that mold. Sennert was even inclined to see the constituent atoms of the various elements as having a kind of chemical identity that persists with little or no change after the formation of mixed compounds. Basso too believed that atoms preserve their individuality in mixed compounds. Van Melsen portrayed the position of these atomists, especially Sennert's, as an effort to blend the atomic theory of Democritus with Aristotle's theory of minimal particles.[2]

Bérigard's "atomism" deserves a special mention for its kinship with Anaxagoras's viewpoint, in that it admits an infinity of qualitatively different corpuscles.

Finally, we must add here the name of Isaac Beeckman (1588–1637), a Dutchman, who apparently influenced Descartes's philosophical thinking. Beeckman's position, somewhat unusual at the time, was patterned after that of some of the philosophers from India: He distinguished four types of atom, each corresponding to one of the four elements, the association of which produced, in his view, all observable substances. He rejected substantial forms, arguing that all properties of substances were explainable on the basis of atoms. Another Dutch scientist of the same period, the renowned physicist Christiaan Huygens (1629–1695), simply adopted Gassendi's views of atoms and void.

For the sake of completeness, we might also mention the English physician Walter Charleton (1619–1707). Inspired by Gassendi, he was an ardent proponent of Christian atomism. He is remembered for having remarked that "while the four ordinary elements might be considered the fathers of all accretions, they are not their grandfathers." He also produced a calculation of the number of atoms in a grain of incense turned into smoke. He estimated that number to be 7.776×10^{16} (in modern scientific notation).[3]

This brief census highlights the diversity of opinions prevailing among the less prominent atomists of the seventeenth century. They frequently deviated from the strictly mechanistic conceptions of the Greek founders of the atomic theory, and seemed eager not to break completely with the other great theories of antiquity, notably those of Empedocles, Plato, and Aristotle. This situation is not surprising since in those days no experimental proof existed, indeed could not even have been foreseen, of the fundamental properties of atoms, let alone of their very existence. Still, the writings of these scholars had the merit to ensure the continuity of atomism through a period of great intellectual ferment, when the foundations of modern ideas about the structure of the world were taking shape.

 PART FOUR

The Advent of
Scientific Atomism:
Nineteenth and
Twentieth Centuries

✺ 18

A Brief Overview

The mind darts from nonsense to nonsense much like a bird flies from branch to branch. It cannot do otherwise. The important point is not to feel steady on any of them.

—Paul Valéry

The theory of atoms was inspired in the fifth century before our era by the spiritual stirrings of ancient Greece in pursuit of a novel form of natural philosophy. For the next twenty-three centuries, it was to remain essentially a vision of the imagination. The many intellectual debates it stirred, however important they may have appeared at the time, were primarily games of the mind.

Not a single piece of empirical evidence, based on either observation or experimentation, existed to prove or disprove the key hypothesis of the theory—the corpuscular structure of matter. This situation continued until the dawn of the nineteenth century. The only "scientific" data produced during the seventeenth century dealt with the existence of void, which was the second postulate of the original atomic theory. And that was certainly not enough to prove the existence of atoms.

This state of affairs was to change suddenly and profoundly at the beginning of the nineteenth century with the birth and consolidation of scientific atomism, in the modern sense of the phrase. Yet neither its birth nor its growth proved smooth. On the contrary, the development of scientific atomism was a convoluted process. It experienced many vicissitudes in its painstaking accumulation of empirical data, which oftentimes suffered from the imperfection of research tools and the difficult maturation of theoretical concepts (for example, the distinction between atoms and molecules). It was further inhibited by the hostility of certain philosophical movements inspired by positivism, which were opposed on principle to any effort to unveil the underlying reality of directly measurable or observable experimental data. This caused a surprisingly large number of prominent scientists, particularly chemists, to adopt a timid, cautious, and restrictive attitude toward the theory. Pierre Thuillier is entirely justified to talk about "the resistible rise of the atomic theory."[1] This atmosphere persisted through the entire first half of the nineteenth century, and the antiatomists continued rearguard battles until the beginning of the twentieth century. Those who defended and nurtured the atomic doctrine in the face of such antagonism, ultimately leading it to victory, deserve all the more credit.

This eventual triumph did not, however, signify a complete victory for the classical theory in the form enounced by its distant Greek authors. A number of stunning experimental surprises in the twentieth century, as well as the bold theories proposed along with, and sometimes even before, them touched off renewed controversy about the nature of atomic reality. For instance, in the early twentieth century, as the atomic structure of matter was poised to become universally accepted, the atom suddenly lost its essential classical property of indivisibility. Experiments revealed that it was not, after all, the smallest particle of matter, but that it had in fact a complex structure. A quarter century later, an even harder blow struck the fate of the atom with the discovery of its dual nature: It was found to behave at once like a particle and a wave (and wavelike properties suggested that it was continuous). These great discoveries, and others, were to turn our entire philosophical vision of the world, its "reality," and even our day-to-day lives literally upside down.

The intensity of the "philosophical" debates that had revolved around the atomic doctrine from its outset continued unabated even after the emergence of scientific atomism. The discussions actually took on a new dimension, as new questions were raised about the nature of the universe and its meaning. They became embroiled in religious issues, although theological considerations, which had so strongly permeated the feuds about the atomic theory during the Renaissance and the Age of Enlightenment, had all but disappeared by the nineteenth century.

The central topic of this book is to retrace how philosophical and religious thought concerning the atomic structure of the world evolved during the course of the centuries. It is therefore not our intention, nor is it even possible given the limited space available, to give here a detailed historical account of the scientific development of the atomic theory, which often involved contentious arguments of a highly technical nature. The specifics of these arguments are now somewhat outdated and likely to be of interest only to scholars of the history of science. The debates centered primarily on the distinction between atoms and molecules, already mentioned several times, the determination of a scale of atomic weights, the significance of "equivalents," the development of a chemical notation, the classification of the elements (which culminated in the periodic table), and many other perhaps narrower but nonetheless important issues. There exists an excellent literature covering, either in condensed form or in great detail, the slow and often confused progress of knowledge in these areas.[2]

On the other hand, in order to put these developments in their proper philosophical and religious perspective, it seems indispensable at least to outline the chronology of the principal steps of the rise of scientific atomism. Such an approach will help us to grasp the significance of the

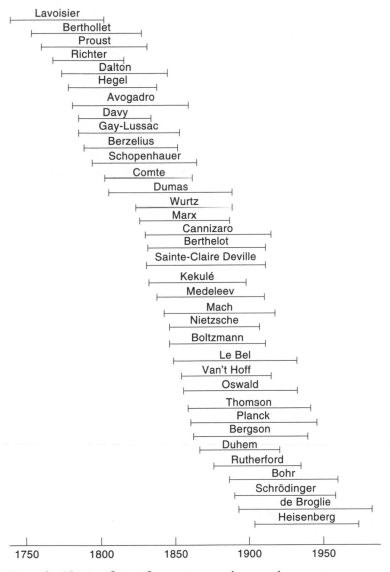

Figure 8. *The rise of scientific atomism: contributors and opponents.*

many spiritual and philosophical struggles regnant at the time. Toward that goal, we will divide this brief historical summary, covering about two centuries, into two great periods, each made of two subperiods. To be sure, divisions of this type are inevitably somewhat subjective, and this one is no exception. The vagaries of discoveries and conceptual innovations do not necessarily fit such a rigid framework. In particular, the dividing lines between the proposed periods are actually rather more blurred, and the chronology of the pertinent advances less clear-cut, than is implied by such a simplistic classification scheme.

With this reservation in mind, and in the interest of simplicity, it is expedient to distinguish the following two great periods. The first corresponds roughly to the nineteenth century. It is a time of relentless drive, over many obstacles, in search of the indivisible atom. The second period unfolds largely in the twentieth century, when the atom becomes a certainty, but is no longer indivisible; it becomes structured and less esoteric, but also more capricious. Figure 8 shows the main contributors to these developments, listing the names of a number of philosophers and scientists directly participating in the debates of the time. The two great periods just sketched form the subject of the remaining part of this book.

 **19**

The Nineteenth Century

In Search of the Invisible and Indivisible Atom

To think is much like to weigh. It is a function of the one who does the weighing, not a function of the balance.

—Alain

BREAKTHROUGH AND CONFUSION (1800–1860)

A major breakthrough occurred at the very beginning of the nineteenth century. It was made possible by two important advances in the latter part of the previous century. Both are due to Antoine-Laurent Lavoisier (1740–1794).

The first is the demise, after a life of two thousand years, of the theory of the four elements. It was brought about by the demonstration of water's compound structure. Lavoisier's contribution in this area was complemented by the work of two English scientists: Henry Cavendish (1731–1810) and Joseph Priestley (1733–1804). Priestley, in particular, is credited with proving the compound nature of air as well.[1] Lavoisier capped his own achievement by reaffirming (in Boyle's footsteps; see chapter 12) the definition of what can properly be referred to as "element": "With the word 'elements' or 'principles of bodies,' we associate the notion of the ultimate entity arrived at by analysis; all substances that we have not yet decomposed by any means, we consider elements," he wrote in 1789 in his *Traité élémentaire de chimie* (Elementary treatise of chemistry). This definition was both rigorous and pragmatic, as it left the door open for future developments.[2]

The second of Lavoisier's contributions, which laid the groundwork for all future atomistic research, was to establish, on a systematic and universal scale, the law of conservation of matter (or mass, in modern language) and to demonstrate the importance of gravimetric studies in chemical analyses of the structure of matter.

The breakthrough announced in the opening paragraph was based on precisely such studies. It took place at the turn of the nineteenth century, a crucial period in the history of scientific atomism. It witnessed, in particular, the discovery of important gravimetric laws pertaining to combinations of chemical elements: Joseph-Louis Proust's (1754–1826) law of constant proportions in 1806, and John Dalton's (1766–1844) law of simple multiple proportions in 1802–1804. The first states that two elements always combine in invariant ratios of weight, and the second specifies that when two elements can combine in several different ways, their relative weights in the resulting combinations form simple ratios, that is to say, rational numbers.[3]

A third law, known as the law of proportional numbers or of reciprocal ratios—chronologically the oldest, as it was enounced as early as 1791 by the German chemist Jeremias Benjamin Richter (1762–1807)—also played an important role in developing the notion of chemical "equivalents," or equivalent weights, which was to prove a highly popular crutch for scores of chemists, even a few prominent ones, anxious to avoid having anything to do with atoms. The law states that the weights of two elements a and b that combine with the same weight of a third element c form a simple ratio with the weights of a and b that combine directly.

This set of gravimetric laws became the impetus for the formulation of a new and quantitative corpuscular theory of the structure of matter. Credit for this goes to John Dalton, whose *A New System of Chemical Philosophy*, published in 1808, heralds the advent of modern atomic theory. Dalton reaffirmed that atoms are indivisible and indestructible and are the ultimate constituents of matter. He further pronounced a number of fundamental propositions: All atoms of a given element are similar and have the same weight; in contrast, atoms of different elements have different properties, specifically different weights; the "elementary particles" of compound bodies are formed by the union of a definite number of constituent atoms; and the weight of these compound particles is the sum of the weights of the atoms that make them up. No creation or destruction of matter can take place during chemical transformations. He wrote: "It is as preposterous to try to introduce a new planet into the solar system or to annihilate an existing one as to try to create or destroy a hydrogen particle." The most important conceptual advance, which will define the primary thrust of research aimed at validating the atomic concept, was to characterize the constituent atoms of all elements, however many they may be, by a specific property—their weight.[4] This way of looking at things introduced into the world of atoms a new type of discontinuity. It paved the way for a systematic rationalization of chemistry.

The propositions set out by Dalton are capable of elegantly accounting for all the gravimetric laws enumerated above. The constant ratios, expressible in terms of integers, of the weights of the constituents in composite bodies could be construed as evidence on a macroscopic scale of interactions at the microscopic level between basic units with fixed weights. For Dalton, this agreement strongly suggested a corpuscular structure of matter, even though it did not constitute definite proof. Yet, as we shall see, the plausibility of such an "interpretation" will by no means be universally accepted.

The atomic theory also owes Dalton credit for two other important developments: He established a scale of relative weights for a number of elements—a primitive scale, to be sure, which will be refined and extended by others, but the first one to be proposed—and he devised a system of symbols assigned to simple bodies (elements) as well as composite bodies to represent both their identity and their weight characteristics. The symbols he chose for atoms were essentially circles with distinctive markings identifying each element. Their only value today is historical. The modern notation relies on the first letter of the Latin name of each element followed by a second letter to avoid any possible duplication. Compound substances are identified by numerical subscripts defining how many weight units of each element they contain. This system was devised by the Swede Jöns Jakob Berzelius (1779–1848), one of the leading and most influential chemists of his time. We shall resort exclusively to this notation, even when discussing Dalton's ideas.

This two-pronged advance did not take place without posing a number of problems. For instance—and this is a key point—attributing relative weights to elements implies a knowledge of (or in the absence of such a knowledge, as was the case in Dalton's days, an assumption about) the composition of the relevant substance. On this topic, Dalton enounced a number of conventions and rules:

First the conventions: Considering two bodies, A and B, apt to combine, the order in which the combinations can arise, starting with the simplest, are as follows:

1 atom of A + 1 atom of B = 1 binary atom of C
1 atom of A + 2 atoms of B = 1 ternary atom of D
2 atoms of A + 1 atom of B = 1 ternary atom of E
1 atom of A + 3 atoms of B = 1 quaternary atom of F
3 atoms of A + 1 atom of B = 1 quaternary atom of G
etc.

Turning now to the rules:

1. When a single combination between two bodies is obtained, one must presume, in the absence of any indication to the contrary, that it is binary.
2. When two different combinations are obtained, one must presume that one is binary and the other ternary.
3. When three combinations are obtained, one should expect that one is binary and the other two ternary.
4. When four combinations are observed, one should expect one to be binary, two ternary, and one quaternary, etc.

Note that Dalton was unconcerned with the constitution of simple substances like hydrogen or oxygen, so obvious was it to him that they were necessarily monoatomic in their free state, an error soon to be corrected by Avogadro. Moreover, Dalton believed that these combinations of atoms occurred by simple juxtaposition, without any internal change.

Because of their arbitrary simplicity, the use of Dalton's rules will present a number of practical obstacles to the emergence of structural chemistry and the development of a scale of atomic weights. Let us take a specific example: The only compound involving oxygen and hydrogen known at the time of the publication of Dalton's work was water, which would then automatically acquire the structure HO in his system. Gravimetric analyses of water show that its composition involves 8 grams of oxygen for 1 gram of hydrogen. If the weight of the hydrogen atom is taken as reference, the atomic weight of oxygen would, under these conditions, be fixed at 8. Hydrogen peroxide, discovered ten years later by Louis-Jacques Thénard (1777–1857), had a composition involving 16 grams of oxygen for 1 gram of hydrogen, which in the same system suggests the structure HO_2. If, however, keeping the weight of hydrogen as unity, we were to choose hydrogen peroxide rather than water as the reference substance, the weight of oxygen would be 16, the formula of hydrogen peroxide would be HO, and that of water H_2O. This example illustrates that if the gravimetric ratios of the constituents of compound substances are provided by measurements within limits of experimental accuracy, what they tell about the atomic composition remains ambiguous without supplementary data. In other words, in a coherent reference system, the symbolic notation we have just used, based on letters of the alphabet, may correctly describe the weight *ratios* of the elements involved, but it is inherently insufficient and indeterminate in terms of figuring out the true structure of the combinations in question. Needless to say, the greater the number of elements involved, the more ambiguous things become.

In short, from the time the quantitative laws of combinations were established, their results could be interpreted from two different points of view. Either they were to be taken literally as the actual weights involved in specific combinations, or they only indicated equivalent ratios describing how the constituent elements of substances unite or replace one another. This notion of equivalence was first proposed by Richter in 1792 in his book *Stoichiometry*, even though the term "equivalent entities" was not formally introduced until 1814 by William Wollaston (1766–1828) in his work *Synoptic Scale of Chemical Elements*. It was based on Richter's own observations that neutrality is preserved in exchanges of acids and bases during reactions of double decomposition between salts, reactions that are symbolized by

$$AB + A'B' \longrightarrow AB' + A'B$$

The entities A and A', on the one hand, or B and B', on the other, are said to be equivalent inasmuch as they can replace each other in the two salts.

However, from a theoretical perspective—perhaps more abstract on the surface, but far more concrete and intriguing in its implications—these gravimetric laws of combinations can also be interpreted as evidence on a macroscopic level of an underlying specific structural organization of matter that is describable by a system of atomic weights. As we know today, it is this second, far bolder view that was to be vindicated. In the meantime, a given chemical formula remained subject to a dual interpretation. This difficulty must be borne in mind if one is to understand the significance of the debates that for so long engulfed chemists and, to a lesser degree, physicists.

Another difficulty and source of misunderstanding has to do with the use of a uniform atomic nomenclature that applied to both elements and compound substances. Dalton spoke of atoms as single, double, triple, and so on, which would correspond to elements, binary compounds, ternary compounds, etc. This he did quite deliberately, as he attributed to the term *atom* a broader meaning than conventionally understood. In his lexicon, an atom is the constituent particle of *any* substance, whether simple or compound. That is to say, the atom is the smallest particle of a substance that still preserves the properties characteristic of that substance. For instance, while discussing the "atom" of carbon dioxide, he declares: "Now, even though this atom can be divided, it then ceases to be carbon dioxide, being decomposed into carbon and oxygen. I therefore see no inconsistency in speaking of compound atoms." This definition encompasses both atoms and molecules in modern terminology. Beyond a simple semantic problem, it inevitably contains the seed of misunderstandings, which will not fail to materialize.

A similar overuse of the word *atom* can be found in the work of J. J. Berzelius, who, to make matters worse, used a different nomenclature. He divided atoms into two categories: elementary atoms and compound atoms. In turn, the latter was of three distinct types: (1) atoms resulting from the union of two elementary substances, which he called compound atoms of the first order; (2) atoms formed by the union of more than two elementary substances, which he believed were to be found only in organic compounds, and which he accordingly called organic atoms; and (3) atoms made up of the union of two or more compound atoms (salts, for example), which he called compound atoms of the second order. This proliferation of nomenclatures was obviously not likely to shed much light on the essence of the problem, nor did it simplify the debates.

Significant progress toward a solution to at least one aspect of the problem, namely, the distinction between atoms and molecules, was provided by the volumetric laws of Joseph-Louis Gay-Lussac (1778–1850), and by Amédée Avogadro's (1776–1850) hypothesis. The laws of Gay-Lussac play somewhat the same role as the gravimetric laws of Proust and Dalton in that they establish certain numerical ratios governing the interaction and association of simple substances in their gaseous state. Formulated in 1809, they specify that the combinations of gases always occur in very simple ratios of volume, and that when the product of the reaction is itself a gas, its volume too forms a very simple ratio with that of its components. For instance, 2 volumes of hydrogen added to 1 volume of oxygen produce 1 volume of water vapor.

These laws, augmented by physical observations demonstrating that all gases behave according to the same laws of expansion and compressibility, led Avogadro to propose in 1811 his famous hypothesis—which was to become a full-fledged law a few decades later—that "equal volumes of all gases in the same conditions of temperature and pressure contain the same number of molecules."[5] More precisely, they contain the same number of "integral molecules," as, unlike Dalton, who spoke only of atoms, Avogadro dealt exclusively with molecules. That forced him to distinguish at least two types: elementary molecules (our atoms) and compound or integral atoms (our molecules). Avogadro showed that his law was in agreement with Gay-Lussac's observations, provided one accepted that "the molecules forming any simple gas, that is to say, those that maintain between themselves a distance sufficient to preclude any mutual interaction, are not made up of a single elementary molecule, but result, instead, from a certain number of these molecules united into one by attraction, and that when molecules of another substance must join these to form compound molecules, the resulting molecule that would be produced splits in two or more parts, or resulting molecules, made up of one half, one quarter, etc., the number of elementary molecules that formed the molecule of the first substance, combined with one half, one quarter, etc., the number of molecules of the other substance which would combine with the total molecule, or, which amounts to the same, with a number equal to the number of half molecules, quarter molecules, etc., of this second substance; thus, the number of resulting molecules of the compound becomes double, quadruple, etc., what it would otherwise be without this splitting, so as is required to fit the volume of the resulting gas."[6]

The main conclusion to be drawn from this absurdly long and awkward sentence is Avogadro's recognition that in the gaseous state the elements themselves can exist in a diatomic or polyatomic form. That is a crucial advance. The synthesis of water, which Dalton and Gay-Lussac would write as $H + O \longrightarrow HO$ and $2H + O \longrightarrow H_2O$, respectively (in modern notation), is described according to Avogadro's law as $2H_2 + O_2 \longrightarrow 2H_2O$. In this manner, the problem of the atomic weight of oxygen is resolved. Since one atom of oxygen combines with two of hydrogen, the atomic weight of oxygen is not 8 but 16 if that of hydrogen is unity (keeping in mind that 8 grams of oxygen combine with 1 gram of hydrogen). Dalton also disposed of the problem (to which we will return later) of valency (or atomicity, in the language of the time) of oxygen: One oxygen atom combining with two of hydrogen must possess a valency of 2, or must be bivalent.[7]

Significant as this development was, it did not signal the grand entrance of modern chemistry just yet. The quantitative theories developed by these pioneers would have to wait another half century to enjoy wide support and a full century to be universally accepted.[8] As a matter of fact, the confusion between atoms and molecules was further aggravated by the introduction of yet another linguistic variation. Jean-Baptiste Dumas (1800–1884), in his *Leçons de philosophie chimique* (Lectures in chemical philosophy, 1837), and Justus Liebig (1803–1873), in his *Chemical Letters* (1844), both spoke in terms of physical atoms and chemical atoms.[9] We defer to Dumas the task of defining what is meant by these terms, particularly since he did it in an especially picturesque, almost facetious, fashion. He wrote: "I find in Lewis the story of a demon who abducts a young lady and who, hoping to win her good graces, pledges to obey her first three orders: 'Show me,' said she, 'the most sincere of all lovers.' He complied promptly. 'Very well,' she continued, 'but now show me an even more sincere one.' The demon was taken aback. 'What would have happened had the lady been under the control of he who can show us an atom, and then divide it in two? He would have experienced no qualm producing the most sincere lover, followed by a more sincere still.' Mr. Griffins failed to understand that I had carefully distinguished between atoms relative to physical forces and atoms relative to chemical forces; that is to say, indivisible masses for the former, and other indivisible masses for the latter. It is thus possible to divide with one type of force that which resists the other type. In the case of chlorine and hydrogen, chemistry dissociated atoms that physics could not dissociate. That sums it all up."[10]

Unfortunately, all was apparently not that simple. The clash with the atomic theory occurred when Dumas set out to compare the values of the atomic weights of nitrogen, phosphorus, and arsenic, among others, determined from gravimetric laws, with those derived from the density of their vapors in accordance with Avogadro's law. To make the results agree would have required assuming two atoms in a molecule of nitrogen gas, but four in a "molecule" of phosphorus or arsenic (and, incidentally, a single one in a "molecule" of mercury and cadmium). But this state of affairs "completely confounded Dumas's ideas."[11] Given this confusion, he wrote in 1836 in his *Leçons de philosophie chimique*: "It becomes necessary to either renounce the most pleasing chemical analogies . . . or agree that equal volumes of phosphorus, arsenic, and nitrogen do not contain the same number of atoms." To agree would indeed have been sufficient, as Daumas and Jacques remarked judiciously.[12] Unfortunately, Dumas made the worst possible choice on this issue. He went on to say: "What is left of the ambitious exploration we began into the realm of atoms? Nothing firm, it seems. What does remain is the conviction that chemistry strays, as always, when it abandons experiments and decides to proceed through the unknown. Guided by experiments, you will find the equivalents of Wenzel [a precursor of Richter's], the equivalents of Mitscherlich, but to no avail would you search for those atoms such as your imagination dreamed when it bestowed on this word, unfortunately anointed in the language of chemists, an undeserved credibility. If I were the master, I would outlaw the word 'atom' from science, convinced as I am that it goes far beyond experiments."

The passion, almost recklessness, with which he wants to banish atoms from the world is astounding. "How can an experimental chemist of such caliber arrive at this epistemologically disastrous conclusion?" asked J. Pétrel.[13] His tentative answer is that it is due to "an erroneous reflex," common, according to him, in the processing of information by "our reasoning machine." Be that as it may, it is not beyond reason that philosophers should have, on occasion, ridiculed atoms. That theologians should have tried to eradicate them from the world is equally understandable. But that a scientist of stature should wish to dismiss a scientific theory is both astonishing and shocking. While Dumas luckily was not the master that he longed to be, one of his successors in the line of French chemists who shared his views would turn out to have enough administrative and political power to actually implement, half a century later (!), Dumas's threat. As it turned out, Dumas's attack was but the opening salvo of a broad offensive launched by scholars, foremost among them a number of French chemists, against the atomic theory. The offensive did not let up until the turn of the century. Other players became embroiled in this battle. Some were no less prominent than Dumas and just as virulent, and all the more culpable because they came still later in the century.

The key reason for the stance adopted by Dumas and his followers in this battle (and a veritable battle it was, producing its own casualties) comes through in his writings, in which he trumpets his "refusal of any speculation too far removed from observable facts." He viewed atomic theories as "sterile conceptions that could only throw the study of chemical phenomena into deplorable confusion," and he proclaimed that "whether ancient or modern, chemists insist on seeing with the eyes of the body, not with those of the mind; they are interested in developing theories of facts, not searching for facts supporting preconceived theories."

Dumas the scientist was turning partly philosopher, and, most unfortunately for him, he chose—whether he did so consciously or not is difficult to assess, and opinions on this subject continue to differ—a particular brand of philosophy that was very fashionable at the time but would prove grievously flawed. We are referring, of course, to Auguste Comte's positivism. In any event, from this moment on until the end of the century, the atomists no longer would be opposed just by philosophers and theologians, but even by some of their own colleagues. From triangular, the shape of the battlefield changed to square.

Under these conditions, it is gratifying to note that, in spite of occasional errors, experimental knowledge of the structure of matter nevertheless continued to move forward during this long period of groping. Even though the path was often bumpy and jarred by conceptual zigzags, gradual advances led inexorably to an accumulation of data whose atomic significance was becoming increasingly clear and compelling, at least to those who were willing to keep an objective mind.

Several important advances along this path during the first half of the nineteenth century must be mentioned. One is the law of Dulong (Pierre Louis, 1785-1838) and Petit (Alexis, 1791-1820), enounced in 1819. It states that the product of the atomic weight and the heat capacity of an element is a constant, independent of the nature of that element and approximately equal to 6.4 calories (thus demonstrating the power of rationalization afforded by injecting the atomic concept into what was otherwise a chaotic morass of empirical data). Other advances included the discovery, also in 1819, of isomorphism by Eilhard Mitscherlich (1794-1863); the increasingly accurate determination of the atomic and equivalent weights of new elements, thanks to the work of Berzelius, Auguste Laurent (1807-1853), and Charles Gerhardt (1816-1856); and the gradually expanding list of elements, eventually leading to the periodic table devised in 1869 by Dmitri Ivanovitch Mendeleev (1834–1907), of whom we will speak in more detail later in this chapter.

Also noteworthy is the development of electrochemistry, principally through the work of Alessandro Volta (1745–1827), Sir Humphrey Davy (1778–1829), and, later, Michael Faraday (1791–1867). Discoveries in the field of electrochemistry strongly suggested that chemical reactions could result from electrical phenomena. This inspired Berzelius to propose in 1819 an electrochemical theory of chemical combinations, in which each atom, whether simple or compound (using his own terminology), exhibits two poles charged with electricity of opposite sign. However, one of the charges predominates and gives atoms a "specific unipolarity," which gives rise to a classification of elements into two categories: electropositive and electronegative. Berzelius even arranged the elements according to an electrochemical scale, from the most electronegative (oxygen) to the most electropositive (sodium). The affinity between atoms would then reflect these opposite properties. According to this theory, the formation of chemical compounds results from neutralization of opposite charges, a process in which the "intensity of polarity" and the capacity for "induced polarization" also play important roles. The theory further introduced the notion of "residual polarity," with which Berzelius sought to explain the stability of compounds in which charges are fully neutralized so as to cancel the forces of attraction between their components.

Berzelius was actually not the first to propose that electrical forces play a role in the formation of chemical compounds or to look in these forces for a complement to, or even a substitute for, the purely mechanical interactions envisioned by the ancient atomists (not to mention a few more recent ones). Newton had already explicitly suggested the possibility that electrical forces might be involved. However, while the principal pioneers of electrochemistry either kept their distance from the atomic theory, as did Faraday, or rejected it outright, as did Davy, Berzelius systematically brought the electrochemical theory and the atomic theory together.[14] As such, he qualifies as a precursor of the modern theory of chemical bonds, at least for one type of chemical compounds. Still, his theory was beset by a number of difficulties. For instance, it did not explain how similar atoms combine, and in this respect it was at odds with Avogadro's propositions. Even half a century later, Berthelot would continue to consider the union of identical atoms "a mystical conception." In spite of these problems, the dualistic aspect of the theory was quite innovative in its explicit identification of a specific physical factor as a possible cause for the association of atoms or groups of atoms into more complex structures. It certainly caught the attention of a number of philosophers of the day, including Hegel (see discussion below).

Having been exposed to this new phenomenon of disagreement between scientists about the significance and validity of the atomic theory, it seems fitting to also examine the positions adopted by the philosophers of the time, particularly since one of them would develop theses that injected a far broader philosophical dimension into the debate between antiatomists—especially the equivalentists—and atomists.

Before proceeding, however, we wish to justify why we have chosen to divide the nineteenth century into unequal periods, the year 1860 marking the transition. The reason has to do with a very unusual event for the time: the gathering in Karlsruhe, Germany, of an international congress of chemists, the very first in history. Its commendable purpose was precisely to attempt to bring some semblance of order in the then-prevailing chaos of concepts and terminology. While the goal was unfortunately only partially met, the fairly wide consensus reached on formulating a distinction between atoms and molecules does rank among its most significant achievements.[15] That success was due in large measure to Stanislao Cannizzaro (1816–1910), who tirelessly reaffirmed the importance of Avogadro's law as the basis for determining atomic weights, condemned the proliferation of nomenclatures, and managed to keep the debates focused on the issue of atoms and molecules. He stated, for instance: "The various quantities of a particular element involved in the constitution of different molecules are integral mutiples of a fundamental quantity that always manifests itself as an indivisible entity and which must properly be named atom." Roscoe summed up Cannizzaro's definitions in these words: "A molecule is a group of atoms forming the smallest unit of a chemical species, whether simple or compound, that can be isolated or that can exist by itself: it is the smallest quantity of a substance that can enter in a reaction or can be generated by it; an atom is the smallest unit of an element that can exist in a compound substance as an indivisible chemical mass."[16]

What amounted to the first-ever synod attended by chemists did not, of course, instantly resolve their conceptual differences or smooth out the real difficulties they were struggling with in their efforts to describe the structure of matter. Nonetheless, the debates and exchanges of views during this event had a very positive effect and, further bolstered by progress in experimental research, did pave the way for the emergence of modern theories. In that sense, 1860 was a watershed year.

THE PERSPECTIVE OF THE PHILOSOPHERS

Three important philosophers, active mostly during the first half of the nineteenth century, explicitly presented their views on the atomic theory. They are Hegel, Schopenhauer, and Comte.

Hegel

When it comes to philosophers, one should never fear
not to understand. Indeed, the gravest danger is to
understand.

—Paul Valéry

It is widely accepted that Georg Wilhelm Friedrich Hegel's (1770–1831) philosophy is exceedingly complicated and difficult to understand. Indeed, at least in Bertrand Russell's opinion, it is actually the most abstruse of all great philosophies: "He [Hegel] tries to incorporate in his philosophy a great many intuitions and to make them coherent, with the result that his attempts to reconcile opposing points of view produce obscurities and contradictions."[17] "What is more," argues D. Collinson, "these difficulties are compounded by the differing interpretations of his philosophy by those who analyzed it."[18]

Since I have no ambition to join the already extensive line of interpreters, I will simply outline here the essential features of Hegel's philosophy in order to try to define his position vis-à-vis atomism.

Hegel's philosophy is probably one of the most monistic and holistic ever to have been conceived. The basis of his vision of the world is his refusal to attribute reality to anything that is "separate." What appears as the persistence of separate objects is but an illusion. The only reality, and the only truth, is embodied in the "whole," conceived not as a simple, homogeneous entity, but as a complex system, which he calls the "absolute." Actually, he does not altogether deny the existence of separate objects, but argues that they are real only to the extent that they participate in the absolute. Moreover, the real is rational, while the absolute is spiritual. It follows that mind is the ultimate reality. Finally, according to Hegel, the whole changes continuously through a process referred to as "dialectic," involving three stages: thesis, which formulates a proposition or a particular point of view; antithesis, which considers the opposite point of view; and synthesis, which reconciles both positions and becomes itself the basis for a new thesis. The cycle is repeated endlessly toward ever-increasing rationality, which, all things considered, constitutes a rather optimistic and reassuring view of things. The ultimate conclusion of this dialectical evolution is the "absolute idea," the meaning of which is somewhat obscure. Bertrand Russell writes that "the absolute idea is pure thought thinking about pure thought," which is precisely the definition of God proposed in the Middle Ages by the Jewish Aristotelian philosopher Levi ben Gershom, better known as Gersonides (1288–1344).

It is easy to appreciate that in such a philosophical framework the notion of innumerable atoms, in constant motion, separated by void, did not arouse much enthusiasm. As a matter of fact, Hegel's position toward atomism is generally negative. In his *Science of Logic*, he emphasizes the "inconsistency [a word which is used repeatedly in his pronouncements on the corpuscular theory] of the categories on which the corpuscular theory rests, be it the ancient version, or that which aspires to be modern."[19]

In spite of a generally hostile stance, his criticisms of the ancient atomists are (somewhat) more tempered than those aimed at their modern counterparts, his contemporaries. We will try next to define successively his own conception of an atomic doctrine, his opinion of the beliefs of the ancient atomists, and his view of contemporary atomism—the latter two being, of course, closely intertwined.

Definition of atomism. Perhaps the easiest definition to grasp is as follows: "It is the view of the atomistic philosophy that the absolute is determined as an entity per se, as one or several ones. Repulsion is recognized as the principal force manifesting itself in the notion of the one; however, it is not attraction that unites them, but blind chance. The one having been fixed as one, the union of these ones must obviously be viewed as something entirely external. Void, which is accepted as the second principle, on the same level as atoms, is repulsion itself, described as nothingness existing between atoms."[20]

The purely conceptual character of the atom is underlined in a passage containing a number of phrases that are characteristically biting, in which Hegel exposes the metaphysical nature of atomism and criticizes it as erroneous metaphysics: "The atomistic doctrine constitutes an essential step in the historical development of the Idea, and the general principle of this philosophy is the thing-in-itself in the image of the Several. Since even today atomism is favored by those physicists who refuse to deal in metaphysics, it must be reminded here that one cannot escape metaphysics or, more specifically, the reduction of nature to thoughts, by throwing oneself into the arms of atomism, since the atom is itself actually a thought, and consequently the apprehension of matter as comprised of atoms is a metaphysical apprehension. It is true that Newton expressly warned physics to beware of metaphysics; nevertheless, it must be noted much to his credit that he himself failed to act in accordance with his own warning. Besides, only animals can be pure and simple physicists, given that they do not think, while man, as a thinking being, is born metaphysicist. The only thing that matters then is whether the chosen metaphysics has the proper nature, and particularly whether, rather than the concrete logical Idea, it is not the result of unilateral elaborations of the thought, tainted by our reasoning, that produce what we hold to be true and constitute the fundamental basis for our theoretical and practical endeavors. That is a legitimate criticism of the atomistic philosophy."[21]

Of the numerous commentaries on this definition of atomism, I choose to reproduce here that of A. Lécrivain, a prominent expert in Hegelian philosophy. I consider his commentary to be one of the most comprehensible to the nonspecialist: "Hegel shows the two-sided nature [of atomism]: both positive and insufficient. He discerns the positive in the ancient conception of Leucippus, and to some extent of Democritus, seeing in it a specific speculative aspect inasmuch as atoms and void are posited not as empirical and tangible realities, but rather for what they are, namely, visualization of thought and pure concepts. As such, the atom is the representative expression of the logical category of the thing-in-itself, which itself is the culmination of a process of qualitative determination, the fulcrum beyond which the process tilts toward a quantitative determination. As for void, conceived as principle of movement, it corresponds to the logical moment of genuine negativity, that is to say, of the infinite. In light of this intrinsically speculative significance, there is little doubt that, for Hegel, modern atomism provides of that concept a somewhat adulterated picture."[22]

Opinion on the ancient atomists. This is where Hegel's position is the most balanced, although it remains fundamentally critical. A few excerpts from his *Lectures on the History of Philosophy* highlight his qualms about the atomic theory. He expresses a major objection to the teachings of Leucippus and Democritus: "If, starting from a broader and richer view of nature, we were to demand that it also be made more comprehensible on the basis of the atomic theory, then we would soon find ourselves dissatisfied and see what is inconsistent and unsatisfactory with proceeding with such a principle." His reason is that "the word atom can be translated as 'individual,' which immediately conjures up the image of a singular, concrete entity. These principles must be held in high regard, as they constitute progress: but their shortcomings become apparent as soon as one considers their implications. The representation of all that is concrete and real is developed as follows: the filled is anything but simple; on the contrary, it is infinitely multiple. These infinite multiples move about in void; for void simply is. Their union [their encounter] produces the coming into being—that is to say, the birth of an existing entity, to be perceived by the senses—while their dissolution and separation produce the passing away. All other categories fit in this framework. Activity and passivity consist of convergences of atoms. But their contact does not cause them to become one: for what is truly—abstractly—one cannot beget a multitude, not any more than what is truly—abstractly—multiple can beget one. Thus atoms are separated from one another by void, even in the case of their phenomenalistic union in what we call things. Void is also the principle of motion, as atoms move through it; and this movement is, so to speak, an invitation on their part to fill void and negate it. Such are the tenets of the atomists.

"Having so defined the void and the filled, nothing could be more congenial than to have atoms float around in this pervasive infinitude, where they are now separated and now united, in such a way that their union is but a superficial association, a synthesis independent of the nature of what is united, but where, in the end, these entities being in themselves and for themselves, remain separated, with no relation to one another specific to them, and no particularity.

"Given this completely external relationship, independent beings bind with other independent beings, all the while remaining independent; we are thus dealing with nothing more than a mechanical coupling. Any living, spiritual, etc., reality is therefore simply contrived; changes, generation, creation, are all just a union. And so is the weakness of this conception exposed. This representation of atoms has been revived in the modern age, particularly by Gassendi. Still, and this is the essential point, as soon as these atoms, these molecules, these tiny particles, are allowed to retain their independence, their union can only happen mechanically; once united, these elements remain external to one another, their very linkage is merely external. And that is a contrivance."[23]

The meaning of this fundamental criticism is unambiguous. Hegel's objection against the atomic theory centers on its inability to reveal the latent properties of the complex compounds they are supposed to form on the basis of ordinary contacts between atoms moving through void. The whole is obviously more than the sum of its parts, but the mere juxtaposition of atoms could never account for it. Hegel was not the first to voice this criticism, which had dogged the atomic theory since its inception. Aristotle and Cicero had already articulated a similar objection in antiquity, as did Newton and others at the dawn of modern times. Nor will Hegel be the last. As a matter of fact, the problem of the real nature of combinations liable to arise from interactions between atoms was to remain the bane of the atomic theory until the definitive formulation of the modern concept of molecule and an understanding of the forces responsible for their cohesion and individuality. The challenge was to find a mechanism capable of transforming the whole into a superstructure endowed with new and specific properties. Success did not come about until the advent of quantum mechanics, whose concepts and methodology finally overcame the almost insurmountable obstacle that the notion of the impenetrability of atoms had put in the way of progress on that front.

Hegel amplified his criticism in the pages of his *Lectures* devoted to Epicurus.[24] His target was the issue of primary and secondary qualities, or, in Hegel's own words, "the opposition between fundamental properties (specifically, weight, shape, and size) and derivative qualities, perceptible only relative to us." His denunciation of Epicurus's propositions on this topic is particularly scathing: "It would then be crucial to clarify the relation between atoms, which are the essence, and sensible phenomena. But here Epicurus gets mired in murky developments, devoid of sense. On this point, we discern in Epicurus and in other physicists nothing but a confused jumble of concepts, abstractions, and realities lacking any rational foundation. All specific structures, all things, all objects, even light, color, etc., and the soul itself, are nothing but particular orderings and arrangements of atoms. Locke also made that claim. Molecules ordered in space form the basis of everything. Those are hollow words." Here again is the same objection, namely, the obvious inability of atoms, whatever the degree and complexity of their superficial unions, to produce the substances and phenomena of nature: "The determination that atoms are formed in one way or another becomes a completely arbitrary fiction. As for the transition to concrete phenomena and material bodies, either Epicurus failed to deal with it at all, or what he alleges is pure nonsense. . . . His ideas about particular aspects of nature are ludicrous; they are an irresponsible hodgepodge of assertions of all kinds: such thoughts are therefore of no interest whatsoever."

Concerning Epicurus's atomic description of the soul, even if it involves the most subtle and perfectly round atoms, Hegel is of the opinion that "we will waste no time with such babbling; the words are hollow. We can have no consideration for Epicurus' philosophical thoughts, if they deserve to be called thoughts."

Yet a tribute or two to Epicurus can be uncovered in Hegel's writings. On one occasion he states: "One often hears praises of Epicurus's physics. If one considers that the mission of physics is to adhere to immediate experience and, when that is impossible, to apply the results of that experience to what is beyond the reach of experience using similarity as one's guide [by analogy], then Epicurus does indeed qualify if not as the originator, at least as the principal champion of this type of thinking, as the one who affirmed that it constitutes knowledge. . . . It can thus be claimed that Epicurus is the inventor of the empirical science of nature, or empirical psychology." For all his misgivings about Epicurus's science, Hegel did admire his moral philosophy, writing: "Epicurus's moral ethics is the most decried part of his doctrine (hence the most interesting); it can also be considered the best." It is a refreshing change from the hordes of scientists, philosophers, and theologians who, instead, admired his science but had little appreciation for his moral philosophy, or from those who liked neither. I have yet to meet anyone who liked both.

Assessment of the modern atomic theory. The main target of Hegel's criticism was the brand of atomic theory developed by the prominent Swedish chemist Berzelius, who was quite famous at the time and whose *Theory of Chemical Propositions*, published in 1818, was widely praised. Although in broad agreement with Dalton's work, this theory differed from it in a number of respects. For instance, Berzelius did not accept that atoms could have different shapes. He believed, instead, that they were all spherical and of uniform dimension. Hegel refers only seldom to Dalton in his own writings and, when he does, it is invariably in pejorative terms, claiming that he "shrouded his thinking in the worst form of atomistic metaphysics."

The backdrop of the squabble between Hegel and Berzelius was the work of another prominent chemist of the day, Claude Louis Berthollet (1748–1822), whose principal writing, *La Statique chimique* (Chemical equilibrium), published in 1803, also enjoyed wide recognition. He was opposed to Dalton's views, refusing to admit the existence of entities that are invisible and inaccessible to experiments, and dismissing the law of simple multiple proportions. Berthollet's main contribution was the concept of selective chemical affinity, which he proposed to explain on the basis of the masses of the reagents involved (and of the physical conditions surrounding the reaction), that is, in terms of observable and measurable data.

The polemic involved technical considerations of little current significance. Still, the overall nature of Hegel's objections remains of interest. His harshest criticism of Berzelius is that he "rehashed Berthollet's conceptions, repeated them word for word, simply adding the trappings of a metaphysics characterized by a lack of critical thinking; the categories involved become the only things deserving of close examination. Theory oversteps the bounds of experiment. On one hand, it contrives descriptions unsubstantiated by experiments, and on the other, it relies on conceptual images. Either way, the theory is vulnerable to logical critique." He later refers to Berzelius's propositions as "metaphysics devoid of any foundation, having nothing whatsoever to do with the propositions of metalogical saturation." Concerning the laws of proportions in chemical reactions, particularly Richter's law, he takes the view that "if experimentation was a sure-footed guide, the contrast was all the greater when one mixes these great discoveries with the barren desert, disconnected from the paths of experiment, of the so-called corpuscular theory."

These excerpts reflect Hegel's fundamentally negative attitude toward the atomic theory and its concepts. From his own perspective, the theory is not necessary for a description of the observable world. It is not compelling and must therefore be superfluous, if not downright false. As expressed by A. Doz in his commentaries on *Theory of Measures*, Hegel's view is that "the categories of concrete measurements constitute a level of intelligibility that is self-sufficient. It is therefore unnecessary, and it may even be harmful, to insist on propping them up with extraneous hypotheses. . . . For Hegel, the discontinuity characteristic of combinations is a direct reflection of the overlap between the qualitative and quantitative natures of measurements. One can—and since one can, indeed one must—dispense with the atomic hypothesis."[25]

Hegel's other criticism of Berzelius was his dualistic electrochemical theory of chemical combinations. This subject will prompt recurring debates.[26] Interestingly, Hegel adopted a favorable, although somewhat restrictive, view of the role of electrical phenomena in chemistry. As A. Doz stressed, the principal unknown, accepted as such by Berzelius himself, was that while electricity seemed to explain the formation of chemical compounds by the attraction of opposite charges, it was powerless to account for their stability, since "these charges cancel out." In Hegel's words, "If theory gives the insight that electricity is the cause of chemical affinity, electricity, on the other hand, sheds no light on just what is chemical in a chemical process." He asks the question: "Is the chemical not, in its corporeality, something different from electricity even today?" Doz interprets the phrase "even today" to mean that Hegel "does not contest the possibility that electrical phenomena may be involved in chemical processes, but he views their character as subordinate." The question was remarkably insightful for the time. It would be another century before it found an answer in the quantum theory of chemical bonds.

Schopenhauer

That which sees is incompatible with what is being
seen, although it is not always evident.

—Paul Valéry

Arthur Schopenhauer (1788–1860) could be called the "philosopher of will." Some may even claim that he was obsessed by it. The title of his most famous book, *The World as Will and Representation*, published in 1818 and reprinted in 1844, summarizes the essence of his doctrine.[27]

He was strongly influenced by Kant and adopted his distinction between the noumenon, or the thing-in-itself, and its image or "representation," which constitutes our experience of the world and is conditioned by a set of *a priori* criteria of human understanding resulting from the structure of our brain. However, and therein lies Schopenhauer's originality, he knew—or thought he knew—the nature of the thing-in-itself: will, of which the world of phenomena is merely a manifestation. He wrote: "Will is the thing in itself, the ultimate foundation, the essence of the universe, whereas life, the visible world, and phenomena are but the mirror of will." For him, every object and every individual in the world are simply a fragmentary manifestation of will, an image the only reality of which is will. Will alone is the source of everything and the tendency of things to manifest their nature; will alone determines the march of the world, both animate and inanimate. "Will, will devoid of intelligence (the only way it can be), a blind and irresistible desire, such as it manifests itself in the raw and vegetal world and in the laws that govern it, as well as in the vegetal part of our own bodies, that will, I maintain, through the perceived world which pledges subservience to it and evolves only to serve it, manages to know that it wants and to know what it wants; it is in this very world and in life itself that will is realized."

This cosmic will happens to be bad and evil. As essential as this aspect of Schopenhauer's profoundly pessimistic philosophy may be, it is not directly relevant to our discussion and we will not dwell on it any further.

"The world as representation" consists of, according to Schopenhauer, two components, "two poles: first, the knowing subject pure and simple, stripped of the forms of its knowledge, and, second, raw matter devoid of shapes and qualities." He held that "the fundamental flaw of all philosophical systems has been to ignore this truth, that intellect and matter are complementary, that is to say, they do not exist without one another, that they prop each other up and are solidary, that each is but a reflection of the other, in short, that they are to be properly regarded as one and the same viewed from two different angles; and that this unity constitutes the phenomenon of will or of the thing in itself; and that, as a consequence, they are both secondary; it further follows that the origin of the world is to be found neither in one nor in the other."

He goes on to say: "Matter and intellect are two interwoven and complementary entities; they exist only for one another and relative to one another. Matter is the representation of the intellect; the intellect is the only thing, and it is in its representation that matter exists. United, they constitute the world as representation, or Kant's phenomenon, in other words, something secondary. The primary thing is that which manifests itself, the thing in itself, in which we shall learn to recognize will."

One aspect of Schopenhauer's philosophy that is of particular interest to us is his belief that matter exists only in its own world as representation. It is in that sense that he considered himself an idealist. "It is healthy to maintain an idealistic perspective so as to counterbalance a materialistic perspective. Any controversy about the Real and the Ideal can be viewed as dealing with the existence of matter; for in the final analysis it truly is the reality or the ideality of matter that is being debated. Does matter exist in our representation, or is it independent of any representation? In the latter case, matter would be the thing in itself, and anyone assuming that matter exists by itself must, to be consistent, declare himself a materialist, that is to say, consider matter the principle explaining all things.

On the contrary, he who denies that matter is the thing in itself inherent-ly becomes an idealist."

This point of view put Schopenhauer at odds with the atomistic thinkers and explains his critical opinion of them: "It is indeed true that 'The world is my representation,' the fundamental axiom of subjective philosophy, can just as reasonably be countered by the credo of objective philosophy, whose paradigm is 'The world is matter,' or 'Only matter is' (because it alone is immune from death or change), or even 'All that exists is matter.' That is the fundamental axiom of Democritus, Leucip-pus, and Epicurus. But upon closer examination, it can be truly benefi-cial to look for the premise of a philosophical system not outside but within the subject itself; it affords one to remain one step ahead, and quite justifiably. For conscience is the only immediate datum, and we overlook it if we go straight to matter and adopt it as our premise."

His opposition to atomism, which at times could be acerbic, assumed several forms. First, he proclaimed that the infinite divisibility of matter is an integral part of the built-in context of our knowledge. It is one of our *predicabilia a priori* (premises), part of the core of "fundamental truths rooted in our intuitive *a priori* knowledge and viewed as basic principles." For instance, his fifth and sixth *predicabilia* declare, respectively, that "mat-ter is infinitely divisible" and that "matter is homogeneous and continu-ous, in other words, it is not made up of parts that were originally diff-erent [the *ominomeres*] or separate [atoms]; consequently, it is not an aggregation of parts separated essentially by something alien to matter." This premise implicitly rejects the concept of void.

Against this backdrop of antagonism on principle, Schopenhauer developed an explicit criticism of the atomic theory, which, like Hegel's, included both the ancient theory developed by Leucippus, Democritus, and Epicurus and its modern version, at least in the form he was familar with at the beginning of the nineteenth century. He articulated his criti-cism in the context of a more general condemnation of reductionism, which is not overly surprising given the holistic character of his doctrine: "But throughout the ages an etiology forgetful of its true purpose has attempted to reduce any organic life to chemistry or electricity; in turn, chemistry, which professes to be the science of qualities, appealed to mechanics [the action of atoms]; in due course, mechanics turned partly to phoronomy, which concerns itself with time and space united so as to make movement possible, and partly to pure geometry, in other words, simply position in space; finally, geometry can be reduced to arithmetic, which by virtue of the dimensional unit, is the easiest form of rational principle to grasp, understand, and explain in its totality. Would specific examples of the process we just sketched help? Just consider Democritus' atom, Descartes' vortex, or Lesage's mechanical physics. . . . This process

resurfaces later, even in the middle of the nineteenth century, in the form
of a crude materialism whose self-perception of originality is matched
only by its shallowness; disguised as a vital force, which is nothing more
than a foolish sham, it pretends to explain manifestations of life by
means of physical and chemical forces, to cause them to come from certain mechanical actions of matter, such as position, shape, and motion
of atoms in space; it purports to reduce all forces in nature to action and
reaction, which would become the 'things in themselves.' These preposterous, unwieldy and awkward theories, concocted by the likes of Democritus, are worthy of those who, fifty years after the publication of
Goethe's theory of colors, still believe in Newton's theory of homogeneous light and are not ashamed to admit it. They should be told that
what is tolerable in a child [Democritus] is unforgivable in a grown man.
They will bring shame onto themselves, but they will all manage to
abscond and feign ignorance."

The attack continues in subsequent pages: "This path leads it [materi-
alism] fatally toward the fiction of atoms, on which it purports to build
the mysterious manifestations of all primitive forces. . . . And the
inevitable fate of atomism is sealed: What had already happened to it during its childhood, at the time of Leucippus and Democritus, replays itself
now due to geriatric infantilism, because Kantian philosophy was ignored
in France and forgotten in Germany. And the confusion is even greater in
this stage of second childhood: not only are solid bodies presumed to be
made of atoms, but even liquids like water, or gases like air. But the theory
of atomism goes even farther astray. All these atoms are supposedly
endowed with various and incessant motions of rotation, vibration, and
such, according to their specific function; also, every atom would have its
own ether atmosphere, or some special property, and other hallucinations of the same ilk. At least the fantasies of the natural philosophy
devised by Schelling and his followers had for the most part a spiritual
nature, a certain loftiness imparted by their boldness, and a degree of
cleverness; the present ones, on the contrary, are ponderous, shallow,
inept, and clumsy; they are the product of brains incapable of conceiving,
first of all, of the existence of a matter with qualities not invented by
themselves, of a genuine absolute object, that is to say, an object without
subject, and second, any cause different from motion and collision: those
are the only two principles they understand, to which they pretend *a priori* to reduce everything, for such is their concept of the thing in itself."

The reader could not have missed the pointed criticism leveled at the
French for not being familiar with Kant's philosophy, and at the Germans
for having forgotten it. This short review is but a glimpse of the systematic
assault launched against the atomists in these two countries, with a particular virulence toward the French: "The French theories claiming to form

light by means of molecules and atoms are especially absurd and revolting. Their fallacies, as well as those of the entire theory of atomism, are
laid bare for all to see in a dissertation on light and heat published by
Ampère, an otherwise highly competent individual, in the April, 1835,
issue of the *Annals of Physics and Chemistry*. All solids, liquids, and gases
are, he claims, formed of similar atoms, and the way these atoms aggregate is what determines their differences: what is more, while space is infinitely divisible, matter is not; for once division is pushed to the level of
atoms, any subsequent division must occur in the spaces between atoms.
Light and heat are then described as vibrations of atoms, and sound
would be due to the vibration of molecules composed of atoms. In truth,
French scientists are obsessed with atoms, and they sound as though they
have actually seen them. It is amazing that a nation as keen on empiricism, as pragmatic and 'matter-of-fact' as the French should be so enamored of such an utterly arcane hypothesis, far removed from any possible
experiment, and use it with false confidence as a basis to concoct theories out of thin air. That is simply the result of a chronic neglect and backward state of metaphysics in their country; for, despite V. Cousin's best
intentions, his lack of depth and poor judgement could not do it justice.
Down deep, because of the past influence of Condillac, the French have
simply continued to blindly follow Locke. For them, the thing in itself is
literally matter, whose intrinsic qualities of impenetrability, shape, hardness, and other 'primary qualities,' must afford the ultimate explanation
of all things in this world: It is impossible to drive this notion out of their
mind, and they implicitly assume that matter can be moved only by
mechanical forces. In Germany, Kant's doctrine has long ago exposed the
absurdities of atomism and of mechanical physics in general; yet, at the
present time, these views enjoy wide acceptance even here, as a result of
the platitude, crudeness, and ignorance promoted by Hegel."[28]

Schopenhauer's wrath did not even spare the Swede Berzelius and his
notion of "chemical atoms which amount to nothing more than the
expression of ratios involved in chemical combinations, in other words,
pure arithmetical quantities that are, in the end, nothing more than
number games." He almost sounds like a French "equivalentist." Schopenhauer's critique of the French is a bit strange in that he appears to por-
tray them all as rabid and belligerent atomists. Alas, that was hardly the
case. As a matter of fact, Ampère was a rare exception in France in supporting the atomic theory. Many, if not most, French chemists of the time,
including, sadly, the most famous, were at least as adamantly opposed to
atomism as Schopenhauer. Ampère, of course, was himself a physicist.

Finally, the third front of Schopenhauer's attack against atomism followed a line similar to the one used to defend teleology, a concept rejected by the ancient atomists (although, as we have seen, some Christian

atomists of the Renaissance and the Age of Enlightenment did accept it). Lucretius finds himself under fire, an honor he shares with Bacon and Spinoza. Schopenhauer's opinion was that all three had an aversion toward teleology because "they believe teleology to be mired in speculative theology so abhorrent to them that they went out of their way to literally bury it." In fact, Schopenhauer was hostile mainly to Spinoza for judging "Lucretius's polemic against teleology so pedestrian and crude that it refutes itself and proves the opposite thesis." Schopenhauer was himself an advocate of Aristotle's final causes, which he considered "the true principle of the study of nature and especially of the organic [or living] world." Schopenhauer was equally combative toward the English antiteleologists of his time. The criticisms he hurled at them were no less fiery and insulting than the ones he reserved for the French. He argued that while the ancients can be forgiven for their outrageous rejection of teleology, because they did not know Kant, his own contemporaries had no such excuse. As it happens, Schopenhauer voiced many hostile opinions; some made rather bizarre associations, such as "Jews are worse than Hegelians."

Auguste Comte

No more than you, sir, am I in a position to do full justice to Mr. Comte. I cannot, however, help but feel emotional when I see so many honorable men in France, England, and America follow his name as they would a banner. Based on my experience in matters of the human spirit, I would predict that Mr. Comte will have his own label in the future and will occupy a prominent place in the history of philosophy. It will be a mistake, I must point out; but the future will commit so many other errors! Mankind demands names to serve as rallying points and beacons; it does not exercise much discernment in its choices.

—Ernest Renan

But you don't seriously believe that none but observable magnitudes must go into a physical theory.

—Albert Einstein

To be wrong is quite common in philosophy.

—Donald Davidson

Given the phenomenal rise of science in the nineteenth century and its veritable explosion in the twentieth, it is difficult to view the theory concocted by Auguste Comte (1798–1857) as more than an amusing footnote. His philosophical doctrine was hostile toward probability calculations, criticized any effort to learn the physical constitution of celestial bodies, rejected the very idea of the unity of matter, and condemned any research aimed at determining its structure, to the point of prohibiting the use of microscopes. He denounced all physical research conducted outside "usual" conditions, all theories on the evolution of biological species, all investigations on the origin of societies, and much more. If the theory of relativity or that of the big bang, quantum mechanics, psychoanalysis, and other fundamental disciplines of modern science are not explicitly included in this list of excommunicated subjects, it is simply because they did not yet exist in Comte's day. I say "explicitly" because some of these disciplines are implicitly condemned by anticipation.

Such is the case, for instance, of the theory of relativity and its explanation of universal gravitation. On this particular subject, Comte wrote: "As for determining the essence of this attraction and gravity, those are questions that we consider insoluble and outside the realm of positivist philosophy; we rightfully relinquish them to the imagination of theologians or to the contemplations of metaphysicists. The obvious proof that it is impossible to work out a solution to this problem is that whenever the greatest minds have tried to make a rational pronouncement on the subject, they became trapped in a circular argument by defining one principle in terms of the other: they claimed that attraction is nothing more than a universal gravity, while gravity consists simply in terrestrial attraction."[29] These comments highlight one of the greatest weaknesses, indeed the fatal flaw, of Comtian logic: It refused to consider that fundamental science looks toward tomorrow and that past failures do not preclude future success. We will return later to this odd conception of scientific progress.

Faced with this situation, some scholars have suggested that the best strategy is to give Ernest Renan a bit of posthumous satisfaction by simply sweeping this entire episode under the rug. For instance, not a single mention of Auguste Comte is to be found in Bertrand Russell's captivating *History of Western Philosophy*.[30] Equally silent is the well-known analysis of the history of science by W. Dampier, which is somewhat surprising considering that the title of his book advertises consideration of science's "relations with philosophy and religion."[31] Perhaps the explanation is that both Russell and Dampier are British. If I take the risk of appearing more indulgent than they by devoting a few pages to this infamous French philosopher, it is in part because I consider it an obligation to present the views on atomism of all well-known thinkers, whatever the merits of their beliefs, and, more important, because Comte's propositions appealed to a number of influential scientists of the time who were active participants in the controversy surrounding the atomic theory.

Auguste Comte was a fierce antiatomist. Actually, he was "anti-many things," as we have seen. This attitude stemmed directly from the fundamental principles on which he had built his philosophy, known as positivism.[32] Within the limited scope of this book, I will argue that the primary characteristic of this doctrine is a dual set of constraints imposed on the development of science, one utilitarian, the other conceptual.

The first constraint is related to the very purpose Comte assigned to the exact sciences, which he conceived as a tool for the development of a social science that, supposedly based on a firm rational foundation, would have as its mission to organize or, rather, reorganize society. The actual nature of this social reform is of little concern here. But what is immediately apparent is the potential danger of limiting the scope of scientific research—despite an avowed declaration of complete freedom—by subordinating it to utilitarian criteria, no matter how commendable their intended practicality. Armed with this principle, Comte proceeded to classify the sciences according to a hierarchy described by Michel Serres as "an encyclopedia of exact sciences that was dead on arrival."[33] But this fundamental strategy explains in large part the exclusions mentioned at the beginning of this chapter. What benefit to social organization, asks Comte, could possibly result, for instance, from experimental research in the fields of very high or very low pressures? The question may be candid, but it cannot hide the flawed thinking behind it. It is not hard to anticipate how detrimental to the development of science, including the social sciences themselves, such a restrictive and short-sighted stance will prove to be.

The second constraint, which hits at the core of Comte's intellectual construct, is more conceptual. It deals with restrictions imposed *on principle* on the scope and depth of knowledge and, by extension, on the description of nature and even on acceptable methodologies. The linchpin of this construct is the assertion that science must "restrict itself to the study of immutable relations that effectively constitute the laws of all observable events." Any hope to gain "insight into the intimate nature of any entity or into the fundamental way phenomena are produced" is summarily dismissed. Hypotheses "concerning the identification of generic agents which might account for various kinds of natural effects... are in the realm of chimeras, indeed are antiscientific, and, far from helping it, can only drastically hinder the real progress of physics." The sole purpose of science is, then, to establish phenomenological laws, that is to say, constant relationships between measurable quantities. Any knowledge about the nature of their substratum remains locked in the province of metaphysics. The French philosopher Gaston Bachelard will call this doctrine "systematic phenomenism" or "eviscerated phenomenology."[34] As if that were not enough, even though he advocated the supremacy of laws, Comte did not hesitate to dismiss those that did not fit his purpose. For instance, as a true-blue antiatomist, he rejected Proust's law of constant proportions, under the pretext that it does not make any prediction: It may explain the proportions of the constituents of a compound substance, but it fails to predict whether that compound will form or not. In fact, he was more favorably disposed toward Berthollet, a prominent chemist of his time, who rejected the law of constant proportions because his own work on alloys, glasses, and solutions led him to believe that the proportions of constituents in chemical combinations could vary over a rather large range. We might note in passing that this conflict could have been easily resolved by studying the structure of the compounds involved. But of course, for Comte, such a study belonged in metaphysics. He was even leery of perfecting laws to excess, too much precision being, in a strangely twisted view, incompatible with their very existence.[35] Had it been left up to Comte, he probably would have allowed Mercury's perihelion to continue to defy Newton's law until the end of time, with all the damage that such myopia would have inflicted on physics as a whole. Claude Bernard appears eminently justified when he commented in his famous *Introduction à l'étude de la médecine expérimentale* (Introduction to the study of experimental medicine), published in 1865: "It is better not to know anything than to cling to preconceived ideas based on theories that are confirmed only at the cost of ignoring anything that does not fit."

It is in the context of these twin constraints and the restrictions they imply that we must analyze Auguste Comte's position not only vis-à-vis the theory of atomism, but toward any research aimed at increasing our knowledge of the structure of matter. His position can be summed up in one word: rejection. Strictly speaking, Comte was willing to consider an occasional use of the notion of atoms, but only to the extent that "speculations" based on such an "artifice" be "useful," and under the strict reservation that it be limited to the spheres of physics and chemistry, to the specific exclusion of biology. In Auguste Comte's own words: "We must recognize that . . . each of our real sciences, after it has been appropriately focused on a search for effectual laws, and only on that, still contains important natural questions that the human mind will never be able to resolve, and yet deservedly qualify as positivist. Only a proper and often delicate appreciation for the true essence of each science must be allowed to preside over the choice of the relevant artifices, in order that the use of such speculative freedom aid, rather than obstruct, the growth of effective knowledge. In this respect, the hypothesis concerning the molecular structure of material bodies, commonly adopted in physics, is a good model, provided that one refrain from attributing to it a dubious reality and that one abstain from extending it to inappropriate subjects, such as biology. These two conditions are all too seldom met nowadays. . . . Physics is also the main impetus behind the corpuscular or atomic theory, which is in the process of firming up its logical structure. The concept is as appropriate in physics as inertia is in mechanics. But our tendency to endow our subjective constructs with an objective existence corrupts both these notions by giving the illusion of an exact description of underlying reality. . . . Only blind generalization transposed it in the realm of biology, where it proved directly contrary to the profoundly synthetic character of elementary notions in that field."[36]

He emphasized on several occasions the impossibility, in his opinion, of discovering the "intimate" constitution of substances and the "real" mode of aggregation of their elementary constituents. For this elusive knowledge, he substituted the simplistic concept of "binary" constitution. For instance, he wrote: "In order to simplify its fundamental notions, today's chemistry would be well advised to take more judicious advantage of the inevitable degree of uncertainty in its research on the nature of the intimate constitution of substances. The real mode of accretion of their elementary particles is fundamentally beyond our reach. Since it cannot possibly constitute the true object of any chemical study, we are left with the only rational alternative, in the restricted sphere of our positivist research, to conceive of the immediate composition of any substance as being binary. . . . Once any hope of knowing the intimate constitution, at once impenetrable and inconsequential, of substances is discarded, one can confidently accept that chemistry will always be entitled to regard any combination as binary."[37]

My own view is that the set of restrictions positivism imposed on itself is precisely what led it into an impasse. While no one can take issue with a doctrine claiming that we will never succeed in uncovering the "reality" of things—Comte was neither the first nor the last to make that assertion—the real issue is to clarify what is meant by "reality." If it refers to what some people sometimes call "independent reality," implying that it "does not proceed from the human mind," it leaves out another, much more directly accessible kind of reality often described as "empirical reality," which encompasses all the perceptions we humans have of independent reality.[38] It is as though Comte failed to see clearly the weight of empirical reality, made up of many successive layers of knowledge, or to recognize where the boundary between these two realities lies. In turn, this failure caused in him a propensity to relegate too hastily to the realm of metaphysics what at the time did not belong—or appear to him to belong—in physics. He dismissed the important possibility of expanding our knowledge through technological progress (as evidenced by his ban on the use of the microscope). He put an ideological brake on the inquisitive power of reasoning, hypotheses, and daring theories. For him, "genuine philosophical hypotheses must constantly display the characteristics of simple predictions about what experiments or reasoning could have uncovered immediately, had the problem come up in more favorable cir-cumstances." He virtually froze the world's science and its empirical reali-ty in the nineteenth century without apparently ever realizing that mod-ern science was then just in its infancy. As it turns out, "the childhood of science is long or, more precisely, eternal," as d'Alembert stated, and rarely do children know what they will be capable of when they become adults. To have doubts that one will ever reach the summit is one thing. But to quit at the bottom of the trail is quite another.

Today's metaphysics may well turn out to be tomorrow's physics, pro-vided that it can grow unfettered. The atom happens to be a perfect example. Of all of Comte's lunatic pronouncements, one of the most out-rageous is his prohibition of the microscope, harking back to some of Galileo's contemporaries and colleagues who refused to look through his telescope. What a contrast between this timid and regressive stance and Francis Bacon's infectious enthusiasm two centuries earlier, when he extolled the microscope and predicted that this new instrument would some day reveal Democritus's invisible atoms. He was proven right, of course, three and a half centuries later, with the advent of the scanning tunneling microscope (see chapter 20). Under the circumstances, it is understandable that Michel Serres considers Auguste Comte "not much more than a collector of tombstones, in the exact sciences anyway," and that in his estimation "the *Treatise [on Positivist Philosophy]* is, scientifically speaking, a graveyard full of fossils."[39]

The unthinkably narrow-minded imposition of such a rigid straitjack-et on the progress of scientific research doomed positivism to inevitable failure. While many scientific advances in a number of fields contributed to its decline and eventual demise, no factor was more decisive than the relentless accumulation of evidence in support of the existence of atoms. In this regard, Comte's timing in unsealing his indictment against the corpuscular theory was particularly poor. Jean Ullmo commented: "The ultimate test of all positivistic theories . . . was the debate about the theo-ry of atomism and its eventual triumph. . . . The advent of this scientific entity called 'atom' and its remarkable fertility sealed the fate of positivism and its narrow-mindedness." Ullmo elaborates on the significance of this victory: "Why are scientific objects so persuasive in research? In what way are they more compelling than the laws from which they are issued? These questions touch on a fundamental point: Objects are more powerful than laws because they suggest structures. . . . To propose a hypothetical struc-ture in order to explain known laws is to describe a new object; it amounts to crafting a new explanatory theory. . . . From the perspective of scientif-ic progress, relationships are merely the starting point, but interpretations are the essential tool."[40] The remainder of this book demonstrates how correct this analysis is, especially in the case of atoms.

In the end, the failure of positivism came about through an accumulation of mistakes, not the least of which was its inability to correctly grasp the link between experiment and reasoning in advancing scientific knowledge. The mistake was both logistical, as is always the case when one forsakes, or fails to coordinate, part of the means at one's disposal, and psychological, as any attempt to erect barriers against the yearning of the human spirit to move forward is bound to be. The guideline of "usefulness" could not have compensated for the sacrifice of this yearning, all the more so since nothing is more insidious than to decree *a priori* what is "useful." Had we heeded Comte's admonition, there would today be no atomism; nor would there be any nuclear energy, astrophysics, microbiology, molecular biology, virology, molecular pharmacology, gene therapy, and God knows what else. Michel Serres was not exaggerating when he wrote that "the bottom line is that our entire science violates positivism; it is rooted in what is prohibited in the *Treatise*."

Another grave mistake was to have entrusted to philosophy the task of guiding the direction of scientific research. For that to succeed would have required infinitely more constructive imagination than Auguste Comte possessed. To pronounce as absolute certainty that we will never see the hidden face of the moon is to display less imagination, less prescience, and even less scientific foresight than did Jules Verne.

Comte was hopelessly lacking in good judgment. He failed to make a distinction between the extraordinary and the impossible, and to realize how risky predictions can be, particularly if they deal with the future, as Yogi Berra might have said with his legendary wisdom.

CONSOLIDATION AND CONTROVERSY (1860 UNTIL THE END OF THE CENTURY)

Nature, eager to do serious chemistry, finally created chemists.

—Gaston Bachelard

While science in general experienced a phenomenal growth during the second half of the nineteenth century, it was the rise of physical and organic chemistry that contributed the most to the development and consolidation of the theory of atomism. For its proponents, the theory represented a powerful tool for a uniform and consistent explanation of an increasing number of observations. And yet, even as evidence continued to mount, a direct proof of the existence of atoms remained elusive (until the last decade of the century). The scientific community reacted to this type of situation in three possible ways. One segment affirmed the reality of the atomic world and pursued vigorous research to prove it. Another adopted a position of cautious neutrality, recognizing the intellectual impact and usefulness of the atomic theory, but carefully avoiding taking sides on the issue of the material world it presumed (this was the position of some prominent chemists, even as their own work was adding powerful ammunition toward the victory of that theory). Finally, there were those who rejected *a priori* the atomic hypothesis or, at the very least, the possibility of ever proving its validity. Whether such a hostile position was consciously or subconsciously influenced by the doctrine of positivism, it amounted, in the end, to nothing more than blatant prejudice. It incited some of the most prominent and, sadly, most influential chemists of the day, to wage a fierce rear-guard battle against the atomic theory.

"Did anyone ever see a gas molecule or an atom?" exclaimed Marcellin Berthelot, the flag-bearer of this militant minority, for which the phrase was to become a rallying cry. It is quite evident that Berthelot's own answer to this leading question was negative, not only as far as the past was concerned, but even for all times to come; he clearly implied that any effort to prove otherwise was doomed. We have already pointed out this curious pessimism, underestimating the possibilities of future technology, as the hallmark of Auguste Comte's philosophy. Mendeleev displayed a bit more wisdom when he stated: "From mushrooms to scientific laws, nothing can be discovered without looking and trying." It is important to stress that the battle for or against atoms was being fought among scientists, even if it sometimes assumed the character of a philosophical debate. In fact, the problem was of relatively little interest to philosophers in the second half of the nineteenth century, as they were engrossed in other matters. The few who did touch on the subject did so more out of nostalgia toward the great ideas of the past; they tended to focus on the atomic theory of antiquity, while modern developments left them rather apathetic. The Church, for its part, remained strictly on the sideline.

Among the many remarkable advances in the fields of physical and organic chemistry are two specific developments uniquely significant to our subject matter. As usual, the scope of our discussion will be restricted to these, and rather superficially at that, considering the many contributions, their interconnections, the number of actors involved, and the occasional contradictions brought about by their evolving views over the the course of the years.[41]

The first development is the establishment of the periodic classification of the elements, marking the successful climax of concerted efforts to arrange the chemical properties of elements according to their atomic weight. The second is the emergence of structural chemistry, which ousted what was a simple and primitive verbal description of the elemental composition, be it atomic or equivalentist, of substances and replaced it with a systematic determination of their internal architecture. This process ushers in the modern theories of valency and, ultimately, of chemical bonds.

We now proceed to discuss both developments. The determination of the number of elements or, equivalently, the number of distinct atoms characterized by their individual atomic weight had ranked among the most important problems of the atomic theory since Dalton developed his famous laws. That number had steadily increased from thirty-six in Dalton's days to approximately sixty by the middle of the century.

A decisive step toward recognizing the importance of an atomic description of the world was the realization that atomic weights might carry a deeper significance than that of an empirical parameter useful for classification purposes only. Progress was prompted by the insight of several scientists—including particularly Beguyer de Chancourtois (1819–1886), J. A. Newlands (1838–1898), and Lothar Meyer (1830–1895)—that the chemical properties of elements exhibited a striking periodicity with their atomic weight. The most complete and boldest articulation of this thesis was provided in 1869 by Mendeleev and his celebrated periodic table. Constructed initially with the sixty-three elements known at the time, using atomic weights determined through a combination of their vapor densities and Avogadro's hypothesis, the table demonstrated conclusively, in Mendeleev's own words, "that when arranged according to their atomic weight, the elements exhibited a periodicity of their properties"; in other words, they could be grouped into chemical families. The fundamental reason for the similarity of chemical behavior was unknown—the mystery would be solved half a century later by understanding the electronic structure of atoms—but a decisive step had just been taken by establishing a link between the chemical properties of elements and one of their specific measurable characteristics. Moreover, Mendeleev had the remarkably daring foresight to occasionally invert the order suggested for some elements by their atomic weights as they were known at the time, when he deemed such an inversion necessary to preserve their classification into chemical families, and even to leave some spots in the table vacant when he felt that the corresponding element was missing. As we know today, the future was to completely vindicate both decisions.

Moreover, the periodic table was to prove the conceptual and practical advantage of atomic weights over the notion of equivalent weights, which lacked the power to suggest such an orderly classification scheme. And so it is all the more surprising that the author of a chart so fundamental to atomism, the profound significance of which was about to be confirmed, never dared to "come out" and publicly acknowledge his own belief in the reality of atoms, even though he had written that "the word 'element' implies the notion of atoms." In his mind, the periodic table seemed to represent nothing more than an interesting coincidence, without any fundamental inference about the ultimate corpuscular nature of elements. Mendeleev's attitude of caution—some will call it timidity—appears to have been inspired by his adherence to positivistic principles.

In addition, the extremely wide range in values of atomic weights had raised questions on whether "heavy" elements could eventually be made either from "lighter" elements or from other fundamental entities. I have already mentioned Proust's hypothesis, which envisioned the synthesis of all elements from hydrogen. Mendeleev categorically rejected that hypothesis or any other assertion that elements, whose individuality he strenuously defended, could be made of common fundamental entities.

Mendeleev turned out to be wrong on that point, but he was quite correct in observing that "the arrangement of elements, or groups of elements, according to their atomic weights corresponds to their valencies." Mendeleev is not the only inventor of the concept of valency. Edward Frankland (1825–1899) is generally credited with the formulation of the concept, even though many other individuals were instrumental in defining and refining it.[42] By suggesting a link between valency and the inner structure of elements, Mendeleev had uncovered a fundamental characteristic of the atomic world. It opened the door to the second area of research that would provide overwhelming confirmation of the atomic theory—structural chemistry.

Decisive strides in this area occurred in organic chemistry. Spurred on by a veritable explosion of experiments synthesizing a host of new substances, chemists were busily trying to classify and organize them, and to understand their structure. At the core of these developments is the theory of valency.

The most significant advances in terms of classification and systematization are related to the introduction of the notions of radicals, substitutions, homologous series, types, and so on. We will express these concepts in atomic terms, even though many chemists still preferred in those days to think of them in terms of equivalent weights.

The possibility of transferring groups of atoms conserving their global identity, called "radicals" by Laurent and "residues" by Gerhardt, from one compound to another was demonstrated in the years 1830 to 1840 by several organic chemists, the most famous of whom were Friedrich Wöhler, Justus Liebig, and Jean-Baptiste Dumas. It quickly led to the notion of substitution, as chemists began to see in these radicals an inexhaustible construction set with which they could assemble and remove elements almost at will. The names of Dumas and Laurent are forever associated with the development of this technique. Its success promptly led to the discovery of homologous series of substances, thanks to the instrumental contributions of Laurent, Gerhardt, and, in the 1860s, Marcellin Berthelot. This, in turn, led to the notion of "types" (more appropriately called "prototypes" in modern language), or elementary modules in which the members of different series could be identified through appropriate manipulations of elements and radicals. The most notable contribution in this area is due to Gerhardt, who distinguished four main types: hydrogen, hydrochloric acid, water, and ammonia. In Gerhardt's own notation, these substances were written as:

$$
\left.\begin{matrix} H \\ H \end{matrix}\right\} \qquad
\left.\begin{matrix} H \\ Cl \end{matrix}\right\} \qquad
\left.\begin{matrix} H \\ H \end{matrix}\right\}O \qquad
\left.\begin{matrix} H \\ H \\ H \end{matrix}\right\}N .
$$

This representation reveals the varying ability of different elements to serve as support for interatomic bonds and underscores the importance of the notion of valency.[43] The term "valency" was actually not introduced until 1868 by Wichelhaus. Prior to that time, people frequently used the designations "atomicity," or even "basicity," "units of affinity," and others. Toward the middle of the nineteenth century, a consensus emerged to define the valency of an element as the number of hydrogen atoms with which it could combine. In the "types" discussed previously, for example, hydrogen and chlorine are monovalent, oxygen is bivalent, and nitrogen trivalent. In this context, another seminal advance was the discovery by August Kekulé of the tetravalence of carbon—a vital element in organic chemistry and, a little later, in biochemistry. He also demonstrated the ability of carbon atoms to bond to one another and form large and complex structures, and established that the bonding involved a valency of either one or two. All these properties were magnificently combined into the well-known hexagonal model of benzene, which Kekulé proposed in 1865. As a further refinement, he introduced the concept of "oscillations" of simple and double valencies between adjacent carbons—a harbinger of the modern theory of resonance.

These developments, and those due to chemists such as Archibald Scott Cooper (1831–1892), Alexander Boutlerov (1828–1886), and others, prompted a deeper appreciation for the role of the arrangement of atoms in molecular structures. Regardless of the ability to transfer groups of atoms, whether they be called "radicals" or "residues," from one compound to another during chemical reactions, the makeup of substances is ultimately determined by the arrangement of atoms. Phrased differently, the problem boiled down to the possibility (or impossibility, as the case may be) of replacing "raw" formulas of chemical compounds, which simply indicate the numerical ratios of the constituent atoms, with "developed" formulas that describe the way individual atoms are bonded.

From a historical perspective, this evolution may appear quite natural today, but it definitely was not in the nineteenth century, when such transformations could only raise troublesome questions. Whereas there exist today many direct physical methods to determine molecular structures (X-ray characterization being undoubtedly the most popular), no such technique was available in the nineteenth century, a time when studying chemical reactions was practically the only way to attack the problem. Such an approach is inherently indirect, with all the drawbacks that necessarily implies.

Another major difficulty, expressly pointed out by Kekulé, also held back the emergence of developed formulas: It is impossible to truly depict a three-dimensional spatial arrangement of atoms in a two-dimensional representation. This difficulty became even more glaring after the discovery of isomers, particularly optical isomers. As early as 1848, Louis Pasteur (1822–1895) thought them to be related to an asymmetry of the inner structure of molecules of otherwise identical composition. He even explicitly suggested in 1860 that they could result from different arrangements of atoms.

A decisive step toward a solution to the problem of structures was taken in 1874 by Achille Jacques Le Bel (1847–1930) and Jacobus Hendricus Van't Hoff (1852–1911), thanks to their brilliant hypothesis of an "asymmetrical carbon." They proposed that the four carbon valencies pointed toward the corners of a regular tetrahedron, with the carbon atom at its center. The designation "asymmetrical" referred to the ability of this model to account for the existence of optical isomers in polysubstituted derivatives of methane. This proposal took structural chemistry across a critical threshold and propelled it into three-dimensional space. To appreciate it, one need only compare the tetrahedron representing methane in this theory (Figure 9a) with the linear notation (Figure 9b) used previously by Kekulé to describe the same compound.

Le Bel and Van't Hoff would have been entirely justified in using their spatial model as a compelling argument in favor of the existence of atoms. Yet Van't Hoff wrote in a letter to Svante Arrhenius: "The models, the atom, the molecule, their sizes and perhaps even their shapes are all somewhat doubtful, as is the tetrahedron itself."[44] As it turns out, this bold theory failed to meet immediately with wide acceptance; it even aroused sharply negative reactions. For example, Adolf Wilhelm Hermann Kolbe, one of the great organic chemists of the time, wrote: "I commented not long ago that the lack of a good liberal education in many professors is one of the main causes of deterioration in chemical research. The result is a natural philosophy that has the appearance of depth and is spreading like wildfire; but in reality, it is trivial and hollow. This explanation, purged about fifty years ago from the exact natural sciences, is actually

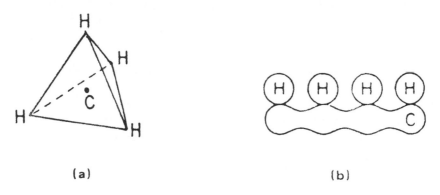

(a)   (b)

Figure 9. *Representation of methane (CH₄) a) after Le Bel and Van't Hoff; b) after Kekulé.*

being salvaged by a gang of scientific charlatans crawling out of the compost of mankind's mistakes. Like an old worn-out prostitute, it is being dressed in a gaudy outfit and fraudulently introduced into high society, where it obviously does not belong. Those who find my criticism too harsh are urged to read, if they can stand it, the recent monograph by a certain Mr. Van't Hoff entitled *The Spatial Arrangement of Atoms*, a book replete with childish nonsense. . . .This young peddler, employed by the College of Cows in Utrecht, has apparently no appreciation for research in rigorous chemistry. He prefers to mount his winged horse, his Pegasus obviously borrowed from the stables of the College of Cows, and, in his insolent flight toward his chemical Mount Parnassus, he claims to have spotted atoms arranged in space. The more sober chemical world has no interest in such hallucinations." This diatribe speaks for itself. For those wondering how a critique as vicious as this could have appeared in a technical journal (*Journal für praktische Chemie*), the answer is that its editor was Kolbe himself. If Kolbe had lived long enough, he certainly would have been surprised to see the first-ever Nobel prize in physics awarded in 1901 to none other than Van't Hoff!

To counterbalance Kolbe's excess of language and narrow vision, I wish to quote here a statement made in 1867 by Kekulé. Aside from hinting at the heavy burden of philosophical preoccupations, it conveys a reasonably positive idea of the kind of atomism envisioned by those chemists of the time who did not insist on shackling scientific discoveries to any preconceived dogmatic framework: "The question of whether or not atoms exist is of relatively little significance as far as chemistry is concerned; that issue belongs more in the realm of metaphysics. In chemistry, all that is relevant is to decide if the hypothesis of atoms is helpful in explaining chemical phenomena. . . . I can unhesitatingly state that, from a philosophical point of view, I do not believe in the actual existence of atoms, inasmuch as this term is to be understood in the literal sense of indivisible particles of matter. But I do hope that someday we will find a mathematico-mechanical explanation for what we have come to call atoms that can account for atomic weight, atomicity [valency], and many other properties of these so-called atoms. As a chemist, however, I consider the hypothesis of atoms not only useful but absolutely essential. I would even go a step further and declare my conviction that chemical atoms exist, with the stipulation that the term designate material particles that no longer undergo any division in chemical transformations. Even if scientific progress were to lead someday to a theory of the constitution of chemical atoms—as important as such an advance would be to a general philosophy of matter—it would make little difference in the field of chemistry. The chemical atom will forever remain a chemical unit."

THE DIE-HARD ANTIATOMISTS

You shall not make for yourself an image. . . .
—Exodus 20:4

It is easy to appreciate the confusion of scientists faced with tantalizing hints, but no irrefutable proof, of the existence of atoms. On the other hand, the often fiercely antiatomistic position of some of the greatest chemists of the day is somewhat puzzling. Indeed, the most acrimonious part of the ongoing controversy about the atomic theory, which was to continue until the beginning of the twentieth century, took place between men of science. But the nature of the arguments bandied about, particularly by the opponents of atomism, often gave it philosophical overtones. Consciously or not, the antiatomist camp often articulated a point of view influenced by positivism, claiming to "reduce science to a systematic phenomenon, to stand guard against assertions steeped in realism and against purely theoretical ideas," and portraying itself as protector of "a code of precautions aimed at warding off errors, rather than a thought process in search of discoveries."[45]

Among the last soldiers in the lost cause of antiatomism, we will make a distinction between the *equivalentists* and the *energeticists*.

The Equivalentists

The most famous and combative equivalentists are Henri Sainte-Claire Deville (1818–1881) and Marcellin Berthelot (1827–1907). Of the first, we will simply reproduce the following pronouncement: "Every time someone has tried to imagine or depict atoms or groups of molecules, I do not believe it ever amounted to more than a crude description of a preconceived idea, a gratuitous hypothesis, in short, a sterile conjecture. Never has any such description inspired a single serious experiment; all it does is seduce, not prove; and these notions, which are so in vogue these days, are for the youths in our school a graver danger than one might think. They lure the imagination and mislead the mind: they disguise themselves as legitimate interpretations of facts and make us overlook our ignorance. . . . I accept neither Avogadro's law, nor atoms, nor molecules, nor forces, nor particular states of matter; I absolutely refuse to believe in what I cannot see and can even less imagine."[46]

Sainte-Claire Deville seems to lack imagination, to put it mildly. As J. B. S. Haldane was to later observe: "Not only is Nature more bizarre than one thinks, but it is even more bizarre than one can imagine."

Berthelot's case is more interesting for two reasons. First, he was a dedicated and militant antiatomist who articulated his objections in minute detail. Second, his role was far more damaging because he had a high-level government position, which gave him the power to interfere with the spread of ideas he disapproved of; his own official edicts practically banned the teaching of the atomic theory in favor of equivalent weights right until 1890. In fact, Berthelot was an unbending equivalentist who limited the objective of science to devising classification schemes and recording relations between observable phenomena, in keeping with the purest positivistic tradition (although he vigorously denied it). The atomic hypothesis was in his view only a "source of confusion" unworthy of being taught in chemistry courses. He wrote: "The definition of equivalent weights is clear and generally amenable to demonstration by precise experiments. Quite different is the definition of the atom. It is founded sometimes on a disguised notion of equivalent weight, sometimes on the notion of gas molecule, which is a circular argument, and sometimes on the notion of specific heat, which is a varying quantity that could not possibly serve as the basis for a rigorous definition. In short, the atom is defined in this new system by three different notions, which often leads to incompatible results and arbitrary choices. Therefore, the definition of atom is itself arbitrary, and it is because of the confusion thrown into science by this ill-defined hypothesis that we refuse to see in it a basis for the teaching of chemistry."[47]

Berthelot defended his position at every opportunity. For instance, he spoke during a famous debate between equivalentists and atomists that took place at the Institute of France every Monday from April 9 to July 25, 1877 (the Academy of Sciences traditionally met then, as it still does today, on Mondays). His main opponent was Charles-Adolphe Wurtz (1817–1884), the most prominent of the French atomists.[48] Berthelot made his case in the following words: "Science must be based on laws and not on hypotheses. Laws can be proposed, debated, and definitively established, after which they become a solid foundation for science to progress steadily according to a methodology and language accepted by all. Unlike physics and astronomy, chemistry has yet to achieve such propitious maturity. To be sure, chemistry does have laws and general truths just as crisp and agreed upon as astronomy and physics. But certain individuals refuse to use these laws as the starting point of science and its only legitimate basis, as Messrs. Dumas and Liebig did forty years ago with universal approval. Today, many chemists less attuned to the precision of physical notions pretend they can replace the strict definitions of the laws themselves with murky descriptions—hypotheses, that is, which change with each generation, each school of thought, or even each individual. To expose the flawed nature of these hypotheses is not tantamount to blocking the progress of science or rejecting established knowledge; rather, by liberating scientists from extraneous baggage, it enables them to proceed with more confidence in their search for the real laws of molecular mechanics. . . . While we are at the moment in broad agreement about the general laws of chemistry, while we can even generally express them in a common language, I would hope that this consensus can someday extend further; by that I mean that we will rid our scientific theories of many obscure speculations, such as latent atomicities, stretched molecules, the precise location of atoms in space, [and] shared saturated atomicities in simple substances."

His objections included the atomic notation. As early as 1860, he wrote in his extensive monograph *La Chimie organique fondée sur la synthèse* (Organic chemistry based on synthesis): "Virtually every single system devised in the last twenty-five years in the field of organic chemistry exhibits the singular and universal characteristic of being based almost entirely on a mixture of symbols and formulas. These are theories of language rather than of facts. As a result, many a chemist has been prone to mistaking the properties of numbers hidden in formulas for the mysterious properties of actual entities: the delusion harks back to the mistake of the Pythagoreans, although it is perhaps less justified because of the nature of experimental science.

"It appears to us superfluous to discuss such conceptions in any detail, as they have no practical bearing on the issue of interest to us. Indeed, they deal only with numerical ratios of elements, rather than with the substances themselves, since they reduce all reactions to a characteristic unit that is necessarily fictitious. In short, they strip phenomena of any real character and substitute for their legitimate description a series of symbolic considerations that the mind finds easier to manipulate than actual reality.

"Chemical symbols can be dangerously seductive: the algebra of combinations is appealing, and the human mind is naturally inclined to substitute for a direct conception of things, which is invariably partly lacking, a simpler view in terms of descriptive symbols that give the appearance of more completeness."

Beyond the question of atoms, Berthelot took issue with the role of hypotheses and theories in the exact sciences. In response to this attack, Wurtz vigorously defended their usefulness, while conceding they had to be distinguished from facts. Addressing the Academy of Sciences during the session of Monday, June 4, 1877, he argued: "The atomic notation is not founded on arbitrarily chosen considerations; it rests on a body of chemical and physical data. With regard to the atomic hypothesis, which is at the basis of that notation, who could ignore the influence it exerted and still exerts today in science? Did it not offer the advantage, among others, of providing a link between the laws I have just reviewed, which would otherwise remain disconnected? . . . In our interpretation of facts, we endeavor to adhere closely to and categorize experimental data, convinced as we are that they must be the unshakable foundation of science, but we do so without absolutely repudiating hypotheses, for no science can do without them, no scientist can eschew them in his analytical efforts to summarize and organize facts by subordinating them to general principles."[49] Wurtz was supported in his defense of atomism by two other prominent chemists: Charles Friedel (1832–1899) and Paul Schutzenberger (1829–1897). Schutzenberger described the system of equivalent weights as "mixed and adulterated."

Be that as it may, Berthelot's actions had disastrous consequences on the teaching of chemistry, on research, and even on industrial development.[50] Bachelard gave a remarkable description of the climate in scientific education in France as he personally experienced it: "In this respect, it would be quite instructive to analyze the spirit presiding over the teaching of chemistry in the early twentieth century in France, even just a decade ago. Most textbooks complied with peculiar governmental decrees by mentioning the atomic hypothesis as an afterthought at the very end of the chapter devoted to chemical laws. Worse, some relegated it to an appendix to emphasize that chemistry had to be taught in an untainted positivistic form, through facts and facts only. Gravimetric laws—so simple, clear, and logical in the framework of atomic intuition— had to be taught without any reference to intuition at all. The trick was never to utter the word 'atom.' Everybody thought about it, but nobody ever said it. A few authors driven by last-minute scruples offered a short history of atomic doctrines, but it always followed a traditional positivistic treatment. Alas, how many of these 'politically correct' books would have made more sense had it been permissible to read them backward!"[51]

The Energeticists/Logical Positivism

The other school of thought opposed to atomism is that of the energeticists. It grew during the second half of the nineteenth century and flourished through the first decade of the twentieth. The most notable representatives of this movement include Ernst Mach (1838–1916) in Austria, Wilhelm Ostwald (1853–1932) in Germany, and Pierre Duhem (1861–1916) in France. Energeticism is a variant of positivism. It advocated a purely phenomenological study of nature, rejected any hypothesis about nature's objective reality, and dismissed any attempt to "explain" its essence. In this view, science need not concern itself with "things-in-themselves," as its mission is narrowly focused on establishing connections between experimental laws and mathematical logic. The hallmark of energeticism is to categorically refuse to appeal to any model, which inevitably implies a refusal to regard as legitimate scientific inquiry the study of the "intimate" structure of substances underlying their phenomenological behavior. Duhem wrote: "We will not discuss whether physical bodies are truly continuous or whether they are formed by disjointed parts separated by void, whether these disjointed parts have finite but very small dimensions, or whether they are simple points. All these questions concerning the real constitution of physical bodies are within the purview not of physics but of metaphysics. . . . Physics only seeks to devise, through concepts borrowed from mathematics, a logical system capable of providing an approximate description of laws relative to physical bodies." A theory was for him merely a "system of mathematical propositions, deduced from a few principles, intended to describe as simply, completely, and exactly as possible a set of experimental laws."[52]

Likewise, Ostwald wrote in his famous paper "The Disarray of Contemporary Atomism," published at about the time atomism was beginning its unstoppable charge toward ultimate victory: "Then, one might ask, if one must renounce atoms and mechanics, what image of reality is left? The answer is that no image or symbol is required. . . . To establish relations between realities, that is to say, tangible and concrete quantities, in such a way that, given the knowledge of some, others can be deduced, that is science's responsibility, and science fails to meet it when it espouses a more or less hypothetical image."[53]

The energeticists believed in the supremacy not just of phenomena but of perceptual data as well, which are the only source of our knowledge of the world. Mach affirmed this principle with particular emphasis.[54] Physical bodies and objects are, in his view, merely the sum of our perceptions as we experience them. These perceptions constitute the only reality accessible to us, which must be described as simply as possible, with the minimum of conceptual effort, in accordance with the criterion of "mental economy" (redolent of the principle of "least expenditure," meaning "requiring the fewest postulates," which Maupertuis used to identify the "best hypotheses").

The principles at the root of the energeticist doctrine are embodied, as we have said, in concepts of mathematical logic on which experimental laws are based. At the center of the doctrine is a belief in the unconditional primacy of energy as the governing agent of the physical world, enabling the comprehension, classification, and unification of all phenomena (mechanical, thermal, and chemical) that constitute the perceptible world. The ultimate expression of this philosophy is contained in the fundamental principles of thermodynamics, particularly the principle of conservation of energy, or principle of Carnot and Clausius. Its goal is to describe and codify the conditions under which matter reacts, without having to make any hypothesis whatsoever about the nature of matter itself. At least, that was the view of those who, like Duhem, were willing to admit that matter actually exists. Ostwald, one the most intransigent champions of energeticism in its purest form, simply denied the existence of matter. After announcing triumphantly that "happiness is our lot; our waning century bequeaths to the next one about to be born a scientific legacy pregnant with hope: the energeticist theory," he added: "Matter is an invention, and an imperfect one at that, that we manufactured to depict what is permanent in the vicissitudes of life. The only true reality, the one that affects us directly, is energy." From this vantage point, the conjunction between the world of perceptions, the only one Ostwald admitted, and ourselves is to be found in the fact that "all our perceptions have a unique and common characteristic: they correspond to a difference in energy between our sensory organs and the surrounding medium."

Such a position appeared to Ostwald to offer several advantages, which also constituted criticisms of a mechanistic approach: "First, natural science would be freed of any hypothesis. Next, there would no longer be any need to be concerned with forces, the existence of which cannot be demonstrated, acting on atoms that cannot be seen. Only quantities of energy involved in the relevant phenomena would matter. Those we can measure, and all we need to know can be expressed in those terms."

Moreover, as did virtually all opponents of mechanism, Ostwald expressly dismissed any prospect of applying the atomic theory to biology, a somewhat backward attitude at the dawn of the twentieth century: "Atomism is a doctrine that has miserably failed in any serious attempt to explain through mechanism all known physical phenomena. It is even less likely to succeed in tackling the incomparably more complex phenomena of organic life. Exploring this route cannot even qualify as an auxiliary hypothesis: it is purely and simply wrong." As Brenner observed: "In order to discard the hypothesis of atoms, Ostwald eliminates matter."[55]

Brenner equates this attitude to a "monistic metaphysics" and contrasts it to the more moderate attitude of Duhem, who did not reject the reality of matter as such.[56] Indeed, Duhem wrote: "We do not subscribe to those theories in which the substantial existence of diverse and massive matter is considered an illusion."[57] However, he did reject any hypothesis concerning the inner structure of matter. For instance, he asserted: "All hypotheses about the inner nature of matter and the structure of mixtures and chemical combinations, especially all atomistic hypotheses, are to be excluded from the sphere of science; no principle derived from these hypotheses shall be used; if an expression makes sense only on condition that one accept these assumptions, either explicitly or implicitly, it shall be rejected mercilessly; or before adopting it, a new definition completely purged of any objectionable doctrine will be agreed upon; in the final analysis, the definitions and propositions of chemical mechanics must involve only quantities representing physical properties that can be measured."[58]

Duhem did not reject the atomic theory simply because it is a hypothesis, as Maiocchi properly noted: "Duhem's loathing of atomism is rooted in his conviction that one must rebuff any theory that is based on figurative models. Although Mach saw in the use of analogies by figuration, particularly in the Maxwellian sense, a guideline for scientific research, he rejected atomic and molecular models because he considered them wanting. Duhem, on the other hand, renounced any figuration at all."[59]

Even though Duhem systematically rejected any atomic description of matter in general, he still felt he had to attack certain specific propositions in the atomic theory or its symbolic notation. He did so at great length in his book *Mixed Substances and Chemical Combinations*.[60] He took issue with two central tenets of atomism: the notion of valency and the significance of gravimetric laws. For instance, he wrote: "The modern chemical notation, based on the notion of valency and so improperly called atomic notation, is a very effective tool for the purpose of classification and discovery as long as it is used only as a symbolic description or as a guideline for various ideas pertaining to chemical substitutions; but should one turn to it for a description of the arrangement of atoms and the structure of molecules, one is likely to be overwhelmed with doubts, inconsistencies, and contradictions."

These arguments are not new. Interestingly, however, Duhem saw "an insoluble contradiction" in the notion of valency viewed "as an intrinsic property of the atom and a consequence of its configuration." The alleged contradiction arises from the fact that the valency of an element can vary from one compound to another. Duhem used the example of nitrogen, which, depending on the circumstances, can be trivalent or pentavalent. This implies that nitrogen has three "atomicities of the first order" and "two atomicities of the second order," which could not possibly be due to "the same cause." He argued that "if the five atomicities were strictly identical, the existence of compounds in which three of them are satisfied and the remaining two are not would be absurd by reasons of symmetry. We are thus forced to admit that in a given atom, in this case nitrogen, there exists a fundamental difference between an atomicity of the first order and one of the second order, regardless of the origin and nature of this difference." Whether such a difference is acceptable is an eminently pertinent question. Duhem's answer is no.[61] As we know today, the answer was wrong, and we even understand why: It is because Duhem failed to incorporate structural considerations in his reasoning. Even Democritus's primitive model, with its undissociable atoms featuring hooks and bumps of various kinds, could account for this multivalency. One need only envision three hooks of a particular type and two more of another to explain the trivalent or pentavalent behavior of a nitrogen atom. This point of view was in fact expressly defended by the French chemist D. Naquet, and the phenomenon itself came to receive a more elegant explanation based on the electronic structure of atoms.[62] Thus, Duhem can blame only himself and his stubborn refusal to consider structural issues for his own inability to conceive of a mechanism for the multivalency of elements. It is impossible to walk through a door after having locked it and thrown away the key.

Another example illustrates the intellectual gyrations Duhem relied on to repudiate the atomic theory. He acknowledged that "the atomistic doctrines rest on a foundation that cannot be ignored, although its firmness should not be exaggerated. That foundation is the law of multiple proportions." Yet he refused to accept its unconditional validity on the grounds that no experimental measurement can ever be precise enough to irrefutably confirm it. He conceded that it could not be disproved, either. As far as he was concerned, the law simply "is beyond the reach of experiment." Even assuming that it could be verified, "would the victory be decisive? For that to happen, not only should the interpretation of the law of multiple proportions, provided by the atomic theory, be plausible and appealing, it would also have to be the only one possible. But who would be willing to guarantee this interpretation and certify that no other explanation could ever be devised? There is more: Upon noting the ease and clarity with which all principles of modern chemistry fall into place in a treatment excluding the name and notion of atom, and the difficulties and contradictions arising the moment one tries to interpret these principles in terms of the atomists' doctrines, one cannot help but think that the unique success of the theory of atoms is an illusory victory without future; that this theory does not enable us to know the true and objective foundation of the law of multiple proportions; that this foundation remains to be discovered; and finally, that all things considered, modern chemistry does not support Epicurean doctrines."

These are stunning errors, coming as they did at the dawn of the twentieth century, and they do not stop there. Not content with rejecting modernism, Duhem reverted to one of the most formal aspects of the beliefs of antiquity. This regression is quite apparent in his comments about the significance of the "new chemical mechanics" and his own views on chemical reactions and the nature of mixed substances. He stated: "Chemical mechanics, in its present form, does not claim to take us to the very heart of what matter is or to reveal to us the *quid proprium* of chemical reactions; its more modest, but also more attainable, goal is to classify and organize the laws experiments enable us to discover; agreement of its corollaries with facts is the only criterion for certainty. As far as mixed substances are concerned, this new science needs only to know their composition, that is to say, the masses of elements that must be destroyed to create this mixture and that can be regenerated by its corruption. Upon the ruins of the notion of mixture crafted by the atomists, it rebuilds the simple and unshakable conception originally formulated by Aristotle.

"One specific and unavoidable question immediately comes to mind: what distinction does the new chemical mechanics make between physical mixtures and chemical combinations? The answer is that it makes no distinction at all between these two concepts; or, more precisely, the principles of thermodynamics on which it is founded do not make any allowance for attributing different meanings to the two designations; they afford no basis in their reasoning or equations to indicate whether a phenomenon constitutes a chemical reaction or a simple change in physical state. According to both Aristotle and modern thermodynamicists, elements no longer subsist in a mixed state; they only exist in potentiality."

Ostwald shared Duhem's Aristotelianism. For instance, in discussing "the idea of matter" and its atomic description, he wrote in 1895: "It is important to recognize that this extension has introduced a host of hypothetical aspects in a concept that at first contained not a trace of hypothesis. In particular, under the influence of this theory, and contrary to any evidence, it came to be accepted that matter undergoing chemical reactions does not disappear to be replaced by another with different properties. What is more, this view forces one to admit that in iron oxide, to take a specific example, iron and oxygen still exist even though all their organoleptic properties have disappeared: they have merely acquired new properties. It has become difficult to remain aware of the strangeness, indeed absurdity, of such a conception, so accustomed have we grown to it. But think about it for a moment: All we can learn about any particular substance are its properties. Is it then not nonsense, or not much better, to pretend that a definite substance continues to exist when it no longer possesses any of its original property? In fact, this hypothesis has but one purpose: to make the general facts of chemistry, particularly the laws of stoichiometry, consistent with the entirely arbitrary notion of inherently unalterable matter."

Need it be stressed that these words were written on the eve of the advent of the theory of chemical bonds, of which those willing to open their eyes could already discern the outline?

In summary, these die-hard antiatomists were dedicated Aristotelians, and they were not even the last ones. As it turns out, Ostwald was the only one to eventually recognize the validity of the atomic theory, in 1908. Mach, in contrast, was still on record as proclaiming his rejection of the theory in 1910. As for Duhem, he wrote in 1913, more than twenty years after the discovery of the electron: "The neo-atomistic school, whose doctrines are centered on the notion of electron, has revived with utmost confidence the method we refuse to follow. It believes that its hypotheses ultimately concern the inner structure of matter and that it reveals the elements as though some extraordinarily powerful microscope could magnify them and make them visible. This confidence we cannot share; we cannot recognize in these hypotheses some divining power bringing to light what is beyond the realm of the perceptible, but we view them merely as models. The time will undoubtedly come when these descriptions and models, because of their increasing complexity, will cease to be of any use to physicists, who will then recognize them as burdens and obstacles. Abandoning these hypothetical mechanisms, they will carefully uncover the experimental laws these mechanisms hinted at; without claiming to explain these laws, they will endeavor to classify them according to the method we just described and to understand them in terms of a modified and generalized energeticism."

It goes without saying that adopting the viewpoint of the energeticists would have been an intellectual catastrophe. The far-reaching progress of science, particularly in chemistry and biology, brought about by the elucidation of the structure of matter—a case in point is the success of molecular biology—is ample testimony to the shortsightedness of these nineteenth-century scientist-philosophers quick to espouse restrictive doctrines. How unjustified was their phobia of models, even as working tools, in light of the now widely accepted technique of computer-aided molecular design, which has become a scientific research tool of unparalleled power.

The deficiencies of the positivist doctrines in general, and of those of the energeticists in particular, came to be recognized long ago. Indeed, they were exposed by a number of the most prominent scientists of the time. Einstein, for one, was quite explicit in criticizing Mach's principle of "mental economy." His comments on the subject are quoted in Heisenberg's memoirs: "Mach's concept of thought economy probably contains part of the truth, but strikes me as being just a bit too trivial. . . . When the child forms the concept 'ball,' does he introduce a purely psychological simplification in that he combines complicated sense impressions by means of this concept, or does this ball really exist? Mach would probably answer that the two statements express one and the same fact. But he would be quite wrong to do so. To begin with, the assertion 'The ball really exists' also contains a number of statements about possible sense impressions that may occur in the future. Now future possibilities and expectations make up a very important part of our reality. Moreover, we ought to remember that inferring concepts and things from sense impressions is one of the basic presuppositions of all our thought. Hence, if we wanted to speak of nothing but sense impressions, we should have to rid ourselves of our language and thought. In other words, Mach rather neglects the fact that the world really exists, that our sense impressions are based on something objective. . . . He pretends that we know perfectly well what the word "observe" means, and thinks this exempts him from having to discriminate between 'objective' and 'subjective' phenomena. No wonder his principle has so suspiciously a commercial name: 'thought economy.'"[63]

Unlike Einstein, Max Planck initially was taken by Mach's view on the exclusive reality of perceptions and was skeptical about the existence of atoms. But he changed his mind at the beginning of the twentieth century and discredited Mach as "a false prophet" in an address delivered in Leyden in 1908.[64]

Others attacked Duhem's Aristotelianism more bluntly. Paul Langevin, for instance, wrote in 1904: "Although necessary, these [energeticist] principles are not sufficient and constitute only one of many trunks from which branches can develop, contrary to the beliefs of those modern evangelists who, having lost sight of the origin of that principle, turn energy into a new idol whose multiple incarnations are supposed to describe everything. . . . These various avatars of energy are at the basis of what Mr. Duhem describes and defends so shrewdly under the name 'qualitative physics'; he invokes Aristotle and replaces measurable quantity by quality, which he nonetheless expresses in numbers, all the while withholding from it any effectiveness. What, then, is the meaning of two numbers describing two different degrees of the same quality? Is there not here a disturbing tendency to limit the scope of investigations, to declare sufficient and definitive what amounts to a general and superficial knowledge of things, to shun a deeper scrutiny because an initial success managed to deliver a few of the more general laws? Who knows what our forward march will yield? Who knows what a microscope might reveal about living organisms? What is there to fear? Why this return to the past, this *ignorabimus* against which our instinct and convictions rebel?"[65]

Likewise, Léon Brillouin lashed out at the idea Duhem and Ostwald had of the "mixed state": "In the current state of knowledge in chemistry, there still are simple substances that cannot be reduced into one another; there are even many of them. It is this experimental notion that prompted the idea that oxygen and iron exist side by side in iron oxide. One may look the other way and avoid talking about it; but as soon as one does talk about it, as soon as one is reminded that every single process used to decompose iron oxide never produced anything but iron and oxygen, it becomes impossible to escape the conviction that oxygen and iron had to have remained distinct. It was not so long ago that this idea finally prevailed and that the transmutation of metals became a chimera. Conservation of matter is an experimental law, and it is more profound than conservation of mass. It is not just the total mass of iron oxide that is equal to the sum of the masses of oxygen and iron; but the individual masses of oxygen and iron that can be extracted are themselves invariant."[66]

Ludwig Boltzmann (1844-1906) adopted a position very much in favor of atoms in the exact sciences. We owe him, among other achievements, the consolidation of the kinetic theory of gases. The theory, which states that the pressure of a gas on the walls of a container is due to the continuous impacts of atoms agitated by incessant movement, dates back to the Swiss atomist Jean Bernoulli (1654-1705) and his son Daniel (1700-1782). Boltzmann is also credited with the development of statistical thermodynamics. He summed up his fundamental view as follows: "Of any extended body of experimental facts, we could never conceive anything more than a mental image, and not a direct description. We must then never follow Ostwald's lead in saying: You shall not make yourselves images. What we should say instead is: Your images will contain as few arbitrary elements as possible."

This criticism really applies to positivism as a whole, and not just energeticism, which is only one of its short-lived offshoots. The fatal flaw in these intellectual movements—a surprising flaw at a time when intense scientific activity and sustained progress were uncovering deeper layers of knowledge—was to have tried to reduce science to pure phenomenalism without recognizing the crucial role played by reasoning in building this knowledge, and to have rejected models for fear that they might be erroneous or fictitious. As Aubert expressed it, such a posture amounted to a throwback to medieval nominalism.[67] Once again, the clearest commentary on the disastrous effect that the grip of positivism had on the way its proponents viewed science was expressed by Einstein: "The aversion of these scientists [Ostwald and Mach] toward the atomic theory can almost certainly be traced to their positivistic philosophical leaning. It is an interesting example in that, even when gifted with a bold mind and sound intuition, scientists can be misguided in their interpretations of facts by philosophical prejudices . . . consistent with their belief that facts themselves can and must produce scientific knowledge unencumbered by any freewheeling conceptual construct."[68]

Jean Perrin, too, displayed considerably more perspicacity when he wrote in 1912 in his book entitled *Atoms* that he had no problem with a succession of structural models stimulating the development of science: "There are cases when hypotheses are actually intuitive and fertile. When we study a machine, we obviously examine as best we can its visible parts, but we also try to guess the hidden mechanisms that explain its workings. To guess the existence or properties of objects beyond the reach of our knowledge, to explain a complex visible with a simple invisible, that is the kind of intuitive intelligence atomism has afforded us, thanks to individuals like Dalton and Boltzmann. . . . A time will perhaps come in the future when atoms can be seen directly and will become as easy to observe as microbes are today. The spirit of present-day atomists will then soar again with those who inherit the power to hypothesize, beyond an expanding experimental reality, about some other hidden structure of the Universe." These words were to prove prophetic.

Finally, the ideas of Mach and his followers aroused reactions outside of the scientific community as well. We would be remiss if we did not mention the opinion of a very special critic: Vladimir Illyich Lenin (1870–1924). Not surprisingly, the great master of dialectical materialism repudiated Mach's theories; he dismissed the doctrine as "antiquated" and its proponents as "dogmatic," preferring to believe in the objective reality of matter. As he wrote in *Materialism and Empirio-Criticism*, published in 1908: "We ask this question: Is it or is it not objective reality when a man sees the color red, feels a hard object, and so on? To answer no is to deny from human knowledge any fundamental grasp, that is to say, to lock oneself in subjectivism and agnosticism. If, on the other hand, one accepts objective truth as attainable, it needs a philosophical concept. Such a concept has been formulated a very long time ago; it is matter. Consequently, to assert that this concept can become 'obsolete' is pure gibberish and amounts to stupidly rehashing the arguments of a trendy reactionary philosophy. Could the struggle of idealism and materialism really have become outmoded in two thousand years of philosophical evolution? And what about the struggle of Plato's or Democritus's concepts? It is absolutely impermissible to confuse, as did Mach's followers, a doctrine about the structure of matter with a gnoseological category, or to confuse the novel properties of new aspects of matter (electrons, for instance) with the old questions of the theory of cognition, the sources of our knowledge, and the existence of an objective truth."[69] He further contended that a theory postulating that the material world is but an organized collection of sensations is inconsistent with established scientific truth. As proof, he argued that there was a time when matter did exist while beings capable of experiencing sensations did not.

This concludes the part of this chapter devoted to the emergence and maturation of scientific atomism during the nineteenth century. If this chapter gives a general impression of confused ideas, of a chaotic and oftentimes conflicting intellectual process, it will have served its purpose well. Rarely in the history of science has a steady and gradual accumulation of experimental data, pointing to an increasingly ineluctable conclusion, aroused more misunderstanding and controversy. It is true that the topic itself, invisible and indivisible atoms, was the perfect breeding ground for such confusion. But it is equally important to stress the inhibiting effect closed-minded and preconceived ideas had on the building of an atomic vision of the world. These ideas either proscribed reasoning and imagination as illegitimate components of science or confined them within restrictive boundaries imposed by philosophical doctrines lacking boldness and foresight. The failure of these attitudes and doctrines, and the contrary triumph of atomism, even if a "reformed" version of atomism, which we will witness in the next chapter, attest to the futility of efforts to curb the scope of scientific exploration.

A FEW NOSTALGIC PHILOSOPHERS

Atomism had attracted the attention of a great many philosophers in the first half of the nineteenth century. They continued to be actively involved in atomic matters as long as the doctrine remained essentially unverifiable, before it began to establish itself as a scientific theory. But in the second half of the century, when its experimental verification was becoming increasingly probable, philosophers seemed to lose interest. The few who did take up the matter in their writings dealt primarily, if not exclusively, with the teachings of their distant predecessors of ancient Greece, and appeared largely unconcerned with modern developments. The most prominent among these nostalgic philosophers are Friedrich Nietzsche, Karl Marx, and Henri Bergson (although Bergson produced the bulk of his writings in the twentieth century, his interest in atomism manifested itself in the early part of his career, toward the end of the nineteenth century and the beginning of the twentieth).

Nietzsche

Greece owes everything to science, and the rest of the
world everything to Greece.

—Jean-Jacques Rousseau

In the first part of this book, we have had several opportunities to mention the opinions of Friedrich Nietzsche (1844–1900) on ancient Greek philosophy, a topic in which this great German thinker had a deep and continuing interest; he also had a well-known fondness for ancient Greek tragedies. The traditional aspects of his philosophy—a particular ethical system epitomized by the notion of a "superman," and marked by harsh criticism of the New Testament (the Old Testament escaped relatively unscathed), contempt for women, and a belief in an "eternal return"—are of peripheral interest to us here. By contrast, his views on the ancient atomists—he does not appear to have had a great deal of interest in modern atomism—merit some discussion.

We do not mean to convey the impression that he had a particular fascination for the ancient theory of atoms. Although he was interested in all the pre-Socratics, his favorite philosopher of the period was Heraclitus, with whose teachings he identified the most. His opinion of Heraclitus tells much about his own conception of the world: "I reserve a special reverence for Heraclitus. While the rest of the philosophical world rejected the testimony of the senses because they suggest plurality and change, he himself rejected the same testimony because it gives to things the appearance of stability and unity. Heraclitus, too, is unfair toward the senses. Their falsehood is not what the Eleatics imagined, nor is it what Heraclitus believed—for they do not lie. Their falsehood is in the way we use their testimony: the senses lie about unity, reality, substance, and constancy. 'Reason' is what causes us to falsify the testimony of senses. To the extent that the senses testify about becoming, destruction, and change, they do not lie. . . . But Heraclitus will forever be justified in saying that being is a hollow fiction. The 'apparent' world is the only world; it is a lie to add the qualifier 'true world.'"[70]

Indeed, one of the primary tenets of Nietzsche's philosophy is to reject any belief in the existence of an objective structure of the world that would be independent of how the mind envisions such structure. For him, "the apparent world, the one we fashioned for ourselves, in which we have accepted substances, lines, surfaces, causes and effects, movement and rest, form and content, is the only one to exist. The notion of a real world is a lie. It would disappear if we were able to eliminate our own interpretations." Our belief in laws of causality and necessity is just a "trick" to enable us to find how we fit in the world, but it teaches us nothing about how things really are. Life itself proves nothing: "To err may well be one of the conditions of life."[71]

From this perspective, Nietzsche's tribute to Democritus is all the more unexpected: "Of all the ancient systems, Democritus' is the most logical: it assumes everywhere the strictest necessity; it contains neither sudden interruptions nor outside interventions in the natural course of things. Only then does thought free itself from all anthropomorphic conceptions of myths; one finally has a scientifically usable hypothesis; this hypothesis of materialism has always been highly fruitful. It is the most down-to-earth conception; it starts from the real qualities of matter and does not from the outset seek to go beyond the simplest forces, unlike the hypothesis of the *nous* or Aristotle's final causes. It is a great conceptual achievement to be able to reduce the entire universe filled with order and exact purpose to the innumerable manifestations of a unique force of a most common type. Matter moving in accordance with the most general laws produces, via a blind mechanism, effects that seem like the designs of a supreme wisdom. . . . To Democritus, we owe many memorial services just to make amends in some small way for the injustices of the past toward him. For rarely did a writer of stature have to endure so many attacks for so many reasons. Theologians and metaphysicists have heaped on his name unending denunciations of materialism. Did the great Plato not consider Democritus' writings so pernicious that he wanted to destroy them in a private auto-da-fé, a plan he abandoned only after realizing that it was too late and that the poison had already been spread too far and wide? Later on, the obscurantists of antiquity took revenge on him by fraudulently introducing under his name their writings on magic and alchemy, saddling the father of all that is rational with the reputation of a crafty magician. Early Christianity finally succeeded in implementing Plato's drastic scheme; and an anticosmic century must undoubtedly have regarded the writings of Democritus and Epicurus as the incarnation of paganism. Lastly, it came to be the dubious distinction of our time to also deny the philosophical greatness of that man and to detect in him sophistic tendencies. These attacks are no longer defensible." What follows is the ultimate compliment: "Democritus, the Humboldt of antiquity."[72]

Karl Marx

Only the universal fame of Karl Marx (1818–1883) justifies the passing mention of his early writings devoted to the atomic theory. He addressed the topic in his doctoral thesis, entitled *Differences Between the Natural Philosophies of Democritus and Epicurus,* and in a posthumously published piece, "Epicurean Philosophy" (*Cahiers d'étude,* 1839–1840).[73] His writings on the subject contain nothing remotely revolutionary, or even mildly penetrating. In other words, nothing can be found that would warrant giving this section the so much more titillating title "Red Atoms." In fact, that Marx chose the atomistic philosophers as the topic of his dissertation does not even imply that he had a particular liking for their ideas. While their general materialism fits well with his future economic and political conceptions, nowhere is an enthusiasm for atomic heroes, so pervasive in Nietzsche's writings, evident with Marx. He was simply conforming to a fashionable trend when he declared Aristotle the greatest philosopher of Greek antiquity. For instance, he wrote: "What must never happen in a good tragedy is precisely what seems to afflict Greek philosophy: a drawn-out denouement. With Aristotle, the Macedonian Alexander of Greek philosophy, it is as though the objective history of philosophy finds its conclusion in Greece. . . . But a very ordinary truth teaches us that to be born, to mature, and to pass away are the path every human endeavor is fated to follow. If so, there is nothing surprising in the fact that, after peaking with Aristotle, Greek philosophy subsequently became decrepit." He did not subscribe to the view of those who consider "the Epicureans, the Stoics, and the Skeptics an almost annoying footnote, without any connection with their illustrious forerunners." On the contrary, he stressed the historical impact of these movements, but refrained from giving Epicureanism any undue importance: "When we examine history, do Epicureanism, Stoicism, and Skepticism qualify as particular phenomena? Are they not, in truth, paradigms of Roman culture, the form under which Greece migrated to Rome? Are they not of such original, profound, and eternal essence that the modern world itself felt compelled to unhesitatingly grant them intellectual legitimacy?"

The fact that these systems manage to completely get around "Platonic and Aristotelian philosophies" puzzled him and incited him to ask the question: "How is it that post-Aristotelian systems find, in a manner of speaking, their foundations ready-made in the past?" His answer is: "It seems to me that while prior systems are more significant and interesting in terms of the essence of Greek philosophy, post-Aristotelian systems, particularly the Epicurean, Stoical, and Skeptical schools, are more significant by reason of their subjective philosophical form and character. But it is precisely the subjective form or spiritual foundation of these philosophical systems that has been until now almost completely neglected in favor of their metaphysical implications."

It is mainly these general considerations, which appear in the introductory section of his doctoral dissertation, that are worth mentioning. The dissertation itself contains an extremely detailed analysis—documented by numerous doxographers—of the differences between Democritus's and Epicurus's atomic views of the world. Marx comes down on

Epicurus's side, in part perhaps because Epicurus heeded some of Aristotle's criticisms when he introduced certain modifications to Democritus's propositions.

On the other hand, Marx's very critical opinion of Gassendi's efforts to rehabilitate Epicureanism is of interest. On this topic, he wrote: "Gassendi, who freed Epicurus from the censure decreed by the fathers of the Church and the entire Middle Ages, a time of institutional intolerance, offered in his commentaries but a single interesting element. He sought to reconcile his Catholic conscience with his pagan knowledge, and Epicurus with the Church; obviously to no avail. It would have been easier to try to clothe the luscious body of a Greek courtesan in a nun's habit. Gassendi was hardly in a position to educate us about Epicurus's philosophy; it is more accurate to say that he learned philosophy from Epicurus. . . . That, incidentally, enables us to see that Pierre Gassendi, who wanted to salvage divine intervention, the perenniality of the soul, and so on, while remaining faithful to Epicureanism, failed to understand Epicurus, and was even less capable of teaching us about his ethics. Gassendi only sought to instruct us through Epicurus and not about him. When he violates iron-clad logic, it is to avoid conflict with his own religious prejudices." Coming from Karl Marx, this judgment is unlikely to surprise anyone.

As evidenced by his correspondence with Friedrich Engels (1820–1895), Marx maintained a keen interest in the development of new theories in chemistry and seems to have had a particular appreciation for the contributions of Laurent and Gerhardt (letter of June 22, 1867), of Kekulé (letter of June 24, 1867), and of Frankland (letter of May 10, 1868). He agreed with Engels (letter of June 16, 1868) that "for all their deficiencies, the modern chemical theories represent a considerable improvement over the ancient atomistic chemistry." He fully appreciated (letter of June 22, 1867) the significance of the "molecular theory," which, incidentally, seemed to confirm one of the fundamental rules of dialectics, Hegel's law "of the sudden commutation from purely quantitative to qualitative change."[74] Indeed, Engels had defined chemistry "as the science of qualitative changes occuring through a modification of quantitative composition."[75] Marx saw a most vivid proof of that concept in the discovery by Mendeleev that the chemical properties of elements are a periodic function of their atomic weights; in other words, that "their quality is determined by the quantity of their atomic weight." In his letter of June 22, 1867, already mentioned, Marx underscored explicitly the parallel between the commutation from quantitative to qualitative changes in molecular chemistry and the transformation of the master craftsman into a capitalist, which he also considered a qualitative change produced by purely quantitative means. As we know, he believed strongly that the laws of dialectics govern the history of human society just as much as they do the laws of nature.

Bergson

Atomism inherently contains a majestic poetry.

—Friedrich Nietzsche

Although his most significant writings place Henri Bergson (1859–1941) squarely as a twentieth-century thinker, his interest in atomism was most evident in one of his early works, entitled *Les Extraits de Lucrèce* (Excerpts from Lucretius), published in 1883, which contains selected poems by Lucretius, along with his own summaries and commentaries. These clearly reflect the admiration that Bergson—a future Nobel laureate in literature—held for Lucretius's poetic worldview.

According to an exhaustive analysis of Bergson's writings by Marie Cariou, what appears to have struck him the most in Lucretius's texts was the complementary participation of the infinitely compact (the atom) and the infinite extent of space (void) in the making of the world. Even though it is merely "the passive element of a mixed entity of which the atom is the active and positive part," void, which is "the necessary reason for the existence of the world of material bodies," owes its importance to the fact that it enables atoms to acquire movement, a property they have had for all eternity by virtue of their own weight. Moreover, although strictly excluded from atoms themselves, void pervades all substances and accounts for the interpenetrability, that is to say, the innermost creative interactions, observed in the world: "Without void, all would be chaos rather than cosmos."[76] Bergson clearly favored movement in all directions over Epicurus's declination, which he rejected as childish fantasy. He did, however, concede to that concept a symbolic value reflecting the "creative spontaneity" of nature and human moral freedom. Bergson seemed fascinated by the endlessly alternating formation and fragmentation of atomic combinations, the random occurrences of which ensure the creation, evolution, and disappearance of worlds. He furnished his own personal definition of disappearance: "The universe will truly perish the day when atomic combinations will have been exhausted." We intentionally use the term *worlds*, as Bergson was a firm believer in a plurality of possible worlds, each comprised of a finite number of atoms, although this number may be infinite when considered collectively. However, unlike Lucretius, who envisioned that other worlds could be quite different from ours, Bergson assumed that they are all more or less alike.

Given the subsequent evolution of Bergson's views, it is not surprising that even in this early work, he was already struggling with "the paradox that life is determined by the sensitivity of atoms that are themselves incapable of feeling." By all appearances, he readily accepted that "life is a particular system of atomic interactions," and that sensitivity can be the result of material organization. In Lucretius's concept of the random formation of initially simple organized beings, of which the only ones to subsist and develop were those that could adapt to the conditions of their environment, Bergson saw the seed of the hypothesis of a creative evolution—the "continuity of springing forth" against a backdrop of atomic discontinuity—that must conform to the requirements of a natural selection process.

As Bergson was becoming more and more the philosopher of duration, intuition, memory, and vital yearning, his attention moved away from the atomic theory. Yet, many of his writings continued to echo themes resonating with that theory. Such was the case, for example, of the perception and reality of the discontinuous, a concept articulated explicitly toward the latter part of the nineteenth century by A. Hannequin in the context of atomism: "Physical atomism is not imposed onto science by reality, but, rather, by our methodology and by the very nature of our knowledge. It would be wrong to believe that atomism necessarily implies the actual discontinuity of matter; it merely implies that we conceive of matter that way in order to understand it, and that mathematics as we know it introduces discontinuity while trying to describe it. In short, atomism has its roots in the ubiquitous use of numbers, which leave their imprint on everything they touch; while numbers are more than a convenient tool for geometry, where they constitute an essential and indispensable method, and while numbers are truly the basis of our entire knowledge of geometry, by the same token one need look no further for the origin of the principles of atomism and for the preeminence of that theory in modern science."[77] This point of view amounts to claiming that atomism came to be because of the very nature of our cognitive processes or, phrased differently, that "the atom was born out of numbers." The contention is that what is continuous is unintelligible, and in order to become intelligible, reality must be resolved into discrete units, for these are the only entities that can be grasped by the human mind in its efforts to analyze and synthesize, in other words, to do science.

Without resorting to the concept of numbers, Bergson similarly justified the intelligibility of the discontinuous itself. He wrote: "Upon examining human intellectual faculties, one realizes that the mind is comfortable and in its element only when it deals with raw matter, particularly with solids. But what is the most general property of raw matter? The answer is that it has extent; it presents us with objects that are distinct from other objects, and within these objects parts that are distinct from other parts. With a view toward subsequent mental manipulations, it is without doubt useful to think of each object as arbitrarily divisible into parts, each such part being further divisible at will, and so on to infinity. But for the present purpose, it is above all necessary to view the real object of interest or the real elements into which the object has been resolved as definitive for the time being and to think of them as so many units. When discussing the continuousness of material extent, we really imply the ability to decompose matter as much and in any way we want; but this continuousness, as can easily be appreciated, boils down to the flexibility we have to choose whatever degree of discontinuousness we wish to find in matter: in the final analysis, once we settle on a particular degree of discontinuousness, it appears to us quite real and captures our attention because that is what our mental process happens to be focussed on. Thus is discontinuousness thought of in its own right and thinkable in its own right; we visualize it through a positive action of the mind, whereas the mental picture of continuousness is rather negative, being in the end simply the tendency of our mind to refuse to view any given system of decomposition as the only one possible. The intellect can conceive clearly only of the discontinuous."[78]

That argument should not preclude nature from actually being discontinuous. In point of fact, Bergson appeared to endorse the notion that matter has an atomic structure.[79] However, he refused to regard mental processes as mechanistic, remaining consistent in this respect with the distinction he made between matter and life.[80] Even though he conceived of mental activity on the basis of an atomistic physico-chemical model, he repudiated any "psycho-physical parallelism."[81] "Assuming . . . that the position, direction, and speed of each atom of cerebral matter are all determined at any given instant, it would not in any way follow that our psychological life should be subject to the same inevitability. For one would first have to prove that to a given cerebral state there corresponds a rigorously predetermined psychological state, and such a proof remains to be established."[82]

Finally, it is appropriate to review Bergson's opinions on an idealistic or realistic conception of matter, on the one hand, and on the nature of interatomic interactions, on the other. On the first point, Bergson proposed to show "that idealism and realism are two equally extremist theses, that it is false to equate matter with the image we have of it, and equally false to make it into something capable of eliciting pictures in us while remaining of a completely different nature than these pictures. Matter is to us a set of images. By image we mean a certain entity that is more than what the idealist calls a representation, but less than what the realist calls a thing—an entity half way between the thing and its representation. This conception of matter is simply based on common sense. . . . In short, we consider matter as it was before idealism and realism dissociated it into something between its existence and its appearance. Now that philosophers have made that dissociation, it is unquestionably difficult to ignore it. Nevertheless, we beg the reader to try to do just that."[83]

On the second point, Bergson's eloquent discourse presages ideas that are to prevail in the twentieth century: "If there exists one truth that science has established beyond any doubt, it is the notion of mutual action of all parts of matter on one another. Forces of attraction and repulsion exert themselves between presumed molecules. Gravitation's influence reaches across interplanetary spaces. Something must therefore exist between atoms. It may be claimed that it is no longer matter but force. One could picture threads stretching between atoms, getting ever thinner until they become invisible, if not downright immaterial. But what good is such a crude picture? The preservation of our own lives probably demands that a distinction be made in our daily experience between inanimate objects and the actions exerted by these objects through space. Just as it serves our purpose to fix the seat of an object at the precise location where one could touch it, its palpable outline becomes to us its genuine boundary, and we proceed to perceive in its action something mysterious that emanates and is different from it. But precisely because any theory of matter purports to rediscover the reality hidden in these customary images, always subjugated to our needs, it must first abstract itself from such images. And we are indeed witnessing a gradual confluence of the concepts of forces and matter, as physicists gain a deeper understanding of their effects. We are seeing forces becoming materialized and atoms idealized, both terms converging toward a common limit, the universe thus regaining its continuousness. People will continue to speak of atoms; atoms will even keep their individuality in our minds intent on isolating them; but the materiality and inertia of atoms are bound to dissolve into either movements or force lines, two concepts whose complementarity and reciprocity will reinstate the continuousness of the universe."[84]

20

The Twentieth Century

From an Invisible and Indivisible Atom to One That Is Divisible and Visible

The atom is a world more profound than the sun.
—Victor Hugo

Nature displays the same unbounded magnificence in an atom as in a nebula. Any new means of study reveals Nature to be greater and ever more diverse, fecund, unpredictable, beautiful and richer, and unfathomably immense.

—Jean Perrin

By the dawn of the twentieth century, the atomic theory seemed almost universally accepted. Compelling evidence was being garnered on several fronts. Chemists were engaged in experiments described by an alphabet of letters symbolizing the constituent atoms of the known elements. Physicists were studying the kinetic properties of gases (Rudolf Clausius [1822-1888], James Clerk Maxwell [1831-1879], Ludwig Boltzmann [1844-1906]), electrochemistry (Svante Arrhenius [1859-1927]), osmotic pressure (Jacobus Van't Hoff [1852-1911]), and, at the turn of the century, Brownian motion (Albert Einstein [1879-1955] and Jean Perrin [1870-1942]). Using a variety of techniques, Jean Perrin managed to determine thirteen different values of Avogadro's number, all roughly consistent with one another, thus demonstrating that atoms could effectively be "counted."[1]

Perrin's success even incited Ostwald to finally recognize the validity of the atomic theory, which he did in the preface to the fourth edition of his work *Grundriss der physikalischen chemie* (Foundations of physical chemistry) (1908). It also prompted Henri Poincaré, long skeptical about the existence of atoms, to publicly admit at a conference sponsored by the French Physical Society shortly before his death that "the atomic hypothesis has recently acquired enough credence to cease being a mere hypothesis. Atoms are no longer just a useful fiction; we can rightfully claim to see them, since we can actually count them." To many scientists, the publication in 1913 of Jean Perrin's book *Les Atomes* (Atoms) marked the decisive victory of the atomic theory.[2] At this stage in the history of atomism, it looked as though Leucippus and Democritus had finally won the battle, at least on the essential issues.

But the victory of the classical atomic theory proved short-lived, if it even had ever truly materialized. At just about the time when the existence of atoms as fundamental constituents of matter was becoming practically irrefutable, new discoveries opened up unexpected perspectives. A new chapter in the history of atomism began with the discovery that these individual building blocks are themselves made of a number of smaller components. While they still remained the basic units of the chemical elements, atoms turned out to be themselves composite entities. This meant, however, that they had lost their traditional and fundamental characteristic of indivisibility. Even though they retained their role as distinct and individual constituents of all observable substances, they could no longer be touted as the most fundamental particles of matter. And so the triumph of the atomic concept was immediately marred by an etymological crisis. Only through a semantic stay of execution—in deference to the conservatism of scientists, in short because of a long-established tradition—was the atom allowed to keep its ancient name. The notion of indivisibility had to be pushed back to a level beyond atoms, toward some of their smaller constituents.

In view of the numerous discoveries and many scientists involved in this new chapter of the atomic saga, we can only afford a condensed and selective summary. We will deliberately concentrate on some specific aspects of twentieth-century atomism that raised troubling implications about "the anxieties we harbor under the name of philosophy" and religion.[3]

The twentieth century was marked by a quick succession of astounding developments, not just in the atomic theory proper but in all scientific fields. A yearly event was instituted to recognize the latest advances in a formal and sometimes dramatic setting. We are referring, of course, to the Nobel prize, the first of which was awarded in 1901. Numerous authors of discoveries of interest to us were to receive this high scientific distinction. This achievement will be noted by adding to the dates of birth and death of the relevant individuals the symbol *NP 19xx*, indicating the year in which they were awarded the prize. So as not to penalize some scientists already mentioned at the beginning of this chapter, the following updated list is provided: Svante Arrhenius (NP 1903), Jacobus Van't Hoff (NP 1901), Albert Einstein (NP 1921), Jean Perrin (NP 1926), and Wilhelm Ostwald (NP 1909). As far as we know, no one has succeeded in determining the extent to which the very existence of the Nobel prize may have accelerated the pace of scientific progress.

THE MULTICORPUSCULAR ATOM

From the Discovery of the Electron to Rutherford's Planetary Model of the Atom

In 1911, Rutherford introduced the most profound change in our conception of matter since the time of Democritus.

—Sir Arthur Eddington

The crucial discovery of the electron brought the hypothesis of indivisible atoms to an end and ushered in a period of intense scientific study of their complex structure. Although much of the work that led to that discovery was done by several scientists over a number of decades, the key step was the demonstration in 1897 by Joseph John Thomson (1856–1940, NP 1906) that cathode rays (which result from electrical discharges in tubes containing rarefied gases) involved particles carrying a negative electrical charge. This proved that electricity had a granular structure. Moreover, Thomson determined the ratio m/e (mass/electrical charge) of these particles: He measured a value on the order of 10^{-11} kg/C (kilograms/coulomb), or more than a thousand times smaller than the value of $\sim 10^{-8}$ kg/C characteristic of the hydrogen ion.[4] The work of Thomson, followed by the more precise measurements of Robert Millikan (1868–1953, NP 1923) and subsequent developments, enabled the individual quantities entering the ratio to be determined separately: The currently accepted values are 1.6021×10^{-19} coulomb for the electrical charge of the electron and 9.1095×10^{-31} kilogram for its mass, the latter being 1,836 times smaller than the mass of the hydrogen ion. Thomson himself always referred to these particles as "corpuscles." The name *electron* was coined by George Johnstone Stoney (1826–1911), a pioneer of the idea of elementary electrical charge, although the notion actually goes back to Benjamin Franklin (1706–1790).[5]

In a further major advance, Thomson demonstrated that the properties of these electrons remained the same regardless of the type of gas in which they are created. From that observation he correctly deduced that electrons are a constituent of all atoms. In his own historic words: "Since electrons can be produced by all chemical elements, we must conclude that they enter in the constitution of all atoms. We have thus taken our first step toward understanding the structure of the atom."[6] He also realized that the size of an electron must be very much smaller than that of an atom. The existence of electrons as subatomic particles was soon to be corroborated by several other experiments, notably those that elucidated the nature of β particles spontaneously emitted in the radioactive decay of uranium atoms, a phenomenon that was discovered by Henri Becquerel (1852–1908, NP 1903) and studied in marvelous detail by Marie Curie (1867–1934, NP 1903, 1911) and her husband, Pierre (1859–1906, NP 1903). They showed that these β particles are electrons moving at high speed.

The prevalence of electrons as constituents of atoms raised many questions, two of which were of paramount importance. The first concerned the number of electrons likely to be present in a particular atomic species. The answer came from a series of studies involving the diffraction of X rays, which had been discovered in 1895 by Wilhelm Conrad Röntgen (1845–1923, NP 1901), and the scattering of α particles emitted by radioactive substances, a field advanced especially by Antonius van der Broek (1870–1926), Lord Rutherford (1871–1937, NP 1908), and Henry Moseley (1887–1915). These studies led to the conclusion that the number of electrons in an element is roughly proportional to its atomic weight (in fact, not too different from half the atomic weight). Furthermore, the number of electrons was found to correspond to the atomic number of the corresponding element in the periodic table—at least when using an improved definition of that number, no longer in terms of the mass of the atom, as in Mendeleev's classification, but as a function of the positive charge of the nucleus determined by the number of protons it contains (see below).

The second question dealt more specifically with the role of the electron in the structure of the atom. It focused on two main puzzles. The first was the necessity, in order to account for the electrical neutrality of the atom, of determining the origin and nature of the positive electrical charge that had to exist to offset the negative charge of the electron. The second concerned the source of the atomic mass: Since the number of electrons in an atom, which corresponds to its atomic number, was always small—a few tens at most—their masses always added up to something quite negligible compared with the total mass of the atom. Where did the rest come from?

On the first point, Lord Kelvin (known previously as Sir William Thomson, 1824–1907) and J. J. Thomson proposed almost simultaneously (with some minor differences) the first structural theory of the atom conceived as a composite entity. The model proposed in 1902 by Lord Kelvin featured a uniformly distributed positive electrical cloud in which electrons were inserted so as to produce an equilibrium situation when the negative charges were at rest.[7] J. J. Thomson wrote in 1899: "Although electrons behave individually as negative ions, when incorporated in a neutral atom their effect is counterbalanced by something such that the space in which these particles are distributed behaves as though it had a positive charge equal to the sum of the negative charges of these particles." In 1903–1904 he proposed a variant of that model in which electrons were distributed on concentric circles within a sphere of positive electricity and moved along them at high velocity. A peculiar aspect of these models is that they accepted a granular structure for negative electricity but continued to regard positive electricity as continuous. This disparity did not seem to bother their authors.

Curiously, there was a time when J. J. Thomson believed that almost all of the atomic mass was attributable to electrons. "There is enough room for 1,700 electrons in a hydrogen atom," he wrote, implying that, whatever the nature of the positively charged medium, it contributes virtually nothing to the mass. He abandoned this assumption in 1906 when he realized that the number of electrons was only of the order of the atomic mass, but he failed to come up with an alternative origin for the atomic mass.

The shortcomings of the Kelvin-Thomson model were remedied by Lord Rutherford, another giant of modern atomism, who proposed in 1911 an atomic model whose essential features have stood the test of time, despite a series of significant improvements and refinements: it is the solar or planetary model of the atom.[8] The central element of this model is an atomic nucleus smaller than the atom itself, within which both the positive charge and virtually all the mass are concentrated. The key experiments that led Rutherford to this picture were conducted by his collaborators Hans Geiger (1882–1945) and Ernst Marsden (1889–1970). Their objective was to study the scattering of α particles (which Rutherford had shown to be positive helium ions) by atoms.[9] Rutherford correctly ascribed the scattering to the repulsion caused by a positive charge located at the center of the atom, which he called the "nucleus." It quickly became clear that the nucleus had to be considerably smaller than the complete atom. Various experiments suggested that its dimension was of the order of 10^{-8} cm.[10] At the same time, it became evident, and Rutherford was quite aware of it, that the atom was essentially just empty space. Indeed, we know today that a nucleus occupies only about one millionth of a billionth (10^{-15}) of the volume of an atom.

It is difficult to think of the sun without the planets. Likewise, in what is usually called Rutherford's model, the central positive nucleus has electrons gravitating around it much like the planets gravitate around the sun. Since each type of atom is characterized by a specific number of electrons, its nucleus must carry a positive charge equal to the sum of the electronic charges so as to ensure the neutrality of the whole structure. In hydrogen, the simplest of all atoms, the nucleus must carry a charge equal and opposite to the charge of one electron. Since the electron is the fundamental unit of negative charge, it is natural to adopt this nucleus as the unit of positive charge. In 1920 Rutherford gave this nucleus the name of "proton." All nuclei must carry a positive electric charge, which can only be a multiple of this elementary charge. The atomic number of an element then becomes a measure of both the magnitude of the nucleus's positive charge and the number of electrons circling around it.

Assigning the elementary positive charge to the proton, which is almost two thousand times heavier than the electron, its negative counterpart, did perpetuate a gross disparity between elementary particles carrying opposite electrical charges. Some symmetry was restored, however, when Carl Anderson (1905-1990, NP 1936) discovered in 1932 the positron, or positive electron, which was the first particle of antimatter ever to be identified. Its existence had been predicted on theoretical grounds by Paul Dirac (1902-1983, NP 1933). It has since been established that to every type of elementary particle in the universe, there must correspond a matching antiparticle. For instance, the antiproton was discovered in 1955 by Owen Chamberlain (1920-, NP 1959) and Emilio Segrè (1905-1989, NP 1959) at Berkeley. The encounter of a particle and its corresponding antiparticle triggers their mutual annihilation and the release of pure energy.[11]

Although Rutherford is generally given credit for the planetary model of the atom, his own attitude toward it was actually rather lukewarm.[12] In his mind, the key feature of his model was a positively charged nucleus. But he was also acutely aware that a planetary model was not really viable. Planets maintain stable, closed orbits determined by the equilibrium between the centrifugal force of their orbital motion and the Newtonian gravitational force associated with their mass. In atoms, gravitational forces are replaced by coulomb forces of attraction between oppositely charged nucleus and electrons.[13] However, in the framework of classical electrodynamics, an electron moving under these conditions must continually lose energy by emitting electromagnetic radiation as it spirals into the nucleus. Satisfied that his "solar" model provided a satisfactory explanation of his experimental results on the scattering of a particles, Rutherford put off a resolution of the stability issue.

Another flaw of the model was its inability to account for the emission spectra of atoms, which display discrete lines corresponding to specific frequencies characteristic of particular types of atoms. These spectral lines constitute a fingerprint, as it were, of the atoms, as was first demonstrated by Gustav Robert Kirchoff (1824-1887) and Robert Bunsen (1811-1899). In Rutherford's planetary model, the orbital motion of the electron changes continuously as the electron falls in toward the nucleus; accordingly, the frequency of the emitted radiation would have to vary continuously as well.

An answer to these objections—at least a tentative one—was provided by Niels Bohr, who, in the process, made the planetary model much more plausible. But Bohr's model relied on hypotheses that were so unusual and bold that it was perceived at the time as nothing short of dizzying intellectual acrobatics. Before discussing the model in question, it is useful to briefly review two discoveries made in the early part of the twentieth century that were destined to revolutionize our view of the world and, incidentally, directly influenced Bohr's work.

THE BIRTH OF QUANTA

*The introduction of "quanta," or indivisible units
of energy, gave integers a role of utmost importance
in a cosmos that until then had been dominated by
the continuous.*

—Paul Valéry

The first of these discoveries was made in 1900 by Max Planck (1858–1947, NP 1918), while he was studying blackbody radiation, which results from equilibrium between the energies emitted and absorbed by the internal walls of an isothermal cavity, or, stated more plainly, how a heated object changes color as a function of its temperature. To explain the discrepancies between experimental measurements and predictions based on Maxwell's theory concerning the spectral content of light, Planck was forced to reject the classical theory in which energy exchanges can take place in continuous amounts. Instead, he had to postulate that matter emits radiant energy only in discrete chunks that were integral multiples of a fundamental quantity $h\nu$, where ν is the frequency of the radiation. The proportionality factor h has the dimensions of an action (energy x time). It is called the *action constant* or, perhaps more commonly, *Planck's constant*. Its numerical value is 6.626×10^{-27} erg-s. The higher the frequency of the emitted radiation (or, equivalently, the shorter its wavelength), the greater the amount of energy contained in a quantum.

This discovery, which amounted to "discretizing" exchanges of radiant energy, ranks as one of the most profound and revolutionary advances in the history of science. It contributed largely to the advent, two decades later, of quantum mechanics, a new vision that radically altered our conception of the world of the infinitesimal. Planck's constant, "this quantity so minuscule and yet with such momentous consequences in science," is involved in every single atomic and nuclear process.[14] Heisenberg's uncertainty principle, which we will discuss later, puts it at the hub of philosophical debates about what is "real"—indeed about the very concept of determinism. It will turn out to be responsible for the cohesion of atomic structures and even of genes. It makes it fundamentally impossible to know simultaneously and exactly two physical quantities, such as the position and momentum of a particle, although no similar mutual exclusion exists in classical physics. It is also because of its exceedingly small value that the wave nature of matter manifests itself only in the realm of elementary particles, while the macroscopic world corresponding to our daily experience is adequately described by the laws of classical mechanics and dynamics.

So unexpected and upsetting to the prevailing views was the discovery of the granular character of exchanges of thermal and radiative energy that Planck himself proposed his law only with the greatest reluctance. A revolutionary in spite of himself, hesitant to accept radically new ideas, indeed a longtime opponent of the atomic theory, he spent much effort trying to salvage at least the continuous nature of radiation itself.[15] These attempts were unsuccessful and came to an abrupt end when the theory of the photoelectric effect was worked out in 1905 by Einstein. In order to interpret certain troubling aspects of how electrons are ejected from a metal plate under the effect of ultraviolet light, particularly the fact that the energy of the photoelectrons depends solely on the frequency of the incident light and not at all on its intensity (which determines only the number of photoelectrons), Einstein was led to propose that light itself is composed of particles named *photons*, each of which has an energy $h\nu$. Although many physical phenomena, particularly interference effects, were still best explained in terms of the wave nature of light—which thus appeared to exhibit a dual nature—the quantization of radiation in other phenomena (such as the Compton effect, which suggested that X rays are diffracted by electrons as though they were particles with energy $h\nu$), as well as the atomic structure of matter, were all beginning to tilt the balance squarely in favor of a corpuscular view of the universe.[16]

A CONCEPTUAL HYBRID: BOHR'S ATOM

God made integers; everything else is the work of men.

—Leopold Kronecker

Of the two main deficiencies of Rutherford's atomic model, the one that Niels Bohr (1885–1962, NP 1922) considered the more troublesome was, by his own account, the question of stability. At a conference held in Göttingen in 1922, he made the following comments about the genesis of his own model: "I had best begin by telling you a little about the history of this theory. My starting point was not at all the idea that an atom is a small-scale planetary system and as such governed by the laws of astronomy. I never took things as literally as that. My starting point was rather the stability of matter, a pure miracle when considered from the standpoint of classical physics. By 'stability' I mean that the same substances always have the same properties. . . . This cannot be explained by the principles of classical mechanics, certainly not if the atom resembles a planetary system." He added some words of caution, heralding the "philosophical" debates that were prompted by the new atomic models such as his own and, later, the one devised by de Broglie and Schrödinger: "We know from the stability of matter that Newtonian physics does not apply to the interior of the atom; at best it can occasionally offer us a guideline. It follows that there can be no descriptive account of the structure of the atom; all such accounts must necessarily be based in classical concepts which, as we saw, no longer apply. You see that anyone trying to develop such a theory is really trying the impossible. For we intend to say something about the structure of the atom but lack a language in which we can make ourselves understood. We are in much the same position as a sailor, marooned on a remote island where conditions differ radically from anything he has ever known and where, to make things worse, the natives speak a completely alien tongue. He simply must make himself understood, but has no means of doing so. In that sort of situation a theory cannot 'explain' anything in the usual strict scientific sense of the word. All it can hope to do is reveal connections and, for the rest, leave us to grope as best we can."[17]

The second major shortcoming of the Rutherford model was its inability to account for spectral rays specific to each atom. As strange as it may seem, this problem occupied Bohr's attention for only a few short months before he wrote in 1913 his famous trilogy of papers in which he proposed his own model of the atom. He had become acquainted that year with the now-famous Balmer formula (Johan Balmer, 1825–1898), although it had been enounced as early as 1885, the year of Bohr's birth. That event sparked his interest in the spectral aspect of the problem and led him to provide a satisfactory interpretation. The way Balmer's formula was established was a genuine tour de force: It proposed an expression, which turned out to be correct, for an infinite series of rays of the visible spectrum of hydrogen (known as the Balmer series) on the basis of only four frequencies known at the time. The formula is nowadays usually written in the form

$$v = R\left(\frac{1}{m^2} - \frac{1}{n^2}\right)$$

where R is known as the Rydberg constant. The main claim to fame of the formula was to express multiple series of frequencies as a difference between two terms, each involving the square of a series of integers, subject to simple relations. For instance, the first series is obtained for $m = 1$

and $n = 2, 3, 4, \ldots$; the second series corresponds to $m = 2$ and $n = 3, 4,$ $5, \ldots$, and so on. It has been claimed that Balmer was inspired by his Pythagorean conviction that all phenomena ensuring the harmony of the world could be explained by judicious combinations of integers.

This quantitative expression for a series of frequencies pertaining to the hydrogen atom was later extended to other series of spectral lines and other atoms. That led to a generalization known as the "principle of combination" or "Ritz principle," which affirms that for each type of atom, it is possible to find a series of numbers, called spectral terms, such that the frequency of every spectral line can be expressed as the difference between two of these terms. Be that as it may, the formula represented one more challenge for anyone trying to devise a plausible theory of the atomic structure. Bohr himself later recognized that "still, those empirical laws describing the line spectra of elements, laws that were without any explanation up to then, provided us with the first clue that the quantum of action was of crucial importance for the stability of atoms and their spectral characteristics."

Bohr's historical merit is to have recognized that the difficulties confronting atomic physics in his day could be resolved with the help of the ideas of Planck and Einstein. He was aware that, in his own words, "in contradiction with the continuousness characteristic of conventional descriptions of phenomena, the indivisibility of the quantum of action requires the introduction of a fundamental element of discontinuity in the description of atomic phenomena."[18] As some have commented: "Bohr and physics were both fortunate with his date of birth, October 7, 1885, for it enabled Bohr to reach maturity and start his research in 1905, just at the right time when the exciting new concepts of quantum theory and relativity theory were emerging from Planck's observations and Einstein's imagination."[19] Specifically, 1913 was a hallmark year in the history of modern atomism: It was the year when Bohr showed how the introduction of Planck's constant into the planetary model of the hydrogen atom, which he endorsed, led to stable electron orbits and to discrete spectral lines in the emission and absorption of electromagnetic radiation.

The proof rests on two really quite revolutionary propositions, each signaling a radical departure from the laws of classical electrodynamics, which proved inadequate to describe the subatomic world. They were set out in one of Bohr's classical papers: "Every atomic system possesses a multiplicity of possible states, called 'stationary states,' which in general correspond to a discrete series of energy values; these states are characterized by a particular stability such that any change in energy of the atom must be accompanied by its complete 'transition' from one stationary state to another," and "The emission and absorption of radiation by an atom are related to the possible changes in energy, the frequency of the radiation being determined by the 'frequency condition' $h\upsilon = E_1 - E_2$, that is, by the energy difference between the initial and final states of the relevant transition process."[20]

The first hypothesis implied that, contrary to classical theories, which inevitably led to the planetlike electrons plunging into the atomic nucleus because of their constantly changing orbit and continuous loss of energy, these particles were, instead, allowed only certain selected orbits with constant energy, or, equivalently, at fixed distances from the nucleus. These discrete values of energy resulted from the quantization of the electron's angular momentum, which could take on only values equal to integral multiples of $h/2p$, where h is Planck's constant. Explicit calculations of the energy E_n of stable orbits produced the expression

$$E_n = -\frac{2\pi e^4 Z^2 m}{n^2 h^2} \; .$$

It applies to hydrogenlike atoms, that is to say, atoms comprised of an electron in the field of a nucleus with charge $+Ze$ (where Z is the atomic number, so that for hydrogen $Z = 1$), where the stable orbits are assumed to be circular. Aside from universal constants, the expression involves the parameter n, which is the first quantum number introduced in atomic physics, and will eventually acquire the name of "principal quantum number." This quantum number n can take on all integer values from 1 to infinity, $n = 1$ corresponding to the lowest energy, which is the most stable state of the atom, also known as the ground state. Just as the space between two rungs of a ladder cannot provide a stable footing to a climber, so are entire regions of space forbidden to stable orbits. As long as it remains in such a stable orbit with fixed energy, the electron will not emit any radiation.

The second hypothesis ushers in a new era in the field of spectroscopy. It states that any radiative process in an atom, such as the emission or absorption of a photon, is associated with a transition or "jump" between *two* stationary states. Bohr's quantum postulate would turn out to be the key to deciphering the messages encoded in atomic spectra. It is the foundation of the entire field of spectroscopy, which came to be one of our primary tools for studying subatomic structures. From the start, Bohr stressed that any change of state of an atom takes on the attribute of an "individual process," the switch from one stationary state to another being determined strictly by probability considerations. "Generally speaking," Bohr wrote, "it can be said that a particular atom in a stationary state is free to make a choice between many possible transitions to other stationary states."[21] This situation prompted Lederman (NP 1988) to observe that Democritus was quite right when he asserted that "everything existing in nature is the fruit of chance and necessity," because "the various energy states are the necessities, the only conditions that are possible. But we can only predict the probability of the electron being in any one of these possible states. That's a matter of chance." It is easy to glimpse the vast horizons that these new concepts unlocked for heated debates on the limits of causality in natural phenomena.

Interestingly, it is the concept of electronic "jumps" between stationary states that was to draw the most criticism. As Louis de Broglie wrote in discussing what he perceived as the "weaknesses" of Bohr's hybrid model: "It rested on a very peculiar blend of concepts and formulas of classical dynamics and methods of quantum theory. The first step was to assimilate the intra-atomic electron with a material point describing quite regularly its orbit under the influence of coulomb forces. In this description so far entirely consistent with classical ideas, conditions of quantization are suddenly introduced—artificially, so to speak—by insisting that the only stable and physically realizable trajectories from among the infinitude predicted by classical mechanics are those that comply with the requirements of quantization. From that moment on, changes in atomic state could only consist of sudden transitions accompanied by a radiative loss of energy, without any reason to justify these sudden transitions within the classical framework of space and time. Between transitions, the atom is in a stable state—one of Bohr's stationary states—in which it seems to completely ignore the outside world, since it does not radiate any energy despite the precise strictures of electromagnetic theory; then it suddenly jumps from its stationary state to another one by means of a transition that is impossible to describe and explain in space. After using classical concepts as a starting point, we end up miles away from them indeed."[22]

Schrödinger criticized this concept of electronic "jumps" even more bluntly during a debate in 1926: "Surely," he told Bohr, "you realize that the whole idea of quantum jumps is bound to end in nonsense. You claim first of all that if an atom is in a stationary state, the electron revolves periodically but does not admit light, when, according to Maxwell's theory, it must. Next, the electron is said to jump from one orbit to the next and to emit radiation. Is this jump supposed to be gradual or sudden? If it is gradual, the orbital frequency and energy of the electron must change gradually as well. But, in that case, how do you explain the persistence of fine spectral lines? On the other hand, if the jump is sudden, Einstein's idea of light quanta will admittedly lead us to the right wave number, but then we must ask ourselves how precisely the electron behaves during the jump. Why does it not emit a continuous spectrum, as electromagnetic theory demands? And what laws govern its motion during the jump? In other words, the whole idea of quantum jumps is sheer fantasy." He even exclaimed: "If all this damned quantum jumping were really here to stay, I should be sorry I ever got involved with quantum theory." To which Bohr replied: "What you say is absolutely correct. But it does not prove there are no quantum jumps. It only proves that we cannot imagine them."[23]

Given the state of knowledge at the time, Bohr's two hypotheses were nothing short of revolutionary. But the success of a revolution can be measured only by its acceptance, regardless of its ideological foundation. In that respect, Bohr's propositions proved successful, in spite of the fact that they were derived only from abstract intuitions, no matter how sound and profound.[24] "Ideological" justifications will only appear *a posteriori* in the framework of quantum mechanics.

At any rate, Bohr's propositions led to a number of achievements. They made it possible to calculate the radius of the hydrogen atom in its ground state (with a result of 0.53×10^{-8} cm, in excellent agreement with the observed value) and its ionization potential (also quite accurately determined to be 13.54 eV). They also could reproduce the correct frequencies of Balmer's spectral series (and other series determined subsequently) and enabled a calculation of Rydberg's constant

$$R = \frac{2\pi^2 e^4 Z^2 m}{h^3 c}$$

(which gave a numerical value of 109,675 cm^{-1}, in perfect agreement with the observed spectrum). A. Pais, one of Bohr's great biographers, regards this expression for R—the oldest equation in the quantum theory of atomic structure—as the most fundamental equation Bohr ever derived. He wrote: "It represented a triumph over logic. Never mind that discrete orbits and a stable ground state violated the laws of physics which up to then were held basic. Nature had told Bohr that he was right anyway, which, of course, was not to say that logic should now be abandoned, but rather that a new logic was called for."[25]

We will have more to say about this "new logic" subsequently. For now, though, we will briefly note that Bohr's model inspired a flurry of refinements aimed at improving the quantum description of subatomic structures. These refinements shed new light on certain fundamental problems discussed in this book. Specifically, new parameters acquired unexpected significance in terms of the periodic classification of elements and provided fresh perspectives about why atoms and their chemical properties can be organized neatly in categories.

As already pointed out, Bohr had assumed in his historic 1913 model that electrons describe circular orbits. He was aware that this was only an approximation, as the force acting on electrons depends on the square of the inverse of their distance to the nucleus, and it was well known that this implied elliptical orbits. Credit for substituting elliptical orbits for circular ones goes to Arnold Sommerfeld (1868–1951). The result showed that, because of relativistic changes of the electron's mass in such orbits, quantization of energy required not one but two quantum numbers. Bohr's principal quantum number was supplemented as early as 1916 by the so-called azimuthal quantum number; shortly thereafter, a third quantum number, called "magnetic," was introduced to account for the effect of magnetic fields on spectral lines. Roughly speaking, these three numbers are associated with the size, shape, and spatial orientation, respectively, of the corresponding orbit. While Bohr's model is no longer current, the significance of these quantum numbers remains fully valid, even if their interpretation is somewhat different. In modern notation, these quantum numbers are designated n, l, and m. They are all integers, and quantization rules impose on them the following restrictions: n can be any positive integer; for a given value of n, l can take any of n integral values $0, 1, 2, \ldots, n\text{-}1$; in turn, for each value of l, m can take on a total of $2l+1$ values $0, \pm1, \pm2, \ldots, \pm l$. Since the quantum number n is the only one to enter into expressions for the energy, all orbits characterized by a common value of n remain, to a first approximation, grouped together. Such a group of elliptical orbits replaces what was a single circular orbit in Bohr's original model. The introduction of this triple quantization significantly improved the predictive power of Bohr's model. In particular, it explained the fine structure of the hydrogen spectrum (known as the Stark effect).

The success of this procedure, which after all was rather empirical, was astounding. While discussing its ability to describe experimental observations, R. Kronig talks "of what is perhaps the most remarkable numerical coincidence in the history of science." Bohr himself stated in 1926, after the advent of quantum mechanics: "It is difficult to tell whether it was a stroke of good or bad luck that the properties of Keplerian motion could be linked with the spectrum of hydrogen in as simple a way as was believed at the time."[26] Some will even go further, as did Pauli when he commented on Sommerfeld: "He trusted certain numerical relations; it was almost a mystical belief in numbers, not unlike the Pythagoreans of antiquity who were studying the harmonics of vibrating strings. That is why one might be tempted to call all these considerations atomysticism." He tempered his opinion by adding: "But until now, nobody has come up with a better proposal."[27] Admittedly, these words were spoken in 1920. Pauli could not have predicted then that he himself would a few years later be suggesting the need for a fourth quantum number. Interestingly, this time around, the number in question would not be an integer. We will come back to it later.

To sum it up, by the close of the first quarter of the twentieth century, developments in the wake of the Bohr model had led to a proliferation of quantized variables defining the state of an electron in an atom. In the process, it became crucially important to understand the role played by these quantum numbers in determining the electronic structure of atoms other than hydrogen, that is to say, how electrons distribute themselves among the available energy levels. While the single electron in hydrogen must obviously occupy its lowest energy orbit in the ground state, placing electrons in more complex atoms requires, as a start, a knowledge of the maximum number of electrons that can be accommodated in each orbit.

Bohr was very much interested in this problem and made a few contributions toward a solution. He became aware early on that specific orbits could become saturated with electrons, primarily because that was in his estimation the best way to account for the periodicity of the chemical properties of elements. Inspired by an ingenious suggestion made by Walther Kossel (1888–1956) that the chemical stability of inert gases (helium, neon, argon, etc.) might be due to their electronic configurations in "closed shells," and deducing that such a configuration must consist of the number of electrons corresponding to their atomic weight (2, 8, 18, and so on), Bohr established in 1920 his *Aufbauprinzip*, or "construction principle."[28] It proposed that the electronic configuration of atoms was determined by a gradual filling of the available levels, in order of increasing energies, until all available electrons were exhausted. Underlying this process was yet another hypothesis, which he laid out explicitly in 1923 as the "postulate of invariance and permanence of quantum numbers." It specified that the quantum numbers of those electrons already present in an atom were not altered by the addition of another electron. Although these principles were never refuted as such, they proved wanting in that they did not address the question of the number of electrons that could be accommodated in a given orbit. In that sense, Bohr failed to provide a satisfactory explanation for the maximum occupancy of shells by 2, 8, 18, et cetera, electrons. That goal was not to be reached until Pauli's exclusion principle was formulated. Nevertheless, Bohr's efforts in that direction are worth mentioning.[29]

THE CORPUSCULAR WORLD

More than anything else, atomism was the theory of microscopic objects.

—Gaston Bachelard

A few other discoveries brought valuable additions to the multicorpuscular view of the atomic structure, although some of them would have to wait until after the advent of wave mechanics. Reviewing them affords an opportunity to highlight developments at the beginning of the century concerning a number of problems raised at various stages in the history of atomic theory.

The first of these discoveries was that of the neutron, a welcomed advance that resolved certain difficulties besetting atomic models limited to protons and electrons. The major problem was an apparent discrepancy between the measured mass of an atom and what that mass should be if the number of protons were equal to the positive charge of the nucleus. For instance, when compared to a proton, a helium nucleus is four times as massive but has only twice the electrical charge. To be consistent with the original hypothesis, a helium nucleus had to also include two electrons to neutralize the charge of the two positive charges. That model had to be discarded with the birth of quantum mechanics, which showed, among other things, that the wave function of the electron was far too extended for that particle to be contained within the nucleus. The conundrum was finally resolved with the discovery in 1932 by James Chadwick (1891–1974, NP 1935) of a new particle called the *neutron*, with a mass roughly equal to that of a proton, but without any electrical charge. Accepting the neutron as the second fundamental constituent of the nucleus resolved the previous mass inconsistencies.[30]

This discovery also explained the existence of isotopes, which are atoms of a same element exhibiting almost the same chemical and physico-chemical properties, but with different masses. Initially discovered in 1913 by Frederick Soddy (1877–1956, NP 1921) in the series of radioactive elements, and subsequently observed by Francis Aston (1877–1945, NP 1922) in a large number of other elements, isotopes turned out to be made of the same number of protons and electrons, but different numbers of neutrons. The discovery rocked one of the most cherished convictions of the founders of atomism, namely, the immutability of atoms of a given element. Dalton himself had reaffirmed this principle, going so far as to boldly proscribe the existence of deuterium and heavy water. He wrote in 1842: "We have no reason to conceive of a diversity of these particles in a substance such as water. Should such a diversity exist in water, it would have to exist in the constituent elements as well, namely, hydrogen and oxygen. . . . If some particles of water were heavier than others, and if some portion of the liquid were under some circumstances made of these heavier particles, the specific gravity of that mass would be affected, and there is no evidence for this. . . . We can then conclude that the ultimate particles of all homogeneous substances are perfectly identical in weight, shape, etc."[31] Even J. J. Thomson proclaimed in 1914 that "all particles of a given substance have exactly the same mass," and that "all atoms of a given element are identical." These great scientists obviously cannot be faulted for not having had at their disposal the technological means to show that their claims were erroneous. Still, others had a prescience of something akin to isotopes. One such example was William Crookes

(1822–1919), famous for his work on cathode rays and on tubes containing low-pressure gases, who clearly ventured the hypothesis that the atoms of a given element may not all be identical, and that the atomic weight of an element could be an average involving atoms with different weights but similar chemical properties.[32] His was a case of remarkable foresight.

These considerations point to a revival, albeit in a more refined and sophisticated form, of Prout's hypothesis, which contended that all elements must emanate from a primordial atom, which he believed to be hydrogen. The hypothesis had had to be discarded in light of mounting evidence that the atomic weights of many elements were not simply integral multiples of that of the postulated basic unit. Still, the proof that the atoms of all elements were apparently made of the same three fundamental building blocks—protons, neutrons, and electrons—vindicated at least the spirit of Prout's hypothesis. As early as 1914, J. J. Thomson had written: "Since electrons can be extracted from all chemical elements, one must conclude that they are part of the constitution of all atoms. Thus, we have taken the first step toward . . . what numerous chemists have pursued since Prout: to prove that the atoms of chemical elements are all made from the simplest atoms, or primordial atoms, as they were called."[33]

With the benefit of this generalization and adaptation to a new store of knowledge, Prout's hypothesis appeared to be gaining credence in terms of the universal nature of atomic components. On the other hand, his fundamental presumption that the mass of an atom should be just the sum of the masses of the constituents proved lacking, because the classical law of conservation of matter does not hold in the world of elementary particles: The mass of the nucleus can be smaller than the sum of the masses of the constituents. To take a classic example, the mass of a helium atom is 3.2 percent lower than the sum of the masses of the constituents. This "mass deficiency" represents the energy liberated during the formation of the nucleus and is explained on the basis of Einstein's relation $E = mc^2$ (in the previous example, that energy approaches 28×10^6 eV). It is equal to the cohesive energy of the particles in the helium nucleus.

Having just mentioned the stability of the nucleus, we must address its instability as well. This phenomenon, first observed, as we have seen, toward the latter part of the nineteenth century with the discovery of the natural radioactivity of uranium and thorium, was confirmed in 1919 by Rutherford, who successfully transmuted nitrogen atoms into oxygen atoms by bombarding them with α particles. In the process, he realized the age-old dream of alchemists and pioneered the field of nuclear chemistry.[34] The atom's instability became more widely recognized with the discovery of artificial radioactivity in 1934 by Frédéric Joliot-Curie (1900–1958, NP 1935) and his wife, Irène Joliot-Curie (1897–1957, NP 1935), followed by the discovery in 1938–1939 of the fission of uranium by Otto Hahn (1879–1968, NP 1944) and Fritz Strassman (1902–1980), which inaugurated the era of atomic energy.

Bohr, who had been so successful in 1913 in formulating a theory capable of accounting for the stability of atomic nuclei, also correctly interpreted twenty-six years later (in 1939) the mechanism of nuclear fission. The natural instability of some elements and the fact that such an instability can be induced in the nucleus of others confirmed a proposition largely inspired by Bohr, namely, that "the great stability of elements is due to the fact that ordinary physicochemical reactions do not affect the nucleus itself, but only involve the bonds associated with the outer electrons of atoms."[35] On the subject of radioactive decay, Bohr acknowledged that the process "seems to occur without any external excitation," and pointed out that fundamentally "if one considers a number of radium atoms, all one can say is that some fraction of them will decay every second." In this, Bohr saw "a remarkable example of inadequacy of a description based on causality, an inadequacy that is intimately related to certain fundamental characteristics of the way we describe atomic phenomena."

Probability and *causality*—these are two words that were the focus of hotly contested debates prompted by the development of quantum mechanics. Bohr played a central role in the ongoing polemic. Indeed, it can be claimed that it was Bohr's contribution, together with quantum mechanical concepts, that brought an end to the dream of those who had so long believed in the universality of explanations based on mechanics. That dream had been articulated perhaps most forcefully by Pierre Simon de Laplace when he asserted: "An intelligence knowing at a given instant all the forces in nature and the position of each entity in it, if it were powerful enough to submit all this knowledge to analysis, would be able to encompass in a single formula the motions of everything, from the largest physical bodies in the universe to the lightest atoms: nothing would escape its grasp, and the future, just like the past, would at once be present before its eyes."[36] This statement implicitly made the atomic microcosm an exact replica of the planetary macrocosm. That view was beginning to look too simplistic in Bohr's time, and it was to be completely shattered by the advent of quantum mechanics.

THE WAVELIKE ATOM

Hoag: *If there are structures that look like a wave on one occasion and like a particle on the next, then we must obviously come up with new concepts. Perhaps one ought to call such structures "wavicules," and quantum mechanics the mathematical description of their behavior.*

Heisenberg: *No, that solution is a bit too simple for me. After all, we are not dealing with a special property of electrons, but with a property of all matter and of all radiation. Whether we take electrons, light quanta, benzol molecules or stones, we shall always come up against these two characteristics, the corpuscular and the undular. . . . It's just that quantum-mechanical features are far more obvious in atomic structures than in objects of daily experience.*

The de Broglie Relation

There are always at least two truths.

—Maurice Maeterlinck (NP 1911)

If radiation can have a dual nature, wavelike and particlelike, why would matter, which up to now has been regarded only as corpuscular, not also have wavelike properties? That question seems quite natural today. But it took great foresight and boldness to ask it in the 1920s, and sheer genius to answer it. Louis de Broglie (1892–1987, NP 1929) displayed all these qualities and more when, on November 25, 1924, he defended his doctoral thesis, entitled *Recherches sur la théorie des quanta* (Research on the theory of quanta), before a somewhat skeptical and reluctant committee.[37] De Broglie himself self-confidently asserted in the preface to the reissue of his dissertation in 1963 that "it can be considered the foundation of every subsequent development in the theories referred to today as wave mechanics or quantum mechanics."

Clairvoyance and courage were required simply because, as he himself reminisced: "Never before had the electron exhibited clearly wavelike properties similar to those of light in interference and diffraction phenomena. To ascribe wave characteristics to the electron in the absence of any experimental proof seemed a fantasy lacking any scientific basis." Nevertheless, he believed "that one could legitimately wonder if this peculiar wave-particle duality, of which light provides so remarkable and disconcerting an example, does not reflect a profound and hidden nature of the quantum of action and if one should not expect a similar duality everywhere Planck's constant is involved." Moreover, "the role of integers in characterizing the stable states of electrons in atoms seemed almost symptomatic. Integers are frequently implicated, as it turns out, in all branches of physics dealing with waves: elasticity, acoustics, optics. They play a role in phenomena involving standing waves, interference, and resonance. It was then only natural to suspect that quantization rules should reflect some wavelike property of intra-atomic electrons."[38]

Bohr's rather artificial hybrid theory, for all its remarkable success, had some serious deficiencies (it failed to account, for instance, for the intensity of the spectral lines of hydrogen and proved inadequate to quantitatively describe atoms with two or more electrons or even the simplest molecules). Such limitations convinced de Broglie that Bohr's model could correspond only to an intermediate stage in devising a complete and coherent theory of atomic structures.[39]

Historically, it appears that the stimulus behind de Broglie's thinking was the idea of ascribing a mass to photons, inspired by a parallel between Planck's law $E = h\upsilon$ and Einstein's relation $E = mc^2$. But, as a knowledgeable biographer observed, "he realized that the only aspect of his argument that was specific to light quanta was to use the frequency to deduce a mass; at that point, he had the brilliant insight to extend his reasoning to any material particle by working in reverse, that is, to start with the mass of such a particle, which defines an energy $E = mc^2$, and to assign to that particle a wave with frequency υ such that $E = h\upsilon$. Elementary, my dear Mr. Watson, isn't it?"[40]

This idea, bolstered by relativity theory and other physical and mathematical considerations too complex to discuss here (they make use of the analogy noted by Jacobi between Maupertuis's principle of least action and Fermat's principle of least time) led de Broglie to articulate his historic proposition, which completely upset our view of the microscopic world, and is encapsulated in the formula $\lambda = \frac{h}{p}$.[41] Worked out at first specifically for electrons, but later extended to any material particle, this relation associates with that particle a wave whose wavelength λ is related to the momentum p ($p = m\upsilon$, or mass times velocity) of the particle. Once again, the quantity h appearing in the formula is Planck's constant. This relation affirms the dual nature—corpuscular and wavelike—of any material particle. One of the most remarkable consequences of this association is that, when applied to electrons in an atom, it provides an elegant interpretation of the quantization rules. In de Broglie's own words: "These rules express the property that the wave associated with the electron is in resonance [that the peaks coincide with the peaks, and the troughs with the troughs] along its trajectory; in other words, they express that the wave associated with a stationary state of an electron in an atom is itself a standing wave in the sense of conventional wave theory," which means that it corresponds to vibrations with a spatial structure that is independent of time.[42] From a practical point of view, the stable orbits are those for which an integral number of wavelengths can fit along their circumference. And so does quantization lose its mystery; it becomes a natural consequence of the standing wave character of electrons in an atom.

De Broglie's hypothesis promptly stimulated new developments in both experimental and theoretical fields of atomic physics. It was confirmed in spectacular fashion as early as 1927 by the experiments of Clinton Davisson (1881–1958, NP 1937) and Lester Germer (1896–1971), and those of George Paget Thomson (1892–1975, NP 1937), who observed independently the phenomenon of electron diffraction by metallic crystals.[43] Since the momentum of the electron beam was known in each of these cases, the diffraction pattern could be calculated using the wavelength given by de Broglie's relation, and it turned out to agree with experimental results.

Electron diffraction has become a routine modern technique (the electron microscope being the most widely known application), as have neutron and proton diffraction. The diffraction of waves associated with atoms had to wait a bit longer. It was finally demonstrated in 1991 by Jürgen Mlyneck and his associates in Constance, Germany, with helium atoms. Shortly thereafter, D. E. Pritchard and coworkers achieved a similar success at the renowned Massachusetts Institute of Technology, in Cambridge, Massachusetts, with atoms of sodium and, later, calcium.[44] These experiments proved conclusively that atoms can indeed behave like waves. Diffraction effects are considerably more difficult to achieve with atoms than with electrons, primarily because of their much larger mass, which, by virtue of de Broglie's relation, implies much shorter wavelengths, making the development of appropriate optical components that much more challenging. For example, for a helium atom moving at a speed of 1000 m/sec, which is typical for a gas at ordinary temperatures, the wavelength is of the order of 1 Angström, or 10^{-10} m. Still, the growth of a genuine field of atomic optics seems just around the corner, with the likely near-term development of atomic interferometers and microscopes.

Schrödinger's Wave Mechanics: When Orbits Become Orbitals

If only I were more proficient in mathematics!

—Erwin Schrödinger

The impact of de Broglie's formula on the development of a new theoretical arsenal applicable to a description of atomic (and molecular) structures was swift. The most famous, and by far most important, contribution was made by Erwin Schrödinger (1887–1961, NP 1933) barely two years after the publication of de Broglie's work. Schrödinger created what is now known as "wave mechanics," which spelled out the rules governing the behavior of de Broglie's waves. He himself announced the goal and significance of his contribution in the first few paragraphs of his seminal paper on the subject, published in 1926: "In this communication, I wish to show first, using the simplest possible example of the hydrogen atom (without relativity or perturbation) that the usual quantization rules can be replaced by another condition that does not rely on 'integers.' These integers come into the picture in the same natural manner as the integral number of nodes in a vibrating string. This new concept is amenable to broad generalizations, and I believe it touches on the core of the true essence of quantization conditions."[45]

The basis of this new mechanics was the so-called wave equation, also known as Schrödinger's equation. We have no desire to go here into the details of its derivation, other than to say that Schrödinger himself gave it a rather elaborate treatment and that it was inspired by the work of Hamilton, which had established an analogy between the Newtonian mechanics of a particle and the geometrical optics of a light ray.[46] What we will do, instead, is to provide the outline of a less rigorous but quick derivation, restricting ourselves to the simplest form of the equation— time-independent and nonrelativistic (that is, with the mass of the electron considered independent of its speed). The starting point is the traditional wave equation describing the spatial dependence of the wave amplitude ψ of any vibrating system:

$$\nabla^2 \Psi + k^2 \Psi = 0$$

where ∇^2 is the Laplacian operator, which, in a Cartesian coordinate system, reads:

$$\nabla^2 = \frac{\partial^2}{\partial x^2} + \frac{\partial^2}{\partial y^2} + \frac{\partial^2}{\partial z^2}$$

The parameter k is equal to $2\pi/\lambda$, so that:

$$\nabla^2 \Psi + \frac{4\pi^2}{\lambda^2} \Psi = 0$$

Replacing λ by the de Broglie expression, $\lambda = \frac{h}{mv}$, one obtains:

$$\nabla^2 \Psi + \frac{4\pi^2 m^2 v^2}{h^2} \Psi = 0$$

But the total energy E of a particle is the sum of its kinetic energy and potential energy V, or $E = \frac{1}{2}mv^2 + V$, from which one derives $mv^2 = 2\,(E - V)$. Substituting this last expression into the previous equation leads to:

$$\nabla^2 \Psi + \frac{8\pi^2 m}{h^2} (E - V) \Psi = 0$$

That is the standard form of Schrödinger's equation, in which Ψ is known as the wave function of the particle of interest. It is a homogeneous, second-order, partial differential equation. With the restriction that the wave function be subject to the usual conditions consistent with physical problems, namely, that it be continuous (without sudden jumps), uniform (a single solution for a given set of coordinates), and finite (no runaway growth), such an equation has solutions only for certain values of the parameter *E*. These are the eigenvalues of the equation and correspond to the allowed energies of the electron. As pointed out by Schrödinger himself, one of the great achievements of this new mechanics is that the quantization of energy flows naturally from the mathematical form of its fundamental equation, rather than being introduced artificially, as in Bohr's model.

To each allowed value of the energy *E* correspond one or several acceptable functions Ψ, called eigenfunctions, that describe the particle in that energy state. These functions are often called *orbitals*, a term that was to acquire great importance in wave mechanics. While it is evident that one consequence of this new point of view was to depose the previously accepted notion of orbits as precise trajectories of particles within the atom, the significance of these orbitals will turn out to be a topic of considerable debate and controversy, which we will discuss on several occasions in this chapter. Given the importance of the stakes, it behooves us to highlight one essential aspect of these differences of opinion.

The Meaning of the Wave Function: The Advent of a Probabilistic Interpretation

While electrons were useful to understand many other things, we never really understood the electron itself.

—Louis de Broglie

A genuine science must be partly philosophy.

—Max Born

At the core of the debate was the meaning Schrödinger himself attributed to his own results. In the context of the criticisms some had leveled against Bohr's atomic model, we have already mentioned Schrödinger's objection to the very idea of distinct electronic orbits and (gasp!) of "quantum jumps" between them. In his mind, the wave equation was an attempt, and a successful one at that, to avoid Bohr's concept of a discontinuous atomic structure. Following Louis de Broglie's lead, Schrödinger considered the wave function a "real" property of electrons, just as real as acoustic or light waves. He thought of it as representing physical vibrations taking place in all directions of space. He considered stable electronic states to be nodes of such vibrations and interpreted atomic spectral lines as transitions between different vibration modes—transitions that were smooth and continuous and replaced, quite advantageously in his view, Bohr's "fictitious" quantum jumps.

The peculiar properties exhibited by electrons in certain circumstances were for Schrödinger simply a particular manifestation of their wavelike nature, and resulted from the identification of a particle with a "wave packet," a term that describes the superposition of waves in a narrow band of frequencies, interfering constructively in a limited region of space, moving at a speed known as "group velocity," and mimicking the movement of a particle in the classical sense. This point of view eliminated the wave-particle duality: Waves simulated the behavior of particles by collecting into packets.[47]

From the outset, Schrödinger's rather monolithic conception ran into serious difficulties. Among the main objections raised against it were the following: (1) the complex character of the wave function, involving the imaginary quantity $i = \sqrt{-1}$, which made it difficult to view it as representing a physically real oscillation; (2) the fact that, for systems with two or more (say, N) particles, these functions no longer depended on just three spatial coordinates but on a total of 3N coordinates in a 3N-dimensional "configuration space," a notion difficult to visualize in any conventional description of an electron; and (3) the demonstration, initially given by Hendrick Lorentz (1853–1928, NP 1902), that "wave packets" were not a particularly satisfactory choice for describing stable electronic states, since they inherently decay and dissipate rapidly.

A very different interpretation of the meaning of wave functions was proposed in 1926 by Max Born (1882–1970, NP 1954). Initially challenged by a few prominent physicists, but quickly supported by the majority, it is now almost universally accepted. It represented a fundamentally new approach by explicitly introducing a probabilistic point of view into our description of the subatomic universe. Leon Lederman considers it "the most dramatic and profound change in our perception of the world since Newton."[48] Taking into account the complex character of the wave function, Born's principle can be stated as follows: "The square of the module of the wave function Ψ at any given point and any given instant is a measure of the probability that the corresponding particle be found at that point and at that instant."[49] The square of the amplitude of the wave Y represents its intensity in the ordinary sense of wave theory.

This probabilistic interpretation of the wave function forces us to change fundamentally our "normal," indeed "natural," way of thinking of the localization of a particle. From this new vantage point, we can no longer claim that the particle can be found at instant t at a point A of coordinates x_1, y_1, z_1. Instead, all we can say is that there exists some probability of finding it at that point, and that it is either more or less likely to be found at point A than at another point B of coordinates x_2, y_2, z_2, depending on whether the probability function $|\Psi|^2$ at instant t is greater or smaller at A than at B. Since the wave Ψ generally occupies a vast region of space (in principle it can extend all the way to infinity), the particle can be expected to be found anywhere in space. There is no *certainty* about its position. Only the *probability* of its presence at a particular point can be known. That was a radical departure from Schrödinger's interpretation, for while Schrödinger saw in the wave function the description of a real material wave, Born interpreted it only as information about the various possible evolutions of the electronic system it describes. The wave function had become a probability function. The conceptual jump was momentous. Einstein would write: "On this point, the quantum theory of today differs fundamentally from all previous theories of physics, mechanistic as well as field theories. Instead of a model description of actual space-time events, it gives the probability distributions for possible measurements as functions of time."[50]

We will discuss later the position of the great Einstein on this radical proposition. For now, let us focus on Schrödinger's reaction, which was hardly enthusiastic. He wrote in 1953: "A most disconcerting aspect of the probabilistic interpretation was then, and still is now, that the wave function can vary in two entirely distinct ways: as long as an observer does not interfere with the system, it obeys the wave equation; but with each observation, it becomes an eigenfunction of the operator involved in the measurement, the eigenfunction becoming intertwined with the eigenvalue affected. Moreover, the change applies only to the observer seeking the result of the measurement. If you are present but have no information about the result, even if you have detailed knowledge of the wave function prior to the measurement and of the apparatus used in the measurement, the modified wave function is of no concern to you, as it effectively does not exist; at best, there is for you a wave function relative to the measurement apparatus plus the system of interest, but it has no particular relevance to the wave function adopted by the observer who knows the result of the measurement."[51]

These comments illustrate the difficulties encountered when trying to visualize the nature of the process that causes an electron to become localized at a precise position in space, for instance, when we detect an electron by scintillation on a fluorescent screen. These difficulties gave rise to much soul-searching. Such localization implies that the electronic wave function, which was initially spread out more or less evenly in space, suddenly becomes concentrated in a very restricted region; it is as though the wave packet "implodes" or "collapses." The crucial point to appreciate here is that the change in the wave function is the result of our own intervention; it is triggered by the detection apparatus itself, by the very measurement we carry out.

The use of the pronoun "we" is deliberate and takes on a precise meaning: It implies a being endowed with conscience. According to a thesis defended, among others, by the mathematician John von Neumann (1903–1957) and the physicist Eugene Wigner (1902–1994, NP 1963), no apparatus or measurement scheme, no matter how sophisticated and complicated, could by itself ever cause the "collapse" of the wave function. It is only when the result of the measurement is registered and recorded in the mind of a human being—an animal, even a highly developed one, will not do—that the wave function "collapses" into an observable reality.

The connection between probability waves spread out through space and experimental observations of localized particles supposedly associated with them will prove to be a highly contentious issue, halfway between science and philosophy, about the significance of quantum mechanics, the determinism (or, rather, indeterminism) of observable phenomena, the underlying "reality" of the world, and the roles of observer and measurements in how we perceive it.

Electrons in Atoms and the Periodic Table of Elements

It seems that systems with different quantum numbers
attract each other, while systems with identical
quantum numbers repel each other.

—Gaston Bachelard

Schrödinger's method, which he himself applied to the hydrogen atom (by letting $V = \frac{e^2}{r}$ in the wave equation), introduces in a "natural" way the three quantum numbers n, l, and m, as well as their interdependence. It provides an expression for the allowed energies that is identical with Bohr's formula, including the fact that it depends on the principal quantum number n only:

$$E_n = \frac{2\pi^2 m e^4}{n^2 h^2} \; .$$

The value $n = 1$ gives the ground state of the atom, while higher values correspond to various excited states.

To each value of n (giving an eigenvalue of the energy), there corresponds one or more orbitals (or eigenfunctions).[52] In addition to the principal quantum number n, these also depend on the azimuthal quantum number l, which describes the spatial distribution of the electron's wave function.

It is worth pointing out another convenient and useful representation, often preferred by chemists. It depicts an electron as spread out in a kind of charge cloud, the density of which corresponds to the local probability of presence of that electron. It is important to bear in mind that, for all its practical utility, it is only a pictorial representation. For instance, when the statement is made that the cloud of electronic charge is $e/5$ (e being the electronic charge) in some region of space, it only means that things appear as though that region contains one fifth of an electron. In more precise language, one would have to say that there exists one chance in five of finding an electron in that particular region. Needless to say, we are talking here about a whole electron: we never actually see fractions of electrons.

That said, regardless of the type of representation used, atomic orbitals have a certain esthetic beauty. The value $l = 0$ corresponds to orbitals with a spherical electronic distribution around the nucleus: the probability of finding an electron varies with its distance to the nucleus but remains constant along the surface of any sphere centered on the nucleus. Figure 10 shows a schematic representation of this type of orbital, called an *s*-orbital; there exist $1s$, $2s$, $3s$, etc. orbitals, corresponding to $n = 1, 2, 3$, etc. They all have the same spherical symmetry. They differ only by their dimensions and the shape of the radial distribution (defined by the function $D(r) = 4\pi^2 r^2 |\Psi|^2 dr$), which is a measure of the electron's probability of presence as a function of distance from the nucleus (see Figure 11). In the ground state of the hydrogen atom, that

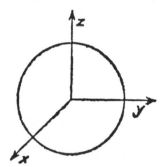

Figure 10. *Shape of s-orbitals.*

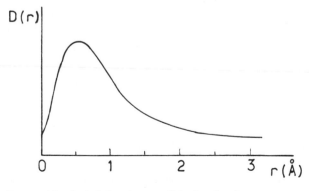

Figure 11. *Radial distribution of the 1s orbital.*

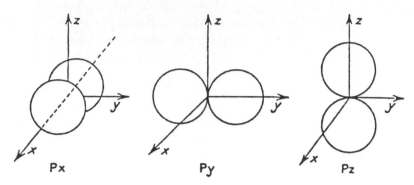

Figure 12. *Shape of p-orbitals.*

function displays a maximum corresponding precisely to Bohr's radius (0.53 Å). However, while in Bohr's model the electron was strictly confined to an orbit with that specific radius, it simply has a maximum probability of being found at that distance in Schrödinger's model (other distances also having a nonzero probability).

In the case of hydrogen, the $2s$, $3s$, et cetera, orbitals are occupied only when the atom is in an excited state. In atoms with several electrons, these orbitals can be filled even in the ground state, as we will see later. Note that the jump of an electron from one orbital to another, for instance via an excitation process, requires a modification of the wave function describing that electron. This, in turn, implies a rearrangement of the distribution of the associated charge cloud.

Figure 12 illustrates the shape of atomic orbitals corresponding to $l=1$, which are called p orbitals. Such orbitals can be described roughly as two equal spheres tangent to each other at the origin; as such, they no longer have spherical symmetry around the nucleus. There are three p orbitals, equivalent but aligned along the three axes of the reference coordinate system; they are denoted p_x, p_y, and p_z. This subdivision corresponds to the existence of the third quantum number m—the magnetic quantum number—which for each value of l can take on a total of $2l+1$ values: $0, \pm1, \pm2, \ldots, \pm l$. The lowest-energy p orbitals correspond to $n=2$ and are designated $2p$ orbitals.

Figure 13 shows schematically the shape of orbitals corresponding to $l=2$, denoted d orbitals. The shape of orbitals corresponding to $l=3$, called f orbitals, is even more intricate. For a given value of n, there are five d orbitals and seven f orbitals, equivalent among themselves except for their orientation in space. The lowest-energy d orbitals correspond to $n=3$, the lowest-energy f orbitals to $n=4$, and so on.

The case of atoms other than hydrogen is, of course, more complex.

Figure 13. *Shape of d-orbitals.*

Overall, however, their electrons occupy atomic orbitals with shapes that are quite analogous to the ones just described for hydrogen. It is evident that orbitals of the same type, $1s$ orbitals, for instance, will have different dimensions in space for different atoms. Generally speaking, the more complex the atom, the closer the inner orbits are to the nucleus. Likewise, the energies of similar types of orbitals will be different in different atoms. Even though their relative values seem to follow a nearly constant order in all elements, that order does differ from the one obtained previously for hydrogen in two respects: (1) Orbitals with the same principal quantum number but different azimuthal quantum numbers no longer have the same energy; (2) The sequence of some high-energy states can be inverted from what would normally be expected. In atoms with several electrons, the ranking of orbitals in order of increasing energy is generally as follows: $1s < 2s < 2p < 3s < 3p < 3d \approx 4s < 4p < 4d \approx 5s < 5p < 6s < 5d < 6p \approx 4f < 7s < 6d < 5f \ldots$.

Table 2 summarizes the classification of atomic wave functions up to $n = 4$. The reader will have noticed that a set of functions corresponding to a given value of n is referred to as an electronic *shell.*

Table 2. *Possible values of the quantum numbers n, l, m, and spectroscopic notations.*

Shell	n	l	Orbital	m	Number of electrons Maximum in orbital	Total
K	1	0	s	0	2	2
L	2	0	s	0	2	8
		1	p	$0, \pm1$	6	
		0	s	0	2	
M	3	1	p	$0, \pm1$	6	18
		2	d	$0, \pm1, \pm2$	10	

cont.

Table 2. *Continued*

Shell	n	l	Orbital	m	Number of electrons Maximum in orbital	Total
		0	s	0	2	
N	4	1	p	0, ±1	6	32
		2	d	0, ±1, ±2	10	
		3	f	0, ±1, ±2, ±3	14	

The rightmost column of the table contains the answer to the crucial question encountered on several occasions in these pages: What factors determine the structure of Mendeleev's table and its relation to the periodicity of the chemical reactivity of the corresponding atoms? From the time the concept of electronic orbits had been developed, the problem amounted to figuring out the maximum number of electrons that could be accommodated in each of these orbits. That problem did not go away when orbits were replaced with orbitals.

The puzzle was resolved by adding to the three quantum numbers n, l, and m a fourth one, known as the *spin*, related to the fact that particles behave as though they are subject to a rotational motion about themselves, thereby giving rise to an angular momentum called spin. The discovery of this property in 1925 by Samuel Goudsmit (1902–1978) and George Uhlenbeck (1900–1988) provided an explanation for the hyperfine structure of spectral lines in a magnetic field (the Zeeman effect). Unlike other quantum numbers, the spin can have only two values: $+\frac{1}{2}$ and $-\frac{1}{2}$ (in units of $\frac{h}{2\pi}$).[53]

Wolfgang Pauli (1900–1958, NP 1945) gets credit for having articulated in 1926 the rules governing the behavior of electrons in terms of the four quantum numbers. He provided the long-sought-after recipe determining the electronic structure of atoms. It is encompassed in his celebrated exclusion principle, which states that no two electrons of a given system can be in the same state, meaning that they cannot be characterized by the same set of four quantum numbers. The most important consequence of this principle is that when two electrons have identical quantum numbers n, l, and m (defining a particular orbital), they must of necessity have different spins. A given atomic orbital can therefore contain at most two electrons, with the restriction that their spin be opposite or antiparallel. Two such electrons are said to be *coupled*.

Completing this all-important principle are the so-called Hund's rules. They specify that, when given the choice of several equivalent free orbitals, electrons always arrange themselves so as to occupy the largest possible number of orbitals; moreover, two lone electrons occupying different orbitals of the same type always have parallel spins.

Armed with this set of rules, we are now in a position to describe the electronic state of any atom. To do so, all we need to know is the number of electrons contained in a given atom and the order in which orbitals are occupied. Orbitals get filled sequentially by the available electrons, starting with the ones with the lowest energy, always subject to Pauli's exclusion principle and Hund's rules. A few examples will illustrate the procedure. Hydrogen has a single electron, which must therefore occupy the 1s orbital in the ground state. The next element, helium, has two electrons, which must both go into a 1s orbital, but with antiparallel spins. In a lithium atom, two electrons will occupy a 1s orbital, while a third will go in a 2s orbital. In a carbon atom, which has a total of 6 electrons, two of them are in the 1s orbital (with antiparallel spins), another two in the 2s orbital (also with antiparallel spins), while the last two must occupy 2p orbitals. In accordance with Hund's rules, the last two will go into two different 2p orbitals, and their spins will be parallel. In oxygen, which has a total of 8 electrons, the occupancy is 2 electrons in 1s, 2 in 2s, 2 (with antiparallel spins) in one of the three possible 2p orbitals, say $2p_z$, 1 on $2p_x$, and 1 in $2p_y$, the latter two with parallel spins. The pattern is now clear.

Figure 14 is a pictorial representation of electronic distributions. Each slot represents an allowed orbital, and arrows symbolize the electrons that occupy them; furthermore, arrows in opposite directions signify

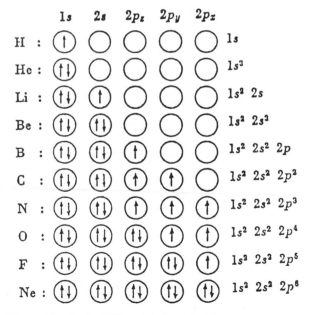

Figure 14. *Gradual filling of orbitals by electrons.*

antiparallel spins, while arrows in the same direction indicate that the spins are parallel. To the right of each row is an abbreviated notation, often used to describe atomic states; its meaning is straightforward, the superscript denoting the number of electrons in the corresponding orbital.

With this procedure, it easy to determine the maximum number of electrons permitted in each shell. It is immediately obvious that the K shell can contain at most two electrons (a single orbital characterized by $n = 1$, $l = 0$, and $m = 0$, but with two possible spins); the L shell contains 2 s electrons and 6 p electrons, for a total of 8, and so on (see the rightmost column of Table 2). It turns out that the maximum number of electrons in a shell n is $2n^2$ (which is known as Stoner's rule).

This scheme provides the key to the periodic classification of the elements. Table 3 depicts the distribution of electrons among successive electronic "shells" for the first twenty elements (and a few additional ones) of the periodic table, chosen so as to highlight the reasons behind the periodicity of chemical properties.

From this vantage point, the most striking feature is that elements that appear chemically related have an identical number of electrons of the same type in the outer, or most peripheral, shell. For instance, all alkaline metals have one s-type electron in their outer shell. It is a $2s$ electron for lithium, a $3s$ electron for sodium, a $4s$ electron for potassium, and so on. But in going from lithium to sodium, one must fill the $2s$ orbital and the three $2p$ orbitals. Consequently, sodium can only appear in the eighth position following lithium in the periodic table. For the same reason, there are seven elements between potassium and sodium. By contrast, since the M shell ($n = 3$) comprises 2 s-electrons, 6 p-electrons, and 10 d-electrons, rubidium comes in the eighteenth position after potassium, and so on.

Table 4 shows the periodic table of all known elements up to element 103 (actually, the most recently discovered element today is number 112). A detailed analysis of the electronic structure of every atom in this table, including the rare earths and the transuranics, reveals relations that may be somewhat more complicated than discussed so far. Nevertheless, the similarities pointed out earlier strongly suggest that the outer-shell electrons do indeed play a determining role in the chemical reactivity of atoms and that they constitute the real structural basis of these similarities, replacing the criterion based on atomic mass that served Mendeleev so well in constructing his table.

Of course, this correlation—and at this stage we can only speak of correlation—demands an explanation for the way these electrons determine the reactivity of atoms and, in particular, how atomic interactions, which

Table 3. *Distribution of electrons in electronic "shells."*

Atomic Number	Element	Shell					
		K	L	M	N	O	P
1	H	1					
2	He	2					
3	Li	2	1				
4	Be	2	2				
5	B	2	3				
6	C	2	4				
7	N	2	5				
8	O	2	6				
9	F	2	7				
10	Ne	2	8				
11	Na	2	8	1			
12	Mg	2	8	2			
13	Al	2	8	3			
14	Si	2	8	4			
15	P	2	8	5			
16	S	2	8	6			
17	Cl	2	8	7			
18	Ar	2	8	8			
19	K	2	8	8	1		
20	Ca	2	8	8	2		
35	Br	2	8	18	7		
36	Kr	2	8	18	8		
37	Rb	2	8	18	8	1	
53	I	2	8	18	18	7	
54	Xe	2	8	18	18	8	
55	Cs	2	8	18	18	8	1
56	Ba	2	8	18	18	8	2

electrons supposedly mediate, can produce cohesive molecular structures with new and specific properties of their own. This is indeed an age-old problem, which Aristotle had capitalized on to criticize the theory of associations by contact proposed by the ancient atomists. In spite of the plethora of qualitative explanations proposed at one time or another, only quantum mechanics will succeed in coming up with a satisfactory answer. We will discuss this point more thoroughly in the section devoted to the association of atoms.

Table 4. *The periodic table of elements.*

Atomic numbers are at upper left in each cell.
Mass numbers are indicated at the bottom; when in parentheses,
they refer to the most stable isotope of the corresponding element.

GROUP

Period	IA	IIA	IIIB	IVB	VB	VIB	VIIB	VIII	VIII	VIII	IB	IIB	IIIA	IVA	VA	VIA	VIIA	0
1	1 **H** 1.008																	2 **He** 4.0026
2	3 **Li** 6.939	4 **Be** 9.012											5 **B** 10.811	6 **C** 12.011	7 **N** 14.007	8 **O** 15.999	9 **F** 18.998	10 **Ne** 20.183
3	11 **Na** 22.990	12 **Mg** 24.312											13 **Al** 26.982	14 **Si** 28.086	15 **P** 30.974	16 **S** 32.064	17 **Cl** 35.453	18 **Ar** 39.948
4	19 **K** 39.102	20 **Ca** 40.08	21 **Sc** 44.956	22 **Ti** 47.90	23 **V** 50.942	24 **Cr** 51.996	25 **Mn** 54.938	26 **Fe** 55.847	27 **Co** 58.933	28 **Ni** 58.71	29 **Cu** 63.54	30 **Zn** 65.37	31 **Ga** 69.72	32 **Ge** 72.59	33 **As** 74.92	34 **Se** 78.96	35 **Br** 79.91	36 **Kr** 83.80
5	37 **Rb** 85.47	38 **Sr** 87.62	39 **Y** 88.905	40 **Zr** 91.22	41 **Nb** 92.906	42 **Mo** 95.94	43 **Tc** (97)	44 **Ru** 101.07	45 **Rh** 102.91	46 **Pd** 106.4	47 **Ag** 107.87	48 **Cd** 112.40	49 **In** 114.82	50 **Sn** 118.69	51 **Sb** 121.75	52 **Te** 127.60	53 **I** 126.90	54 **Xe** 131.30
6	55 **Cs** 132.91	56 **Ba** 137.34	57 * **La** 138.91	72 **Hf** 178.49	73 **Ta** 180.95	74 **W** 183.85	75 **Re** 186.2	76 **Os** 190.2	77 **Ir** 192.2	78 **Pt** 195.09	79 **Au** 196.97	80 **Hg** 200.59	81 **Tl** 204.37	82 **Pb** 207.19	83 **Bi** 208.98	84 **Po** (209)	85 **At** (210)	86 **Rn** (222)
7	87 **Fr** (223)	88 **Ra** (226)	89 ** **Ac** (227)															

* *Lanthanides*	58 **Ce** 140.12	59 **Pr** 140.91	60 **Nd** 144.24	61 **Pm** (147)	62 **Sm** 150.35	63 **Eu** 151.96	64 **Gd** 157.25	65 **Tb** 158.92	66 **Dy** 162.50	67 **Ho** 164.93	68 **Er** 167.26	69 **Tm** 168.93	70 **Yb** 173.04	71 **Lu** 174.97
** *Actinides*	90 **Th** (232)	91 **Pa** (231)	92 **U** (238)	93 **Np** (237)	94 **Pu** (244)	95 **Am** (243)	96 **Cm** (247)	97 **Bk** (247)	98 **Cf** (251)	99 **Es** (254)	100 **Fm** (257)	101 **Md** (256)	102 **No** (254)	103 **Lr** (257)

The Matrix Atom and Heisenberg's Uncertainty Principle

It seems that all the experts have come to an agreement
that virtually everything in the physical world is based
on the arcane formula qp$-$pq$=\frac{ih}{2\pi}$.

—Sir Arthur Eddington

The short period between the introduction of de Broglie's relation (1924) and the initial formulation of wave mechanics by Schrödinger (1926) would seem to leave little time for the emergence of yet another theoretical contribution in the same area of thought, let alone one that could have a comparable influence and importance. Yet that is precisely what happened when Werner Heisenberg (1901–1976, NP 1932) unveiled in 1925 another version of the new mechanics, which he labeled "matrix mechanics."

The methodology proposed by Heisenberg was motivated by a strict requirement for what in his mind must constitute the basis of a description of the subatomic world, at least to the extent that it is accessible to us. The stimulus behind his pursuit was his own conviction, which he expressed with considerable emphasis, that a physical theory apt to describe our world must rely exclusively on directly observable quantities: "If one wants to clarify what is meant by 'the position of an electron,' one has to describe an experiment designed to actually measure 'the position of an electron'; otherwise, the statement is meaningless."[54] Heisenberg noted that atomic orbits, so dear to Bohr, are not observable, and therefore cannot constitute a suitable basis. The only things that are accessible to experimental measurements are spectral lines, and it is therefore on those that any acceptable description of the electronic structure of atoms must be based. However, the characteristics of these lines depend on which pairs of orbits are involved in the pertinent transitions. Consequently, he asserted, all quantitites used to describe electrons in atoms must be related to such pairs.

In order to describe the state of an electron in an atom, Heisenberg made use of matrices, or two-dimensional arrays, comprised of rows and columns of numbers. Actually, Heisenberg did not resort to matrices, for, truth be told, he did not know anything about them. He reinvented them to serve his own purpose. In the process, he also rediscovered the algebraic operations used on such tables, which involve rules that are quite different from those applicable to scalars, that is, ordinary numbers. His first paper on the subject was of questionable clarity, as noted by Weinberg: "I have tried several times to read the paper that Heisenberg wrote, . . . and although I think I understand quantum mechanics, I have never understood Heisenberg's motivations for the mathematical steps in this paper." Added Weinberg: "Theoretical physicists in their most succesful work tend to play one of two roles: they are either *sages* or *magicians*. The sage-physicist reasons in an orderly way about physical problems on the basis of fundamental ideas of the way nature ought to be. . . . Then there are the magician-physicists, who do not seem to be reasoning at all but who jump over all intermediate steps to a new insight about nature. . . . It is usually not difficult to understand the papers of sage-physicists, but the papers of magician-physicists are often incomprehensible. In this sense, Heisenberg's 1925 paper was pure magic."[55] Einstein appeared to have a similar opinion, which he expressed in a letter to Michele Besso: "The most interesting development produced lately by theory is the Heisenberg-Born-Jordan theory of quantum states. The calculation is pure witchcraft, with infinite determinants [read matrices] in place of Cartesian coordinates. It is most ingenious and, owing to its great complexity, safely protected against any attempt to prove it wrong."[56] Einstein's comment is all the more striking because it refers not to Heisenberg's original paper but to an improved version of his theory, done in collaboration with two other scientists.

The situation appears not to have changed much, at least as far as the general public is concerned. As D. Cassidy, one of Heisenberg's excellent biographers, observed: "The new mechanics was—and still is—largely incomprehensible to non-specialists."[57] Such a verdict gives us an excuse for not trying to describe the technique here. The easy way out is further justified on two grounds. First, as was demonstrated initially by Schrödinger and later by Dirac, wave mechanics and matrix mechanics actually turn out to be mathematically equivalent. Second, from a practical standpoint, Schrödinger's formalism was, and remains to this day, far more widely used than Heisenberg's for the purpose of exploring atomic and molecular structures. As a result, the designation "quantum mechanics" as it is commonly meant today almost always refers to Schrödinger's version.

Without dwelling unduly on Heisenberg's formalism, it does, however, seem worthwhile to emphasize an important aspect of his description of the subatomic world. Using matrix notation to describe various physical variables associated with an electron, such as its position or momentum, reduces considerably, if it does not outright eliminate, the possibility of depicting atomic reality in terms of conventional models. In discussing the wave and/or particle nature of the electron, Heisenberg proposed to substitute for such mental pictures, which he deemed deficient, a mathematical formalism that in his view actually constituted the only possible perception of reality, indeed the only revelation of reality itself. Adopting a pure abstraction as the safest, perhaps even the only, manifestation of the ultimate truth was obviously not likely to contribute much to the popularity of his theory.

However, it may be of some interest to stress the relationship established by Heisenberg himself between his own vision of the world and the particular form of atomism developed more than two millenia earlier by Plato (see chapter 4). That relationship also defined his own position with respect to Democritus's classical version of atomism: "In the philosophy of Democritus, all atoms consist of the same substance if the word 'substance' is to be applied here at all. The elementary particles of modern physics carry a mass, in the same limited sense in which they have other properties. Since mass and energy are, according to the theory of relativity, essentially the same concepts, we may say that all elementary particles consist of energy. This could be interpreted as defining energy as the primary substance of the world. It has indeed the essential property belonging to the term 'substance,' that it is conserved." Plato's vision of the world, observed Heisenberg, is quite different: "The elementary particles in Plato's *Timaeus* are finally not substance but mathematical forms. 'All things are numbers' is a sentence attributed to Pythagoras. The only mathematical forms available at that time were such geometric forms as the regular solids or the triangles which form their surface. In modern quantum theory there can be no doubt that the elementary particles will finally also be mathematical forms, but of a much more complicated nature. . . . But the fundamental triangles cannot be considered as matter, since they have no extension in space. It is only when the triangles are put together to form a regular solid that a unit of matter is created. The smallest parts of matter are not fundamental Beings, as in the philosophy of Democritus, but are mathematical forms. Here it is quite evident that the form is more important than the substance of which it is the form."[58]

In a 1963 interview, Heisenberg had this to say: "Waves and corpuscules are, certainly, a way in which we talk. . . , but since classical physics is not true there, why should we stick so much to these concepts? Why not say just that we cannot use these concepts with a high degree of precision? . . . When we get beyond this range of the classical theory we must realize that our words don't fit. They don't really get a hold in the physical reality and therefore a new mathematical scheme is just as good as anything because the new mathematical scheme then tells you what may be there and what may not be there."[59] In the final analysis, matrices did represent physical reality in Heisenberg's view. As noted by J. L. Heilbron: "In Heisenberg's romantic metaphor, modern physics strips atoms of the last vestiges of personality. Democritus had deprived them of color and taste, but at least he left them with extension, position, and velocity. For the atom of modern physics, all these properties are simply derivative quantities."[60] Such a position is related to the problem of assimilating mathematical structures produced by the human brain with the structure of phenomena, a problem that we intend to carefully avoid going into.

In any event, the fierce battle Heisenberg waged to eliminate from the atomic theory any nonobservable quantity produced an important result. It led to his much-celebrated—and quite justifiably so—*uncertainty principle*, which marks a watershed in the evolution of our views about the limitations of accessible data in scientific experiments and, by extension, about the troubling issues of determinism and indeterminism in natural phenomena.

Although the uncertainty principle came about as a purely algebraic consequence of matrix theory, it can be introduced in a more empirical and descriptive way, as indeed was done by Heisenberg himself, by means of a thought experiment often referred to as "Heisenberg's microscope experiment."[61] Assume that we wish to determine as accurately as possible the motion of an electron, including its position and momentum, in an atom. To that end, we decide to use a microscope. It is obviously desirable to resort to a microscope with extremely high resolution, which implies the use of short-wavelength radiation, say, gamma-ray photons. However, we know that the motion of an electron—in other words, its position and momentum—is going to be perturbed by impact with the photon via a phenomenon known as the Compton effect. The more precisely we try to determine the electron's position, for instance by increasing the frequency of the radiation, the more aggressive the photons become (since their energy is given by the formula $E = h\upsilon$), and the greater the perturbation caused by the experiment itself on the "canonically conjugate" variable of momentum. And vice versa, of course. It is a saddening result, perhaps, but one that simply cannot be overcome. It establishes insurmountable limits to the answers we can expect, limits that are imposed by the very manner in which we ask the questions.

Heisenberg's uncertainty principle stipulates these limits by way of the famous relation:

$$\Delta q \cdot \Delta p \geq \frac{h}{2\pi}$$

It states that the product of the uncertainties on the position and on the momentum can be no less than Planck's constant divided by 2π. A similar relation applies to another pair of conjugate variables, namely, energy and time. In quantum physics, knowledge about one variable may diminish or even eliminate the prospect of knowledge about another, in sharp contrast to the situation in classical physics. As Bachelard observed: "Until Heisenberg came along, errors on independent variables had been assumed to be themselves independent. Any variable could by itself be an object of study with increasing precision. . . . Heisenberg's uncertainty principle introduced an objective correlation between errors."[62] No measurement apparatus can ever get around this limitation, which reflects an intrinsic property of the world of elementary particles, including atoms. It is only because of the exceedingly small value of Planck's constant that these effects go unnoticed in the macroscopic world, thus enabling us to determine with apparently unlimited precision the position and speed of a tennis ball in flight or of a high-speed train hurtling down a track.

The uncertainty principle had momentous conceptual consequences and jolted some of the most cherished beliefs that had been for centuries, if not millenia, the very foundation of scientific study of the world. Foremost among these was the conviction that all phenomena of nature are governed by strict determinism. Admittedly, a few philosophers, such as David Hume (1711–1776), had occasionally challenged the notion of causality. But these were philosophers and it seemed only natural that they should engage in this type of intellectual exercise. Scientists, on the other hand, unless they were prepared to contradict themselves, really had no choice but to embrace determinism. "A science that is not deterministic would be no science at all," asserted Henri Poincaré as late as 1912.[63] And now, here was this universal constant, discovered by one from their own ranks and gaining widespread acceptance in the sciences, that injects into the exploration of the subatomic world a "quantum fog" threatening the very foundation of a long deterministic tradition. To be sure, from the beginning of the nineteenth century on, no one could dispute the fact that a global description of at least some observable physical phenomena could be formulated correctly only by resorting to the laws of probability and statistics. But it was always assumed that it was because of the large number and randomness of "elementary phenomena" participating in the "global phenomenon," and that, taken individually, these elementary phenomena would continue to comply rigorously with determinism. The uncertainty principle shattered this pleasant optimism and exiled it from the realm of the infinitely small by introducing a fundamental indeterminism, inherent in the very nature of the universe.

If we can no longer determine simultaneously the position and momentum of a particle at a given instant, what becomes of the dream of Laplace, who had claimed that, knowing these quantities and the forces involved, he could predict with absolute certainty the future motion of that particle until the end of time?[64] The dream simply collapses, but one should stress that "in the rigorous formulation of a causal law, the contentious issue is not so much the conclusion as the premise," in other words, the possibility of knowing with accuracy "the initial conditions" of future events.[65] It is this uncertainty that severs the deterministic link between prior and subsequent states, the present and the future, in the world of the infinitely small. On a quantum scale, certainty must step aside in favor of probability and statistics.

This state of affairs did not satisfy all scientists, including some of the most prominent ones. A few, like Louis de Broglie, refused to accept this intrinsic breakdown of causality in the subatomic world. He commented, "The evolution of the wave function Ψ between two measurements is entirely determined by its initial state and the propagation equation: it therefore does conform to strict determinism, but that does not necessarily imply strict determinism for observable and measurable phenomena as well, since each observation or new measurement brings in new elements and interrupts the normal evolution of Ψ." De Broglie cautioned, however: "But it seems to us entirely plausible that physics will someday return to the fold of determinism, at which point the current state of science will simply be perceived as an interim phase during which the inadequacy of our concepts forced on us a temporary suspension of the notion of rigorous determinism in phenomena on an atomic scale. It is conceivable that our current inability to follow the thread of causality down to the microscopic world is due to our use of concepts such as corpuscle, space, time, and the like; these concepts were developed on the basis of our ordinary macroscopic experience, and we naively transposed them to a description of the microscopic. Nothing assures us—in fact, quite the contrary—that these concepts can be adapted to a description of this new reality."[66] Or, as Holbach had stated two centuries earlier: "We can only judge objects we know nothing about in the light of those we have managed to understand."

At the other extreme, some will point out "that we must not deduce [from the uncertainty principle] that we are unable to simultaneously determine the position and momentum of an electron, but rather that the electron does not possess these two attributes at the same time. These two concepts, when considered together, are meaningless for an electron. What is true for an electron is also true for all other particles, including an atom taken as a whole."[67]

Still others tried to soften the intellectual blow caused by the switch from a deterministic physics to one that is probabilistic. One such individual was Hans Reichenbach, who wrote: "With Heisenberg's uncertainty relation, the physics of material waves has established the principle that nature is governed by the laws of probability, not by causal laws. . . . This conclusion will not be deemed as revolutionary by a philosopher of science as by one steeped in metaphysics. From the time Boltzmann gave a statistical interpretation of the second principle of thermodynamics, a change in conception from causal to statistical has remained a possibility that could not be dismissed. As a matter of fact, a more careful formulation of the principle of causality had led to the realization that the notion of causality has a meaning only in the sense of approaching a limit in probability theory. From this point of view, there was a time when the probability of a predicted event could come as close to the upper limit of

1 as desired. The uncertainty principle has simply imposed on this process of approximation a limit less than 1: that is the logical meaning of Heisenberg's principle. Historically, the indeterminism of quantum theory is much more an evolution than a revolution of the principles of physics."[68]

Finally, a few would rave about the conceptual (as well as practical) superiority of probability over "certainty." For example, L. Rosenfeld wrote: "To be rid of a metaphysical hindrance is to soar in the sphere of rational knowledge. People are too accustomed to view determinism as the epitome of scientific 'certainty,' as opposed to what is metaphysically capricious and arbitrary. Probability does not mean chance without rules, but the exact opposite: it refers to that aspect of chance that is systematic. A statistical law is first and foremost a law, the expression of a pattern, a tool for predictions. Probability calculations are no less effectual than classical methods; in fact, they may be more so, since this type of calculation applies to a host of situations where deterministic laws, if they exist at all, prove impotent. We do not lose anything by replacing a deterministic conception by one, broader and more flexible, based on probability: far from weakening our ability to make scientific predictions, we actually reinforce it."[69] In other words, as Gaston Bachelard observed: "Under no circumstance should one believe that probability and ignorance are synonymous just because probability is called in whenever the causes are unknown."[70]

This review provides a glimpse of the treacherous intellectual battleground faced by the nineteenth-century physicists so mesmerized by the secrets of quantum mechanics. They ventured onto it always with courage, often with determination (if not obstinacy), and sometimes with recklessness and temerity, like knights at the service of the lady of their choice. Our next task is to sketch some of these boisterous jousts.

THE SCIENTIST-PHILOSOPHERS

It was an ancient tradition passed from Egypt to Greece that a god averse to man's quietude invented science.

—Jean-Jacques Rousseau

I tried to explain the interpretation of quantum theory to these philosophers. At the conclusion of my presentation, there was neither opposition nor hard questions; but I must confess that this is precisely what shocked me the most. For if at first you are not horrified by quantum theory, you could not possibly have understood it. My presentation was probably so bad that nobody understood what it was about.

—Niels Bohr

Nobody understands quantum mechanics.

—Richard Feynman

The emergence of quantum mechanics propelled to the foreground of scientific thought a series of propositions that upset many established ideas and shattered the conceptual foundations of classical physics. A number of puzzles challenged traditional views: the wave-particle duality, which mixed two apparently incompatible concepts in the description of a single fundamental entity; the quantum of action (Planck's constant), whose exceedingly small but finite value introduced fundamental differences in the way we can describe the microscopic and macroscopic worlds; the importance, perhaps even supremacy, of a probabilistic approach in predicting the behavior of very small particles, which dealt a blow to the unquestioned validity of the rules of determinism; how an observer inevitably influences the outcome of any experiment in the atomic world by using hardware itself made of atoms—all these observations raised fundamental questions about the nature of "reality," indeed whether it exists at all. Because of the nature of the issues involved, scientists found themselves unwittingly dealing with arguments whose ramifications extended beyond a strictly scientific framework, venturing into areas that had traditionally been the province of philosophers. These scientists were forced in that direction because they were virtually the only ones with enough knowledge of the concepts and methodologies of the pertinent theories to infer with any credibility the appropriate consequences. Einstein, who readily conceded that "it has often been said, and certainly not without justification, that the man of science is a poor philosopher," nevertheless deemed that "at a time like the present, when experience forces us to seek a newer and more solid foundation, the physicist cannot simply surrender to the philosopher the critical contemplation of the theoretical foundations; for, he himself knows best, and feels more surely where the shoe pinches."[71]

Be that as it may, by stepping onto philosophical ground, the controversy was destined to be endless. The ambitious stakes involved would guarantee that it could not be otherwise. As Schrödinger stated: "The real difficulty of philosophy lies in the spatial and temporal multiplicity of individuals doing the observing and the thinking. If everything were confined to a single conscience, it would all be quite simple."[72]

In fact, the debate, which began when the very first concepts of quantum theory were being formulated, remains quite current. Tens of books, hundreds of articles, thousands of pages have been, and continue to be, written about it. Besides the numerous biographical or autobiographical writings featuring the principal protagonists,[73] and a number of classical works of a highly technical nature,[74] several recent books have been entirely devoted to a detailed account of this clash of ideas, chronicling how concepts evolved, almost day by day, sometimes hour by hour.[75] They retrace the sometimes bitter feuds and subtle arguments centered around "thought experiments," each of which seemed to uncover a new paradox. Such accounts tend to be complicated by the fact that some of the participants gradually modified their views over time, and sometimes even switched opinion outright.

Given the abundance of pertinent material and the limited space in this book, it is difficult to do much more than sketch a broad outline of this great debate. For lack of a better solution, we will summarize the positions of the two most prominent protagonists of the play, namely, Niels Bohr and Albert Einstein. These men emerged as the two leading figures around whom the other participants rallied, even if they did not always fully endorse all the views of their respective leader. The debate has often been described as a "struggle between giants," a "battle of titans." Bohr turned out to be the flag-bearer for the new vision of the world inspired by quantum mechanics; this is somewhat surprising since, with all due respect for his role in promoting quantum concepts in the atomic universe, he himself did not contribute to quantum theory to the same degree as, say, Schrödinger or Heisenberg. Einstein, on the other hand, was the defender of more orthodox concepts and advocated a rather conservative posture.

Bohr

It is just as correct to claim that a conscious subject is a product of matter as it is to say that matter is simply a description elaborated by a conscious subject.

—A. Schopenhauer

An observer is as necessary to the creation of the universe as the universe is to the creation of an observer.

—J. Wheeler

In the great debate about the meaning of quantum mechanics, Bohr owes his unique position to the concept of "complementarity," which he fathered and made the underpinning of his overall philosophical vision of the world. In a sense, the entire controversy will revolve around the significance and consequences of that concept. Bohr first proposed the idea in Como, Italy, during an international colloquium held in honor of the hundredth anniversary of Volta's death. He restated it a month later at the famous Fifth Solvay Congress in Brussels.[76] Nevertheless, the core of the doctrine is today commonly referred to as the "Copenhagen interpretation" of quantum mechanics, as a tribute to the cultural influence that Bohr's work imparted to his native city. I only met one individual who objected to that designation. He was Rudolf Peierls. In all fairness, his objection was actually a tribute of sorts, since he wrote: "I object to the term Copenhagen interpretation . . . because it sounds as if there were several interpretations of quantum mechanics. There is only one."[77]

The basis of Bohr's "complementary" approach in his description of the atomic world is a simple observation that, in the words of C. Chevalley, amounts to a paradoxical opposition between two impossibilities, to wit, "the impossibility of relying on the conventional usage of classical physics concepts when dealing with atomic phenomena, and the impossibility of replacing those concepts by others in the hope of creating a quantum language able to convey a satisfactory description of atomic phenomena. This paradox boils down to a shortage of words: quantum mechanics lacks the linguistic tools to express what it is all about. . . . Complementarity attempts to offer a possible solution to the dilemma of describing atomic phenomena while preserving the use of the ordinary language of physics."[78]

For our own purpose, the concept of complementarity takes on its primary significance in the context of the wave-particle duality of the fundamental constituents of nature, indeed of the atom itself. But even from that perspective, the meaning of the word is likely to be misunderstood by the casual reader. The conventional notion of complementarity connotes synergistic congruence—in other words, the compatible union of fragments fitting like pieces of a jigsaw puzzle and producing a unified entity, a synthesized and fused whole. The type of complementarity conceived by Bohr, on the contrary, is characterized by a fundamental and irrevocable contradiction between the elements involved. They remain separated by a mutual incompatibility that at best affords them a "peaceful coexistence," and only so long as they carefully avoid each other. Bohr's complementarity amounts to forced integration but maintains segregation based on qualities.

One of the most lucid discussions of complementarity Bohr ever gave is to be found in his own summary of his Como lecture.[79] It describes his motivation for introducing the concept as well as the numerous and complex problems immediately raised in its wake: "Especially had the great success of Schrödinger's wave mechanics revived the hopes of many physicists of being able to describe atomic phenomena along lines similar to those of classical physical theories without introducing 'irrationalities' of the kind which had thus far been characteristic of quantum theory. In opposition to this view, it is maintained [in my theory] that the fundamental postulate of the indivisibility of the quantum of action is itself, from the classical point of view, an irrational element which inevitably requires us to forego a causal mode of description and which, because of the coupling between phenomena and their observation, forces us to adopt a new mode of description designated as *complementary* in the sense that any given application of classical concepts precludes the simultaneous use of other classical concepts which in a different connection are equally necessary for the elucidation of the phenomena. . . . The main purpose [of my lecture] is to show that this feature of complementarity is essential for a consistent interpretation of the quantum-theoretical methods. A very significant contribution to this discussion had been given shortly before by Heisenberg, who had pointed out the close connection between the limited applicability of mechanical concepts and the fact that any measurement which aims at tracing the motions of the elementary particles introduces an unavoidable interference with the course of the phenomena and so includes an element of uncertainty which is determined by the magnitude of the quantum action. . . . This very circumstance carries with it the fact that any observation takes place at the cost of the connection between the past and the future course of phenomena. As already mentioned, *the finite magnitude of the quantum of action prevents altogether a sharp distinction being made between a phenomenon and the agency by which it is observed,* a distinction which underlies the customary concept of observation and, therefore, forms the basis of the classical ideas of motion. With this in view, it is not surprising that the physical content of the quantum-mechanical methods is restricted to a formulation of statistical regularities in the relationships between those results of measurement which characterize the various possible courses of the phenomena"

In terms of the wave-particle duality of subatomic entities such as an electron, the notion of complementarity thus becomes a prescription for how to simultaneously resort to both aspects without becoming vulnerable to internal contradictions. As C. Chevalley asserted: "The preeminent purpose of complementarity is to spare physicists the ridicule once denounced by Aristotle of claiming that opposites belong in the same object at the same time. . . . Complementarity is something that belongs in the realm of discourse, not of things. It addresses a strictly epistemological problem, in the sense that it is not the things themselves that are complementary, but rather their description and how we perceive them. As such, the 'notion of complementarity is simply a characteristic of our perceptions' [according to Bohr]; it is 'an attempt for our ordinary intuition to adapt to experimental facts' [also in Bohr's words]."[80]

To be more explicit, if wave and particle descriptions appear to be mutually exclusive, at least one of them would have to be presumed false in classical physics. Things are quite different in quantum physics: Both descriptions are equally valid, in Bohr's view, provided that one does not insist that they be so at the same time. There is no fundamental problem as long as one does not ask whether an electron *is* a particle or a wave, but simply whether it *behaves* like one or the other in a given situation. In quantum mechanics, the former question is meaningless, according to Bohr, while the latter is easily answered, provided that the experimental conditions be clearly stipulated. Thus, in Bohr's conception, complementarity does not resolve the incompatibility between the wavelike and particlelike behaviors of an electron. It merely acknowledges that both descriptions are necessary to encompass all properties of that entity. The logic of this proposition is based on the argument that because of the uncertainty relations—themselves a consequence of the quantum of action—the two properties can never be found in conflict, as they never exist at the same time. As Louis de Broglie had put it: "People always expect a battle between wave and particle: it never happens because never more than one of the combatants shows up."[81] Bohr's "complementarity" represents the relationship between two contradictory and irreconcilable properties; but it is necessary to invoke both *in turn* to completely describe the behavior of subatomic entities. Bohr himself stated: "All one can ask in a new experimental situation is to resolve all apparent contradictions. For all the differences between atomic phenomena observed under different experimental conditions, they can be declared complementary to the extent that each is well defined and that they cover all that is ostensibly known about the objects considered. The formalism of quantum mechanics, whose sole purpose is to unify observations made in experimental conditions described by simple physical concepts, gives an exhaustive complementary description for a wide range of experiments."

One of the consequences of Bohr's analysis is to expose the inadequacy of our usual concepts drawn from our day-to-day experience of the macroscopic world (concepts that include not only waves and particles, but also causality, reality, objectivity, etc.) in trying to describe atomic phenomena. Since we cannot do without these concepts when trying to express intelligibly the content of our experience, we are compelled to improvise ways to use them, ways that are bound to be unconventional. Viewed from that angle, complementarity only expresses relations that are perfectly obvious, as Bohr was fond of saying. It is a recipe for how to cope with a dilemma. After all, in Paul Valéry's own words: "Science exists to the extent that there is action. It is a collection of recipes that always succeed. Everything else is only literature."[82]

In light of this new concept of complementarity, scientists had to reassess their attitude on the questions of observing and interpreting nature. In particular, they were compelled to recognize the "active" role of experimental instrumentation and, by extension, of the observer himself. In the realm of the infinitely small, we are no longer, indeed could not be, mere passive and objective observers of nature "as it really is," as if it had an independent physical reality preceding (or following) our intervention. As Hans Reichenbach wrote: "To speak of things that remain unchanged when no one is observing them is perfectly reasonable language for the macroscopic world; but it would be dangerous to assume that the same language applies without reservation to the microscopic world. Linguistic rules do not constitute philosophical truth; they result from habits developed in the course of interacting with our environment."[83] The image of nature we can arrive at is set largely by the characteristics of the measuring instrument we choose to use. That choice determines which side of complementarity becomes visible to us. In Bohr's world, the term *phenomenon* must of necessity encompass both observed and observer.

This paradox is the result of an unavoidable perturbation caused by the interaction between the object of interest and the experimental apparatus. Such an interaction obviously exists in the macroscopic world as well. But there the perturbation is negligible and has no practical effect on measurements. One of the consequences of this state of affairs is that the physical laws governing our daily lives appear to be perfectly deterministic. However, the complementary description of the microscopic world, which results from the quantum of action and the uncertainty principle, invalidates strict determinism.[84] The rigor of precise values gives way to the murkiness of probabilistic and statistical information.

We have already indicated in the previous chapter how several leading scientists reacted to this development. Bohr's new insight prompted him to reassess the traditional notions of both causality—"too rigid a framework to encompass the very peculiar laws governing individual atomic processes"—and reality—"a word the meaning of which we have to relearn."[85] Scientists were forced for the first time in history to abandon the comfortable distinction between observer and observed, because of the perturbation caused by their interactions. This situation imposed some restrictions on the significance of the information obtained. The degree of arbitrariness introduced in the characterization of what is considered the "object" defied the classical concept of independent physical reality. Bohr's opinion was that this concept has actually no meaning at all. In his own words: "Our purpose is not to discover the essence of things, the meaning of which would escape us anyway, but merely to develop concepts useful for speaking of natural phenomena in a productive way." The word *phenomena* in this quotation is used according to Bohr's own definition, which takes in both the object of study and the means to observe it. Bohr went on to say: "The quantum world does not actually exist. All we have is an abstract quantum description. It is a mistake to believe that the purpose of physics is to find out how nature is made. Physics is interested only in what we can say about nature." Selleri commented that as far as

Bohr was concerned "physics must deal exclusively with the process of observing, any reference to a supposed elementary reality that is not observed being anathema to scientific reasoning. . . . Only measurements are real." Arthur Eddington articulated this point of view more concretely: "By virtue of its own procedures, physics does not study the inscrutable qualities of the material world but only data provided by instrumentation. To be sure, the data do reflect the fluctuations of the properties of the world; but the only thing we know exactly are the data, and not the properties themselves. Data are to qualities what a telephone number is to a subscriber." D'Alembert was expressing more or less the same idea when he wrote in 1759 in his *Essai sur les éléments de philosophie* (Essay on fundamentals of philosophy): "It is to satisfy our needs and not just our curiosity that we were endowed with sensory feelings; their purpose is to reveal to us the connection between other beings and ourselves, rather than the inner essence of these beings."

Einstein

It is possible that you dream about me, A. tells me. But what is troubling is that I too dream about you. It is also possible that when I dream about you, I feel that you dream about me too. . . . It would be somewhat surprising that so many people would dream at the same time and try to establish contact through so many dreams. Or else such a coordinated network of dreams would surely deserve the name of reality.

—Jean d'Ormesson

The concept of complementarity as it was formulated by Bohr, and its consequences in terms of a quantum description of the world, provoked a lively controversy for decades. Its repercussions are still being felt today. All of the major contributors to the development of quantum mechanics were involved in it, as are their successors to this day. The opinions expressed represent a varied and diverse mosaic. Several thick volumes would be required (a few have recently been published) just to summarize its broad outlines. We will focus here primarily on the position of the most adamant, as well as most illustrious, opponent to Bohr's ideas— Albert Einstein. The debate between these two giants of modern science dealt with more than just the principle of complementarity; it touched on the very meaning of our perception of the world. It often involved subtle arguments and thought experiments whose interpretations would be too convoluted to review here. We will, however, attempt to extract the general ideas.

Einstein's views are summed up in the celebrated phrase "God does not play dice." Indeed, the phrase deserves prominent mention, as it reflects a fundamental aspect of Einstein's position toward the new mechanics: He categorically rejected the notion that a purely probabilistic interpretation could constitute the ultimate stage of our knowledge of the material world. The nondeterministic character of quantum theory thus offended his profound conviction about the nature of laws governing the march of the universe. He was convinced that the statistical aspect of the theory resulted merely from our intellectual inability to uncover the underlying laws. It was simply "an expedient temporary gimmick." Heisenberg, himself in the thick of the debate, wrote in his memoirs: "Einstein was not prepared to accept the essentially statistical character of the new quantum theory. He obviously had no objection against probabilistic predictions in those cases where all the parameters of a given system are not perfectly known. After all, statistical mechanics and thermodynamics are both based on just such predictions. But what Einstein refused to accept was that it could be fundamentally impossible to gain a knowledge of all the parameters necessary to completely determine the relevant processes. 'God does not play dice,' he often insisted during these discussions."[86]

Here again is Einstein's celebrated phrase.[87] While it is natural to ascribe it to his abhorrence of indeterminism, it actually reflects a more far-reaching position, one that addresses the innate existentialist preoccupations of any scientist-philosopher, indeed of any human being. Such preoccupations deal with a sense of "reality" and, as a backdrop, the question of the degree of correlation—or lack of it—between a type of reality that some call "observed" or "studied," "thought," or "empirical," and so on, and the so-called objective reality, to use Einstein's own terminology (sometimes also referred to as "reality in itself," or "independent," "intrinsic," etc.). There is unfortunately no guarantee that these various designations have the same connotation for everyone using them. Other terms often bandied about in this context are worth mentioning, such as "open," "near," or "mathematical reality," "strong" or "weak objectivity," or "underlying reality" (a term that seems particularly appropriate).[88]

Einstein himself had the following to say on the subject in 1953, two years before his death: "I do not doubt for a moment that modern quantum theory (more precisely, 'quantum mechanics') is the best theory consistent with experiments, inasmuch that the treatment is based on material points and potential energy viewed as elementary concepts. But what I find unsatisfactory in the theory is something else, namely, the interpretation of 'the function Ψ.' In any case, it inspired my own thesis, which has been categorically rejected by the greatest contemporary theoretical minds. I nevertheless believe that there is something like the 'real state' of a physical system, existing objectively and independently of any observation or measurement, which in principle can be described with the means of expression of physics. . . . This thesis concerning reality is not a clear statement in itself, by virtue of its 'metaphysical' nature; it only has the character of a blueprint. All of us, including quantum theorists, hold on firmly to this thesis about reality, as long as no one talks about the foundations of quantum theory. . . . No one doubts, for instance, that the moon's center of gravity occupies a precise position at a given time, even in the absence of any observer—whether actual or potential. Should one discard this thesis about reality as being arbitrary and an exercise in pure logic, it would then become extremely difficult not to fall into solipsism. In the sense I just described, I do not feel ashamed to put the concept of 'real state of a system' at the focus of my reflections. . . . Now, there is no doubt that the function Ψ is a way to describe a 'real state.' The question then is whether the character of this description is complete or incomplete. Depending on one's position with respect to that question, one could encounter serious difficulties."[89]

Let us try to define more clearly Einstein's sense of conviction—*faith* might almost be a better term—concerning the existence of an "objective reality." Abraham Pais, to whom we owe a remarkable biography of Bohr and another no less noteworthy of Einstein, is a highly qualified commentator. After explaining Bohr's concept of 'phenomenon,' Pais clarified Einstein's objections thus: "In contrast to the view that the concept of phenomenon irrevocably includes the specifics of the experimental conditions of the observation, Einstein held that one should seek a deeper-lying theoretical framework which permits the description of phenomena independently of these conditions. That is what he meant by the term *objective reality*. After 1933 it was his almost solitary position that quantum mechanics is logically consistent but that it is an incomplete manifestation of an underlying theory in which an objective real description is possible."[90] Einstein maintained that position up until his death.

Heisenberg, who had discussed this subject with Einstein at great length, summarized the dilemma faced by the author of the theory of relativity in these words: "I realized how difficult it is for a physicist to abandon ideas that have up to this point been the basis of his thoughts and scientific work. In Einstein's case, his life-long pursuit had been to analyze the objective world of physical phenomena unfolding in time and space, independently of us, according to fixed laws. In his mind, the mathematical tools of theoretical physics had to describe this objective world and, consequently, to make predictions about its future behavior possible. And now, here were those people trying to convince him that at the atomic level such an objective world did not even exist in time and space, and that the mathematical tools of theoretical physics could only describe what is possible, rather than what is real. Einstein was not prepared to accept the ground being pulled from under his feet—that is exactly the way he must have felt."[91]

Indeed, as early as 1926, Einstein had written in a letter to Bohr: "Quantum mechanics is very impressive. But an inner voice tells me that it is not yet the real thing. The theory produces a good deal but hardly brings us any closer to the secret of the Old One."[92] He would never change his mind.

Let us now return to the terms *complete* and *incomplete* in the context of Einstein's view of quantum mechanics. Einstein readily conceded that quantum mechanical methods account quite well for experimental observations of the atomic world. But at the same time he rejected the interpretations of these observations, objecting in particular that they did not go far enough either in clarifying the causal relations governing phenomena or in characterizing the reality that he firmly believed must exist independently of any observer. Faced with this situation, the only recourse was to blame the "incomplete" character of the formalism in question.

That is precisely the unshakable position Einstein adopted throughout his debate with Bohr, much to Bohr's consternation, for whom the postulate of complementarity was, on the contrary, the most "complete" description of the material world that was possible.

Einstein was obviously not one to just proclaim what he believed to be the truth. He also tried to prove it. His most famous attempt was a paper published in *Physical Review* in May 1935, written in collaboration with Boris Podolski and Nathan Rosen, and entitled "Can Quantum-Mechanical Description of Physical Reality Be Considered Complete?" As was customary in those days, the answer to the question was couched in the form of a thought experiment, widely known as the "EPR paradox," after the authors' initials.[93] It is one of numerous paradoxes that from time to time added drama to the debate and enlivened the controversy surrounding the meaning of quantum mechanics.[94] It is also the only one we will describe, and only superficially at that. At issue is the definition of what the authors call an "element of reality," which they describe as follows: "Every element of reality must have a counterpart in the physical theory. We shall call this the condition of completeness. . . . If, without in any way disturbing a system, we can predict with certainty (i.e. with a probability equal to unity) the value of a physical quantity, then there exists an element of physical reality corresponding to this physical quantity. "

Let us then, together with the authors of the paper, consider the problem of two particles that first interact and subsequently fly apart. While classical physics allows us to determine or at least predict the position and momentum of each particle (q_1, p_1, and q_2, p_2) at any moment by observing the state of one of them, quantum mechanics, by contrast, precludes these quantities from being determined simultaneously for either one of the particles, in accordance with the uncertainty principle. However, it can be shown that it is possible to determine simultaneously and exactly the *difference* between their positions ($q = q_1 - q_2$) and the *sum* of their momenta ($p = p_1 + p_2$). It follows then that if, after the particles have moved apart, we measure p_1, we can also deduce p_2 without introducing any pertubation whatsoever on particle 2. The conclusion, according to the EPR definition, is that p_2 is an element of reality. Similarly, if we measure q_1, we can deduce q_2 without perturbing particle 2. Hence, q_2 is also an element of reality. Therefore both p_2 and q_2 would be elements of reality, whereas quantum mechanics teaches us that they cannot be so simultaneously.

This analysis led Einstein and his coworkers to conclude that quantum mechanics is an "incomplete" theory. They remarked: "This [simultaneous predictability] makes the reality of p_2 and q_2 depend upon the process of measurement carried out on the first system which does not disturb the second system in any way. No reasonable definition of reality could be expected to permit this." For Einstein, this dependence would introduce a mysterious "instantaneous coupling between objects separated in space," which is often discussed in the literature under the more colorful name of "spooky actions at a distance." But since according to special relativity no signal can travel faster than the speed of light, it is generally accepted that if two events are sufficiently far apart in space and close in time, they cannot exert any influence on each other. Some authors have referred to this view of things as Einstein's *principle of separability* or *locality*.[95]

The term *locality*—or, rather, its opposite, *nonlocality*—actually takes us beyond the confines of the Bohr-Einstein debate about the "EPR paradox" and touches on the heart of one of the most vexing problems in quantum mechanics. The debate had the beneficial effect of training the spotlight on that problem. The EPR article had an extraordinary impact and enjoyed great longevity: d'Espagnat and Klein are of the opinion that "it has probably been the most often cited and heatedly debated paper in the entire history of physics."[96] It may not be the most cited, but it surely was the most thoroughly discussed. An entire volume would be needed to chronicle the literature it inspired. We will simply allude to Bohr's response. Needless to say, he did not buy Einstein's argument. After first criticizing the "ambiguity" of the phrase "without in any way disturbing a system" offered as a definition of "elements of reality" in the EPR paper—to which he opposed the clarity and "completeness" of complementarity—Bohr eventually articulated his strongest objection by pointing out that in quantum mechanics it was in fact not possible to conceptually isolate the two interacting particles; one always deals with a complete system, lumping both particles together. As surprising, indeed shocking, as this assertion may have seemed at the time, it zeroes in on one of the fundamental properties of the subatomic world and epitomizes one of the trickiest concepts of quantum theory; it came to be known as *quantum inseparability* or *nonlocality*.

It is widely accepted today that most states of a composite quantum system consisting of, say, two particles are "nonseparable." This implies that any attempt to characterize the quantum state of such a system does not allow for an unequivocal determination of the states of each constitutent, but only a *correlation* between their possible states. For instance, in the case of a system consisting of two particles A and B, it is not possible to claim that A occupies coordinates x_1, y_1, z_1, while B is at coordinates x_1, y_2, z_2, since the converse, namely, A at x_2, y_2, z_2, and B at x_1, y_1, z_1, is equally likely. Any description of the system must combine both possibilities. Rather than to specify the position of each constituent of the system individually, the best we can do is be content with the correlation between two possible outcomes, which in plain words can be expressed as: "If A is at x_1, y_1, z_1, then B is at x_2, y_2, z_2."

Predictably, this counterintuitive concept provoked many stormy theoretical debates—the important contributions of John Bell (of "Bell's inequalities" fame) in this area must be acknowledged. A number of extremely clever experiments, not the least of which were those conducted in France by Alain Aspect in 1982 on the correlation between the polarizations of photons, ultimately brought convincing experimental confirmation of the principle of quantum nonlocality.[97] It has come to be accepted as an integral part of quantum mechanics. It results from the fact that wave functions are spread out in space or, in scientific jargon, are "nonlocal." As a consequence, a measurement performed in one location instantaneously affects the materiality of what resides in other locations, no matter what their distance. In other words, the wave function of a system of two particles does not split into two distinct wave functions when the two particles are separated in space. It only "stretches," in a manner of speaking. The quantum mechanical universe turns out to be intrinsically holistic.

We now return briefly to Einstein and his deep-seated conviction that quantum mechanics is an "incomplete" theory. He believed that, contrary to the assertions of the authors of this formalism, the wave function does not provide all the information necessary for a description of the states of a system. Einstein was to cling to that conviction to the end of his life. He stated in 1953: "A description based on the wave function Ψ is incomplete. One is therefore forced to conclude that there must exist a more complete description. One is also led to the view that the intrinsic laws of nature must involve data pertinent to a complete, rather than incomplete, description. It is furthermore difficult to avoid the suspicion that the statistical character of the theory reflects the incomplete character of the description, and that it may have little to do with things as they truly are."[98]

To the extent that one is unwilling to dismiss quantum mechanics—if for no other reason than there is simply nothing better available—how to reach completeness in describing the microscopic world? One way, which seems natural to some, would be to attach to the wave function some hidden variables (some proposals have even distinguished internal and external hidden variables) whose nature and values would enable a description consistent with the condition of complementarity. Einstein understood this to mean a return to a deterministic framework for all physical phenomena. It should be pointed out that Einstein himself never used the term "hidden variables" in any of his papers or lectures. Others have done so, generally without results, at least up to now. The distinguished mathematician John von Neumann (1903–1957) even derived a famous theorem purporting to prove mathematically that no hidden parameter could ever render quantum mechanics in its traditional form deterministic, much to the rejoicing of Bohr, Born, Pauli, Heisenberg, and other advocates of quantum indeterminism. However, it appears today that the claims of that theorem were somewhat inflated.[99]

Abraham Pais summarized the lesson to be learned from this long controversy as follows: "The dialog between Bohr and Einstein had one positive result: It forced Bohr to articulate the principles of complementarity in ever more precise language." This verdict is all the more significant since it appeared in a book devoted to Einstein!

The Others

Similarity does not exist per se: it is merely a
particular limiting case when the difference goes
to zero.

—Claude Lévi-Strauss

After our discussion of the Bohr-Einstein debate, it may seem somewhat cavalier to speak of "the others" when they have names like Max Planck, Louis de Broglie, Erwin Schrödinger, Werner Heisenberg, or Wolfgang Pauli. The fact remains, however, that regardless of their stature and the hints of individualism in their positions with regard to the subtleties of quantum mechanics, they can be classified, at least at first blush, into two groups rallying around either Bohr or Einstein.

This becomes obvious when one examines Selleri's survey, which in effect asked every one of these scientists three questions, and categorized them according to their answers. Selleri's questions addressed three specific issues:

1. Reality issue: "Do the fundamental entities of atomic physics, such as electrons, photons, etc., really exist, independently of observations made by physicists?"
2. Comprehensibility issue: "If the answer is yes, is it possible to understand the structure and evolution of objects and atomic processes, in the sense that spatial and temporal images can be formed that are consistent with their reality?"
3. Causality issue: "Is it possible to formulate physical laws in such a way that one or several causes can be identified to account for the effects observed?"[100]

Given the context of this book, I cannot resist the temptation to extend Selleri's survey in two directions. First, I would ask all those scientists a fourth question, the answer to which would, I believe, provide useful insights into their respective overall worldview. The question centers on their religious beliefs. It can, of course, be asked in several different ways. I would phrase it as simply and directly as possible:

4. Faith issue: "Do you believe in God? If so, how do you envision him?"

The second direction in which I would propose to widen this survey would be to submit the first three questions to none other than Pope Pius XII. I will indicate my reasons when I analyze his stated position. I trust I will be forgiven for not having had the nerve to submit to him the fourth question.

Let us first review how quantum physicists answered Selleri's first three questions. On the basis of the evidence contained in their original writings, Selleri concludes that, by and large, Planck, de Broglie, and Schrödinger join Einstein in answering yes, while Born, Pauli, Heisenberg and Dirac, together with Bohr, answer no.

To be sure, there are significant nuances within each group, to the point that on certain specific issues, members of one group occasionally cross the dividing line and join with the opposition. There are also opinions that evolve with time and outright about-faces, which place some of the protagonists first in one camp and later in the other.

For instance, among the opponents of "reality" of the external world, Heisenberg took a much stronger position than did either Bohr or Born. He went so far as to assert that "the elementary particles of modern physics are even more abstract than the atoms of the Greeks," that "it is not a material particle in space and time, but only, in some sense, a symbol whose introduction gave the laws of nature a particularly simple form." According to Born, the probabilistic wave function represents "a tendency toward something. It is a quantitative adaptation of the ancient concept of 'potentia' in Aristotle's philosophy. It introduces something halfway between the idea of a phenomenon and the phenomenon itself, between possibility and reality."[101] Duhem would be pleased to know that he effectively received the endorsement of such a prominent scientist.

Heisenberg was also the most intransigent of the indeterminists, embracing extreme opinions on this subject as well. Dirac, on the other hand, although perceived as belonging in the same camp, was to write in 1976: "I think it is quite possible that in the end Einstein was right, for the current version of quantum mechanics cannot be considered definitive. I believe it is very likely that in the future a better version of quantum mechanics will be developed, one marked by a return to determinism."[102]

On the other hand, in the realist camp, de Broglie emphasized clearly the difference he saw between the principle of causality and the more complicated problem of determinism: "I never took a position on determinism because it is a much more general philosophical issue. I am not a philosopher and I do not consider myself qualified to render a verdict on this question. . . . It is a problem quite different from causality in physics, because the principle of causality is simply a framework in which physicists operate."[103]

As far as Schrödinger is concerned, Moore is of the opinion that, even though he appeared to believe in the reality of objects, he went at least partway toward Bohr's and Heisenberg's ideas when he stated: "Our brain, hindered by its limited capability, cannot ask nature questions that would require a continuous series of answers. Observations and individual results of measurements are nature's answers to our discrete questions. Because of that, they do not deal with a pure object, but rather with the relation between object and subject. It is no longer obvious that repeated observations must inevitably lead to a precise knowledge of an object."[104] According to Selleri, Schrödinger's position on the issue of determinism was ambiguous, judging by his answer to the question of whether or not the behavior of a single atom is determined by strict causality: "There is probably no way to resolve this experimentally. Through pure reasoning, it is clearly possible to derive the randomness of laws, or the laws of randomness, if one prefers."[105]

Both de Broglie and Schrödinger changed their positions, in fact more than once, about certain fundamental propositions of quantum mechanics. While they believed in the material reality of the wave function Ψ in the early part of their careers, they both yielded to the pressure of the "general consensus" at the Fifth Solvay Congress and crossed over (Schrödinger more reluctantly than de Broglie) to the probabilistic camp. Interestingly, both were to return to their original position. Schrödinger did so rather quickly, de Broglie only in the 1950s. The intellectual path followed by de Broglie is particularly poignant; the lack of understanding by some of his colleagues about his own position caused him much bitterness.[106] He himself retraced the evolution of his thinking in the preface to the reedited version of his 1924 doctoral dissertation, released in 1963: "The moment I stepped so boldly into completely uncharted territory in 1923, I was convinced that it was necessary to arrive at a veritable synthesis of the notions of wave and particle, all the while preserving the precise images of physical reality that had always been associated with both notions. In the three years following the defense of my thesis, I attempted to construct—somewhat imperfectly, I must admit—a theory consistent with the goal I had set myself. Having shared my recollections on numerous occasions during the last few years, I need not reiterate here how much opposition my theory ran into at the time of the Solvay Congress in October 1927 and how, discouraged by the intense opposition, I switched over—although with some reluctance—to the very different interpretation favored by a number of prominent scientists, which has since then remained the 'orthodox' interpretation of quantum mechanics. . . . Some ten years ago, after a renewed period of deep thinking about this crucial problem of contemporary physics, I concluded that my attempts thirty-five years earlier, as unsatisfactory as they may have seemed, were nevertheless on the right track. A careful examination of the objections against the orthodox theory raised by the distinguished physicists I have just named [Planck, Einstein, Schrödinger] and a few others led me to conclude that this interpretation leads to certain paradoxical and hardly acceptable conclusions hiding under the guise of an elegant and precise mathematical formalism. In my opinion, these paradoxical conclusions all come about as a result of repudiating the clear picture of a particle conceived as a concentration of energy tightly localized in space and time. We abandoned the idea that the wave function is a physical entity truly propagating in space, and it gradually came to be perceived simply as an artificial mathematical tool useful for calculating probabilities. I then resumed my previous efforts to view the particle as a very small region of intense field associated with a spatially extended wave, but I incorporated in my description some new elements which, I believe, brought some degree of improvement."[107]

This is hardly the place to describe the technical details of these "improvements," which came to be embodied in several versions of the so-called double solution theory.[108] We will, however, at least mention the name of David Bohm (1917–1992), a British physicist-philosopher, who substantially influenced de Broglie's ideas and made original contributions of his own to the debate on the implications of quantum mechanics.[109]

This short synopsis of the tug-of-war surrounding the meaning of the quantum revolution, its impact, and its consequences is a testimony

to the richness and complexity of the debate and its continually changing dynamics. The leitmotif of the new mechanics is to continually exhibit apparent cracks, with no one knowing if or when they might be
repaired. As such, it constitutes a marvelous field of epistemological
research. Indeed, some notable studies along this line have been published recently.[110]

The Quantum Scientists and Religion

*In short, our means of investigation are far ahead of
our ability to understand.*

—Paul Valéry

The two great scientific revolutions of the twentieth century—the theory of relativity and quantum mechanics—introduced totally unfamiliar
concepts whose strangeness suggested troubling paradoxes, raised philosophical questions, and stirred metaphysical anxieties. Added to this was
the accelerating pace of scientific progress, which prompted Paul Valéry
to remark that "all the verbal, theoretical, and explicative part of our
knowledge becomes essentially transitory," and that "reality (in the scientific sense) is a kind of function of time."[111]

All these factors probably contributed to the fascination shown by the
principal scientists involved in these dramatic changes for possible connections between science and religion. Planck, Schrödinger, and Heisenberg devoted entire chapters of their writings to this topic, and Einstein
addressed it on numerous occasions.[112] It is obvious that the reflections of
these scientists reached far beyond purely scientific or even philosophical
considerations. The British physicist-philosopher Arthur Eddington
observed: "Physics is presently being formulated in a way that is evidently
only a partial aspect of something far wider. . . . The problem of the scientific world is but a part of a broader problem, namely, that of the entire
human experience. . . . It could be said perhaps that the conclusion to be
drawn from the arguments of modern science is that religion becomes
possible for a rational scientist around the year 1927."[113]

At one time or another, practically every one of the founders of the
new mechanics pondered over the connections between science and religion. They showed as many subtle differences in this area as they did in
their opinions on the scientific and philosophical implications of their
discoveries. Their views run the full gamut, from pure atheism to relatively conventional faith. At one extreme stands Dirac, who, unlike the
majority of his colleagues, displayed a harsh contempt for the traditional
religions of the Western world. "Dirac . . . had little time for tolerance,"
wrote Heisenberg, who related in his memoirs an entire diatribe by Dirac
against religion: "If we are honest—and scientists have to be—we must
admit that religion is a jumble of false assertions, with no basis in reality.

The very idea of 'God' is a product of the human imagination. . . . I can't for the life of me see how the postulate of an Almighty God helps us in any way. What I do see is that this assumption leads to such unproductive questions as why God allows so much misery and injustice, the exploitation of the poor by the rich and all the other horrors He might have prevented. . . . Religion is a kind of opium that allows a nation to lull itself into wishful dreams and so forget the injustices that are being perpetrated against the people." The virulence of this statement, recounts Heisenberg, prompted the following reply from Pauli: "Well, out friend Dirac, too, has a religion, and its guiding principle is: 'There is no God and Dirac is His prophet.'"[114]

Schrödinger's position was more circumspect. All the same, his religious quest is most original for a Western scientist. Incredibly, not once is the word *God* mentioned anywhere in the chapter on science and religion in his book *Mind and Spirit*. As Moore observed: "Schrödinger had no need for God in his vision of the world."[115] He rejected all traditional monotheistic religions, such as Christianity, Judaism, and Islam, but without any virulent recrimination or even open antipathy; he simply found them too naive. He was, however, taken by the mystical writings of India and those of the early Christian Gnostics, both of which provided a system of thought based on the conviction that "the living self and the world are one and are the whole."[116] Eventually he became an adherent of the Vedanta. In the ultimate irony, of all the spiritual systems of ancient India, that is the one that most categorically denied the existence of atoms.

Bohr embodies another type of religious scientist. He broke early on with the Lutheranism to which he was born. "The idea of a personal God is foreign to me," he claimed. At the same time, though, he understood the special nature of religious messages. Perhaps his tolerant attitude reflected his predilection for complementarity and may have had something to do with his refusal to make a clear distinction between "objective" and "subjective" worlds.[117] At least, that is suggested by the following citations: "But we ought to remember that religion uses language in quite a different way from science. The language of religion is more closely related to poetry than to the language of science. True, we are inclined to think that science deals with information about objective facts, and poetry with subjective feelings. . . . But I myself find the division of the world into an objective and a subjective side much too arbitrary. . . . The location of the separation may depend on the way things are looked at; to a certain extent it can be chosen at will. . . . Perhaps we ought to look upon these other forms as complementary descriptions which, though they exclude one another, are needed to convey the rich possibilities flowing from man's relationship with the central order."

The philosophical opposition between Bohr and Einstein (despite their profound and enduring friendship) extended to the religious domain as well. Einstein did not believe any more than Bohr in an anthropomorphic God, writing: "I believe in Spinoza's God who reveals himself in the orderly harmony of what exists, not in a God who concerns himself with the fates and actions of human beings." Also like Bohr, he distanced himself early in his life from the religious culture in which he was born—in his case the Jewish community, even though he would often proclaim his identity with the Jewish people and his kinship with its spiritual heritage and tradition. Yet Einstein was a religious man, even "deeply religious," in A. Pais's estimation.[118] But his religiosity had a "cosmic" dimension, in the sense that "it is grounded in marveling and being in awe at the harmony of the laws of nature revealing an intelligence so superior that all human thoughts and skills can only be exposed for their pitiful shortcomings." Einstein described in moving words his own intimate and edifying experience: "The individual feels the nothingness of human desires and aims and the sublimity and marvelous order which reveal themselves both in nature and in the world of thought. He looks upon individual existence as a sort of prison and wants to experience the universe as a single significant whole. . . ."[119] His hunger is far-reaching: "I want to know how God created this world. I am not interested in this or that phenomenon, in the spectrum of this or that chemical element. I want to know His thoughts, the rest are details." At this high level of aspiration, religion and science are not in conflict but, instead, reinforce each other: "Science without religion is lame, and religion without science is blind," he wrote. Strongly individualistic, the only philosopher with whom he felt a spiritual communion was Baruch Spinoza, for whom he professed great admiration, presumably because of his pantheism and "deterministic" attitude: "Although he lived three hundred years before our time, the spiritual situation with which Spinoza had to cope peculiarly resembles our own. The reason for this is that he was utterly convinced of the causal dependence of all phenomena, at a time when the success accompanying the efforts to achieve a knowledge of the causal relationship of natural phenomena was still quite modest."[120]

No less significant is another of Einstein's comments, in which he evidently alludes to Spinoza's excommunication by the Jewish community of Amsterdam: "There can be no Church whose central teachings are based on it [a cosmic religious feeling]. Hence it is precisely among the heretics of every age that we find men who were filled with the highest kind of religious feeling and were in many cases regarded by their contemporaries as atheists, sometimes also as saints. Looked at in this light, men like Democritus, Francis of Assisi, and Spinoza are closely akin to one another."[121] With these words, Einstein pays implicit homage to Democritus by putting him on the same pedestal as Spinoza. The parallel is somewhat unexpected, since Democritus was a great visionary of atomism, while Spinoza was rather opposed to that doctrine.[122] Einstein was apparently not bothered by Voltaire's opinion that the indivisible unity of a spatially extended substance, which is one of the fundamental principles of Spinoza's philosophy, is such that "the truth of atomism proves the falsity of Spinozism."

Max Planck was another profoundly religious scientist, convinced that science and religion are not in contradiction but, on the contrary, are perfectly compatible and even complementary in the classical sense of the term (he differed from Bohr in this respect). In his own words: "There is nothing to prevent us—indeed our instinctive intellectual drive toward a unified description of the world demands it—from identifying these two ubiquitous, yet mysterious, forces: the order of the universe implied by natural science and the God presumed by religion. Therefore, the divinity that men of religion seek to promote by means of tangible symbols has just as much substance as the power acting in accordance with natural laws for which scientific data provide indisputable evidence." The only difference he saw between these two approaches is one of internal logic: "While religious and scientific endeavors both require a belief in God, He is the starting point in religion and the ultimate objective of any thought process in science. God is the basis of religion and the crowning of any scientific construct aimed at a generalized view of the world."[123] Although Planck's discussions of the relation between science and religion are extensive, nowhere does he offer a hint of his own denominational preference (he came from a Protestant family). Judging from a letter he wrote a few months before his death, his attitude did not necessarily imply a belief "in a personal God, let alone a Christian God."[124]

Finally, Heisenberg too was agreeable to a symbiotic relationship between science and religion, as he saw in all scientific constructs a necessary coexistence of the rigor of mathematical formalism with the blurred images of classical concepts (including the notions of wave and particle). From this perspective, he viewed the advent of quantum mechanics as reinforcing this interdependence: "In science, a central order is implicitly acknowledged by the simple fact that it is acceptable to use metaphors such as 'nature is created according to a particular purpose.' In this respect, my own conception of truth is closely related to the way religions describe it. I believe that this connection can be grasped more clearly now that quantum theory has been developed. For in this theory we can formulate in abstract mathematical language a unified order encompassing very large domains; at the same time, though, when trying to describe in our ordinary language the consequences of that order, we see that we are reduced to using parables, that is to say, complementary interpretations that contain paradoxes and apparent contradictions."[125]

These examples demonstrate how important a role religious considerations played among the most prominent relativity and quantum physicists. The radically new concepts emerging from their research prompted
them to profoundly reevaluate their perception of fundamental problems at the heart of scientific inquiry: reality, comprehensibility, causality.
For the most part, they pursued their reflections outside the context of
traditional religions, blending, in a manner of speaking, their religious
and scientific views, without mutual interference. In this respect, these
modern thinkers differ from many earlier scientists (particularly from the
sixteenth to the nineteenth centuries), who were prone to intermingling
religious arguments with raw scientific data. What emerges in the twentieth century is a religiosity of a higher order, often expressing awed admiration at the beauty and harmonious order of the world, conceived as a
complement to scientific constructs, whether the nature of this complementarity be understood in the Bohrian sense of coexistence of opposites or in the classical sense of a congruous assemblage of parts into a
unifying whole.

In any case, God, whatever he may be, appears to be increasingly
accepted in a world viewed through the prism of modern science. In a
broader context, relativity and quantum physics seem to have stoked an
interest in the "problem of God" among a general public that was opening up to scientific culture and might otherwise have remained more
indifferent. This is a significant new development worth emphasizing.[126]
It has resulted in a steady stream of modern physics books that explore to
various degrees of depth the relation between science and religion and
the possible or presumed—if not certain—role of "God" in the creation
and march of the universe. Predictably, the conclusions are far from
unanimous. An explosion of popular science works on this topic has also
appeared on the scene in recent years. Many are of high quality, written
by experts of impeccable reputation. Among the most noteworthy are:
God and the New Physics, by P. Davies (Penguin Books, 1990); *Other Worlds*,
by the same author (Penguin Books, 1980); *The Cosmic Code*, by H. Pagels
(Penguin Books, 1982); *The Ghost in the Atom*, by P. Davies and J. R. Brown
(Cambridge University Press, 1986); *The Secret Melody*, by Trinh Xuan
Thuan (Oxford University Press, 1995); *Dieu et la science* (God and science), by J. Guitton, G. Bogdanov, and I. Bogdanov (Grasset, 1991); and
others. Of particular interest is *Dreams of a Final Theory*, by S. Weinberg
(Vintage, 1993), in which the author argues that "some people have views
of God that are so broad and flexible that it is inevitable that they will
find God wherever they look for Him," adding that he himself does not
believe we will ever find "any sign of the workings of an interested God in
a final theory." Lastly, given Richard Feynman's popularity and stature, it
may also be relevant to mention here his strictly antireligious position. As

told by James Gleick in *Genius* (New York: Pantheon, 1992), Feynman was particularly averse to the image of an anthropomorphic God, declaring in 1959: "It doesn't seem to me that this fantastically marvelous universe, this tremendous range of time and space and different kinds of animals, and all the different planets, and all these atoms with all their motions, and so on, all this complicated thing can merely be a stage so that God can watch human beings struggle for good and evil—which is the view that religion has. The stage is to big for the drama."

Pius XII

The reader will undoubtedly have noticed that in the previous part of the book devoted to the advent of scientific atomism, no mention was made of the attitude of religious authorities toward this development. Actually, it seems that we have paid far more attention to the attitude of atomic scientists toward religion than the other way around. The reason is quite simple: The rise of the scientific exploration of atoms coincided with a period of distinct relaxation in the antagonism of the Christian Church toward the concept of atoms. The hatchet seemed to have been buried, which is not to say that the Church endorsed all explicit or implicit tenets of the atomic theory, or that it was prepared to accept atoms and molecules as the ultimate word on the nature of the universe and man. Nevertheless, its open antiatomism did recede in favor of a quiet acceptance of a theory increasingly verified by experiments. As an atomic vision of the world was becoming more and more irrefutable, the Church wisely decided to accept it, at least as far as what appeared to be beyond doubt was concerned, namely, the structure of the material world.

In the process, the Church proved itself far more astute than positivism. In fairness, Rome had to confront in the nineteenth century other crises far more perilous to its doctrine than atomism, including discoveries in geological chronology, which suggested the earth was far older than the five thousand to six thousand years decreed in the Bible and, most important, the hypotheses concerning the evolution of species, culminating in Darwin's theory that ostensibly included the human race, squarely challenging the immutability of species and the special creation of man taught by the Church.[127] These more pressing battlefronts tempered any latent opposition to atomism. Actually, certain aspects of the atomic theory in the first quarter of the twentieth century even inspired a more receptive attitude on the part of Church officials. For example, the Vatican came to see in the discovery of the composite structure of the atom and the possibility of its transmutation by natural—and, later, artificial—radioactivity reasons for satisfaction. Pius XII, pontiff from 1939 to 1958, stated as much in an address delivered on November 30, 1941, before the Pontifical Academy of Sciences gathered in plenary session: "Who could have imagined only a hundred years ago the riddles locked up in so tiny a particle as a chemical atom, in a space of a millionth of a millimeter! The atom was then considered a homogeneous globule. The latest physics saw in it a microcosm in the truest sense of the word, hiding in its midst such profound mysteries that, despite the most delicate experiments and the use of the most modern mathematical tools, research is today just beginning its conquest of the structure of atoms and of the fundamental laws governing their energy and motion. By now, the continual change of material things and their transformations, including the chemical atom which had for so long been held immutable and imperishable, are more manifest than ever. One thing alone is immutable and eternal: it is God."[128] The pontiff was to return later to this theme of the "changeability of things" in greater depth in another speech, also before a plenary session of the Pontifical Academy of Sciences on June 7, 1949.

Among all twentieth-century popes, Pius XII was the one who dealt the most extensively with the issue of atomism, particularly with the scientific and philosophical questions raised by the advent of quantum mechanics, questions that could not have left the Church indifferent. He did so in two speeches to the Pontifical Academy of Sciences, the first, already mentioned, on November 30, 1941, and the second on February 21, 1943. They are lengthy and marvelously prepared dissertations that attest to the detailed knowledge the pontiff had of the subject matter. Reading them is not unlike attending a magisterial lecture, as they constitute genuine updates on the state of knowledge at the time.

Of primary interest to our own purpose is Pius XII's position on the philosophical questions that were at the core of the Bohr-Einstein debate, as related in the previous section. Without ever explicitly referring to this debate, Pius XII is implicated in it *de facto* simply by virtue of taking a stand. Incidentally, it should be mentioned that virtually every great scientist who contributed to the triumph of the atomic theory and to the development of quantum mechanics (with the notable exception of Einstein) was elected a member of the Pontifical Academy of Sciences: Bohr, Planck, Schrödinger, and Rutherford in 1936, de Broglie and Heisenberg in 1955, and Dirac in 1961. It gave Pius XII the opportunity to consult with them in person.

If one had to analyze in only a few words the pope's position toward the problems raised in the Bohr-Einstein debate, one would have to say that he was clearly on Einstein's side. Indeed, there is a remarkable convergence of views between the two men on the fundamental issues of reality, comprehensibility of the world, and determinism.

Speaking about reality and comprehensibility, Pius XII proclaimed in a speech in 1943: "Let it not be said that matter is not reality but an abstraction fashioned by physics, that nature is inherently unknowable, that our sensible world is another world in itself in which phenomena reflecting the external world make us dream the reality of the things it hides. No, nature is reality, and reality is knowable. Things may just appear and be silent, but they do have a language that speaks to us and springs forth from their midst like water from an eternal spring . . . impressing onto us an image or a likeliness that is a vehicle for our intellect in its quest for the reality of things. . . . Do not commit the same blunder as those philosophers and scientists who believed that our cognitive faculties are capable of recognizing nothing but their own changes and feelings, which led them to assert that our intellect could only develop a science of appearances emanating from things; hence, only images of things, and not the things themselves, would be the object of our science and of the laws we formulate about nature. What a grievous mistake! . . . Science, which exalts a Copernicus or a Galileo, Kepler or Newton, Volta or Marconi, or other famous and deserving researchers of the physical world that surrounds us on all sides, would be reduced to the beautiful dream of a wakeful mind, a beautiful phantom of physical knowledge; and the affirmation or negation of the same thing would be equally true. No, science is not in dreams, nor in the appearance of things: it does reside in things themselves through the images we perceive of them."

What is more, readers familiar with the history of the periodic table of elements, the cornerstone of the atomic conception of the world, may appreciate the following argument used by Pius XII, in which the way the table came about is interpreted as a testimony in favor of the reality of the world: "When Lothar Meyer and Mendeleev organized the chemical elements according to this simple scheme presented today as a natural system of elements, they were profoundly convinced that they had discovered an orderly classification scheme based on internal properties and trends, one that was suggested by nature itself and whose gradual development promised the most penetrating discoveries about the constitution and state of matter. Indeed, that was the premise of all modern atomic research. At the time of the discovery, the so-called economy of means principle did not come into play because this primitive scheme still displayed numerous empty slots; it could not be a matter of simple convention, for the properties of matter itself mandated such an order. This is simply one example among many others of the most brilliant scientists of the past and present having come to the noble certainty that they were the harbingers of a truth identical and common to all people and generations walking this earth and raising their eyes toward the sky; a truth that is founded in its essence on an *adaequatio rei et intellectus*, a truth that amounts to a grasp, more or less perfect or advanced, by our intellect of the objective reality of natural things; and that is what constitutes the truth of our knowledge."

The identification by Pius XII of "reality" with "truth" is noteworthy, although he might have felt a little more comfortable with the latter term. On December 9, 1939, he made the following pronouncement in a speech to the Pontifical Academy: "Just as we do not create nature, we do not create truth either: neither our doubts nor our opinions, negligences or compromises can alter truth. . . . Between God and ourselves stands nature. . . . Scientific truth is nature's daughter and God's granddaughter."

The parallel with Einstein's position is evident. A similar parallel exists when it comes to Pius XII's views on determinism. Along with Einstein, the pope believed that where, for the time being, purely statistical laws seem to reign, a definitive causal order (which for him is "evidence of God's hand in the universe") is to be someday uncovered: "Being content with statistical laws is a mistake of modern times, as would be ignoring the nature of man's genius, which is his only capacity to learn what he ultimately makes worthy of the intellect. . . . The scientific system, richly connected and well developed in the macroscopic realm, undoubtedly includes many statistical laws; but considering the multitude of elements, atoms, molecules, electrons, photons, etc., they are not manifestly inferior to strictly dynamical laws in terms of certainty and exactitude. At any rate, they are founded on, if not anchored to, the strictly dynamical laws of the microcosm, even though the details of the microcosmic laws may be completely hidden from us at the present time, notwithstanding the great efforts expended in new and bold research aimed at unravelling the mysterious happenings inside atoms. Gradually, though, the veils will be lifted and the seemingly noncausal character of microcosmic phenomena will disappear: a new and marvelous rule of order, encompassing the most minute particles, will be discovered."

Lastly, at the risk of digressing from philosophical considerations in favor of a much more concrete problem (still involving atoms, though), I could not conclude this section without quoting the following excerpt from a speech Pius XII gave on February 21, 1943: "The fascinating transformations of atoms have for many years occupied only researchers in pure science. Without a doubt, the magnitude of the energy released was unexpectedly high, but since atoms are extremely small, nobody thought seriously that they could some day acquire any practical importance. Today, on the other hand, such a possibility has gained renewed attention because of experiments on artificial radioactivity. It has been established that during the decay of a uranium atom bombarded by a neutron, two or three neutrons are liberated, each of which proceeds on its own and can collide with and break apart another uranium atom. In this manner, effects multiply and it is conceivable that the ever increasing number of neutrons colliding with uranium atoms could proportionately increase the number of neutrons liberated in a given unit time, to such an extent that the integrated energy might reach values so enormous as to be unfathomable. . . . But above all, it is essential that such a process not be allowed to proceed in the form of explosions, but that the course of the reaction be slowed by appropriate and careful chemical means. If not, a dangerous catastrophe could ensue not only locally but for the entire planet."

Evidently, at the height of the war, Pius XII was not only familiar with research on nuclear fission, but he was also aware of its potential for building nuclear weapons. Fear of this prospect may have played a role in his ongoing interest in the science of atoms. His successors, John XXIII, Paul VI, and John Paul II, were, for their part, confronted with very important developments in other areas of science, particularly in molecular biology and astrophysics. These developments, which enabled man to decipher the genetic code and go to the moon, have enormous potential implications for the fate of humanity, and redirected the attention of later popes away from nuclear science, toward the sciences of life and of the universe. This is not to say that they had previously neglected these fields or that they subsequently ignored atomism. It quite simply means that because of the dynamics of the evolution of science, each period has its favorite topics.

THE SOCIETY OF ATOMS: MARRIAGE

The greatest torture for me would be to find myself alone in paradise.

—Goethe

More is different.

—P. W. Anderson

Among the most important objections leveled against the ancient atomic theory from the very outset was the apparent difficulty—or outright impossibility, in the opinion of many—to explain how substances with new properties could result from mere juxtaposition of indivisible, rigid, and impenetrable atoms, even if their surfaces featured various types of hooks, grapples, and other projections propitious for mechanical interlocking. Aristotle had already made this criticism his primary argument against the theory (see chapter 5). Cicero was following suit when he wrote: "Given that in the study of nature there are two questions to consider, namely, from what matter is each thing produced, and what force causes it to be produced, they [the atomists] commented at length on matter, but they ignored the force and efficient cause."

This problem was to remain the weakest point of the theory practically until the twentieth century, indeed until the arrival of quantum mechanics. Molecules came to be accepted as associations of atoms long before any understanding of the nature of the forces ensuring their cohesion or of the unavoidable structural transformations necessary to account for unified entities endowed with new characteristics. Even in the simple case of diatomic structures, the whole proves to be far more than the sum of its parts, and mere addition of basic units fails to explain the result. From any angle other than weight, a hydrogen molecule ($H + H = H_2$) is quite different from two separate hydrogen atoms ($2H$). When our distant ancestors began primitive attempts at mixing simple substances, they were already instinctively looking for something with new properties. As time went on, man experimented with materials made of larger numbers of atoms, producing an explosion of substances with ever more varied and novel qualities. This trend is the key to understanding the evolution and increasing complexity of our world.

The shift from a qualitative notion of union to one—more quantitative and structural—of chemical bonding required an understanding of the mechanisms responsible for the formation and cohesion of these combinations. An early fruitful step in this direction is to be found in the work of Newton, who was the first to explicitly propose replacing mechanical interactions by the effect of physical forces. Remarkably, having discovered the universal law of gravitational attraction, Newton realized that it was not sufficient to explain chemical interactions and that for this purpose it would be necessary to supplement gravitational forces with electric and magnetic forces, not to mention forces of repulsion (see chapter 12). His pupil, Boscovitch, then went on to develop the thesis of equilibrium between forces of attraction and repulsion (see chapter 14).

From merely *possible*, the role of electric forces surmised by Newton became *probable* in the nineteenth century, as a result of research on electric phenomena. Volta's invention of the galvanic battery in 1800 deserves a prominent mention in this effort. It provided the first decisive proof that chemical reactions could indeed result from such phenomena.[129] In spite of its many shortcomings, Berzelius's electrochemical theory of combinations (see chapter 19) managed to shape into a concrete form the idea of electrical charges participating in these combinations.

Finally, from merely *probable*, the decisive role of electrical forces in the formation of interatomic associations became *certain* around the turn of the nineteenth century with the discovery of the electron, and later of the planetary structure of the atom (already discussed in this chapter). J. J. Thomson foresaw as early as 1904 a mechanism for the formation of what today is called an *ionic bond*, which consists of an interaction between oppositely charged ions resulting from the transfer of electrons from "electropositive" to "electronegative" atoms.[130] Subsequent work brought to light the specific advantages of such transfers when the atoms involved can acquire through this mechanism outer electron shells containing

two, or more commonly eight, electrons. Such a configuration is characteristic of the noble gases, which exhibit considerable chemical inertia
and are especially stable (see discussion later in this chapter). This purely
electrostatic model proved quite suitable, at least to a first approximation,
for describing interatomic bonds in numerous compounds common in
mineral chemistry. Such compounds often involve strongly electronegative or electropositive elements, in which the outer electron shell (which
J. J. Thomson and Bohr had already identified as being primarily, if not
exclusively, implicated in interatomic interactions) is not very different
from that of the noble gases, differing from them by the deficiency or
excess of at most one or two electrons.

On the other hand, the theory had very little success in the field of
organic chemistry. Walter Kossel (1888–1956) did propose in 1916 structures involving intramolecular multivalent ions to describe some fundamental carbon derivative compounds such as methane (CH_4) and carbon
tetrachloride (CCl_4), which looked like:

$$
\begin{array}{ccc}
H^+ & & Cl^- \\
+ & & + \\
H^+ - C - H^+ \text{ for } CH_4 & \text{and} & Cl^- + C + Cl^- \text{ for } CCl_4 . \\
+ & & + \\
H^+ & & Cl^-
\end{array}
$$

But it quickly became apparent that such a formalism was inadequate
to account for the properties of organic substances, in which C^{4+} and C^{4-}
ions are almost never observed. It was soon concluded that the chemical
bonds in the vast majority of organic compounds had to be of a different
type. Since nobody had a clear idea of what the nature of that bond might
be, it became traditional to symbolize it simply by a line connecting the
linked atoms, which did not go much beyond a crude pictorial description and explained nothing. Incidentally, organic compounds were not
the only ones affected by this state of ignorance; the same problem existed for diatomic molecules formed of two identical atoms (what today we
call "homonuclear" molecules), which clearly could not have an ionic
structure either.

The discovery of the role of electrons in ionic bonds strongly suggested that they had to be implicated in other types of bonds as well. The key
question obviously was how. As early as 1913, Bohr had proposed a model
of the hydrogen molecule in which the bond between the two atoms was
provided by their two electrons circling around together in an orbit perpendicular to the axis of the molecule, midway between the two nuclei.
The model had the merit of suggesting explicitly an electronic origin for
the bond but, as it turned out, it was incorrect.

A significant development along the same line of thought came from
Gilbert Newton Lewis (1875–1946), who was the first to propose replac-

ing the line traditionally used to depict chemical bonds (even in nonpolar or weakly polar compounds) by two dots symbolizing the participation of the two electrons involved. As modest as this pictorial improvement might seem, it had great impact. This new representation highlighted the fact that this type of bond was due to the sharing in internuclear space of electrons supplied by the bonded atoms. As a result, the terms *electronic pair* or *doublet* became permanent fixtures in the language of chemists. Soon thereafter, Irving Langmuir (1881–1957, NP 1932) gave such bonds the name *covalent,* and it stuck.

Lewis applied his concept of electronic pairs to explain the configuration of various types of compounds, often with a view toward satisfying the requirement of eight electrons in the outer shell of atoms. This in turn led him to a cubic model of atoms (replaced in the case of carbon compounds by a tetrahedron, with bonding electrons located at the four vertices). The same type of model was developed independently by Langmuir, although historical accounts often fail to acknowledge that fact. Thanks to Lewis, the fundamental idea that chemical bonds are due to interactions between the valence electrons of atoms gained wide currency in chemistry. Lewis also deserves recognition for asserting that "it is not necessary to consider the two types of chemical bonds, corresponding to highly or weakly polar compounds, different in nature but, instead, different in degree."[131] He symbolized that conviction by displacing the two dots depicting the bonding electronic doublet toward one or the other atom, according to the presumed direction and strength of the polarity. This intuition could be construed as a premonition of the notion of the "partially ionic character of covalent bonds," which was to prove so useful later on.

By the 1920s the idea that chemical bonds, including covalent bonds, were electronic in nature had won wide acceptance. What was still completely missing was an understanding of how two negatively charged particles, placed between two other positively charged ones, could produce entities as stable as molecular structures, given the mix of forces of attraction and repulsion at work.

The answer to this crucial question came from quantum mechanics toward the end of the 1920s and the beginning of the 1930s. Two main methods were proposed to handle this difficult problem. They are known as the *molecular orbital* method and the *valence bond* method. The latter actually came first chronologically—it was introduced in 1927 by Walter Heitler (1904–1981) and Fritz London (1900–1954), who used it in their famous work on the structure of the hydrogen molecule—and enjoyed immense success.[132] We will, however, outline the method of molecular orbitals only. This is for three reasons: (1) it is the most straightforward extension to molecules of the principles described previously in connection with atoms; (2) it is much simpler than the method of valence bonds and is used today almost exclusively for quantitative work; and (3) the two approaches lead to quite similar results when it comes to the issue of interest to us, which is the nature of the forces imparting stability to chemical bonds.

The method of atomic orbitals was developed by a great many scientists—its most important contributors include Robert Mulliken (1896-1986, NP 1966) and Friedrich Hund (1896-1988).[133] It is based on the fundamental idea that, just as one can describe an electron in an atom by its *atomic orbital*, it is possible to do likewise for an electron in a molecule with the help of another function called its *molecular orbital*. The only difference between these two types of orbital is that the first is mono-centric, while the second must be multicentric, since each electron is sub-ject to the simultaneous action of several nuclei. The objective of the method is to determine the shape of the orbital Ψ for each electron in the molecule and, of course, its energy. As in the case of an atom, the quantity $|\Psi|^2$ represents the probability of finding the corresponding electron at a given point in space, or, in a more visual picture, the shape of the associated electronic cloud. Also as in the case of an atom, each orbital can accommodate only two electrons, each with a different spin; in the ground state of the molecule, the various energy levels are gradu-ally filled from the lowest one on up, until all available electrons are used up; in response to excitation processes, one or more electrons jump from an initially occupied orbital to one or more others that are empty.

Furthermore, in the most commonly adopted approach, which is the standard approximation of the method, the molecular orbital of an elec-tron is assumed to be a linear combination of the orbitals characteristic of that electron in isolated atoms. The technique is routinely referred to as the LCAO approximation (for linear combination of atomic orbitals). Its principles are best illustrated by taking the particularly simple exam-ple of a hydrogen molecule.

Let A and B be two hydrogen atoms, and let Ψ_A and Ψ_B be the atomic wave functions of each of their single electrons when the two atoms are far apart. Under these circumstances, Ψ_A and Ψ_B are each simply identical to the $1s$ orbital of a hydrogen atom. The LCAO approximation consists in writing the general form of the molecular orbital Ψ_{AB} of an electron in the hydrogen molecule as a linear combination of atomic orbitals, or:

$$\Psi_{AB} = \Psi_A + \lambda\Psi_B$$

The square of the coefficient λ is a measure of the relative contri-bution from each of the two individual atomic orbitals. In the case of symmetric diatomic molecules, such as H_2, one can obviously have only $\lambda = \pm 1$.

Assuming that a function of this form must be a solution of Schrödinger's equation applied to the case of a hydrogen molecule leads to the following results:

○ There are two allowed energy levels for the electrons in the molecule. If E_0 designates the energy of a $1s$ electron in an isolated hydrogen atom, one of these levels has an energy E_+ less than $2E_0$, and the other has an energy E_- greater than $2E_0$. The subscript indicates whether the coefficient λ is equal to $+1$ or -1.
○ The level E_+ corresponds to a molecular orbital $\Psi_+ = \Psi_A + \Psi_B$, while E_- corresponds to $\Psi_- = \Psi_A - \Psi_B$.
○ In the ground state of the molecule, both electrons occupy the E_+ orbital. In accordance with Pauli's exclusion principle, their spins are antiparallel.
○ The E_- orbital can also be partially or completely occupied, but only in an excited state of the molecule, which we will not discuss here.

Although the calculations of energies and associated wave functions differ somewhat in various implementations of the method, the general characteristics of the results always come out the same. They are schematically illustrated in Figures 15 and 16.

Figure 15 shows the dependence of the potential energy of the composite system on the distance between the two atoms. Recalling that potential energies corresponding to attraction are counted as negative, it is apparent from the graph that attraction predominates at large distances. This attraction increases as the distance is reduced, until an energy minimum (corresponding to maximum attraction) is reached. For

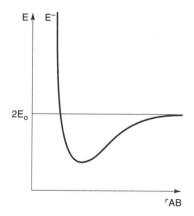

Figure 15. *Dependence of the potential energy on interatomic distance, explaining the formation of a stable hydrogen molecule.*

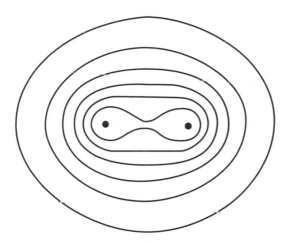

Figure 16. *Isoelectronic density curves of the* Ψ_+ *orbital of a hydrogen molecule.*

shorter distances still, attraction quickly gives way to a strong repulsion when the nuclei are getting too close to each other. The existence of the minimum is the signature of a stable compound. With more refined calculations, the technique predicts with remarkable accuracy the experimental values characteristic of the minimum—namely, the equilibrium distance (0.74 Å) and the dissociation energy of the molecule (3.6 eV).

Figure 16 shows contour lines for several constant values of the density of the electronic cloud associated with the Ψ_+ orbital. It illustrates how electrons tend to concentrate in the region between the two nuclei. Indeed, that is what holds the atoms together and ensures the cohesion of the molecule in its ground state. In the absence of this internuclear cement, the two atoms would inexorably repel each other, which is why the Ψ_+ orbital is said to be *bonding*.

Figure 17 illustrates especially vividly the situation in the ground state of H_2. It shows how the electronic density changes when one subtracts from its value in a hydrogen *molecule* the sum of the corresponding values in two independent hydrogen *atoms* placed at the same distance. Solid curves depict an increase in density, while dotted curves correspond to a decrease. The resulting picture displays convincingly the role of internuclear glue played by electrons as they concentrate in the middle region.

One should, of course, not forget that the electronic distributions depicted in Figures 16 and 17 are actually three-dimensional. They must have a symmetry of revolution about the axis of the molecule, at least in a hydrogen molecule. Bonds with this type of symmetry are called σ-*bonds*. They are characteristic of all single bonds in ordinary chemistry. In addition, the electronic cloud in a H_2 molecule displays a symmetry about a

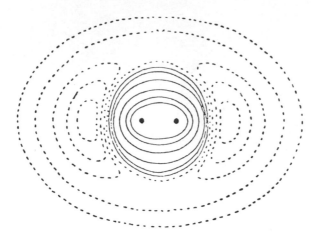

Figure 17. *Differential isoelectronic density curves for the ground state of a hydrogen molecule. The solid curves represent a net density gain when compared to the sum of the densities of two individual atoms separated by the same distance; the dotted curves represent a net density loss.*

plane perpendicular and equidistant to the two nuclei. As expected, this symmetry is lost in diatomic molecules formed of atoms that are different ("heteronuclear" molecules). In such molecules, the electronic cloud is skewed toward the most electronegative atom (the coefficient λ in the formula $\Psi_{AB} = \Psi_A + \lambda \Psi_B$ is then different from 1). This situation describes the partially polar character of the majority of covalent bonds, which, as pointed out above, had been foreseen by Lewis.

The H_2 molecule is evidently the simplest case. A similar analysis of somewhat more complex molecules brings forth a wealth of additional information, including the following:

○ The existence of other types of bonds, such as π-bonds, resulting from the lateral coupling of two p-type atomic orbitals (their axial coupling gives rise to a σ-bond). π-bonds no longer have a symmetry of revolution about the intranuclear axis. Instead, they are characterized by a localization of the electronic cloud in two oblong volumes located on either side of a plane containing the molecular axis (see Figure 18), which correspond to multiple (double and triple) bonds so prevalent in organic chemistry.

○ The formation of chemical bonds (particularly in polyatomic compounds) involving not simple orbitals of type s, p, etc., but weighted linear combinations of those. Such so-called hybrid orbitals are highly directional, as shown in Figure 19. As such, they constitute a kind of

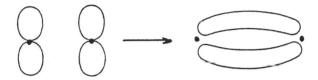

Figure 18. *Schematic representation of the formation of a bonding p-orbital by overlap of two p-orbitals.*

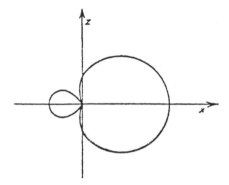

Figure 19. *Schematic representation of a hybrid orbital due to superposition of s- and p-orbitals.*

prenuptial preparation of atoms, as it were. They play a key role in the stereochemistry of molecular structures. Organic chemistry, to take one example, owes its richness in large part to the possibility of a carbon atom to assume three different hybridization states, as depicted in Figure 20.

○ The distinction between *localized bonds*, in which the coupling involves basically only two atomic orbitals, one from each atom, and *delocalized bonds*, resulting from the admixture into a single continuum of several atomic orbitals. Such bonds are generally π-type and occur in molecules with double bonds on adjacent atoms. Benzene, shown in Figure 21, is the classic example in that category.

○ The existence of bonds involving only one electron (as in H_2^+) or three (as in boron hydride, BH_3).

○ The particular case of metals, in which bonds are no longer localized at all in any particular direction. Instead, electrons roam around freely throughout the lattice of constituent atoms. Discrete electronic energy levels, characteristic of molecules, are replaced by continuous bands.

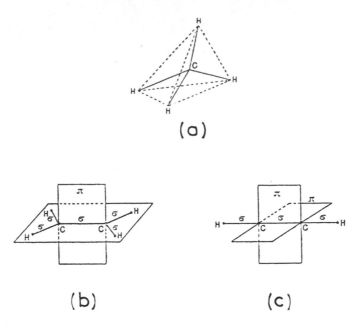

Figure 20. a) *Carbon in a tetrahedral hybrid state, denoted* sp³, *resulting from the mixing of three 2p-orbitals and one 2s-orbital. The configuration is characteristic of methane and derivative compounds. The four CH bonds are of s-type; the angle between them is 109°28'.* b) *Carbon in a trihedral hybrid state, denoted* sp², *due to the mixing of two 2p-orbitals and one 2s-orbital. It is characteristic of ethylene and related compounds. It consists of three s-type hybrid bonds in a common plane, making an angle of 120°, and one p-bond localized in a second plane perpendicular to the previous one. The p-bond results from the lateral overlap of two 2p-orbitals.* c) *Carbon in a dihedral hybrid state, denoted sp. It is characteristic of acetylene and its derivatives. It features two colinear s-type hybrid bonds and two p-bonds localized in two perpendicular planes.*

The multifaceted character of chemical bonds is certainly a fascinating topic and deserves more than a casual mention. But a more detailed treatment would distract us from our immediate purpose, which is to answer a question born with the ancient atomic theory, a mystery that ever since has preoccupied its advocates and opponents alike: What factors control the interactions between atoms and give rise to polyatomic compounds with properties of their own? Given what we already know about ionic bonds, a mechanism that is fairly straightforward, the basic answer to the question emerges from studies of covalent bonds in the hydrogen molecule.

It is evident that all chemical bonds are electronic in nature. The forces experienced by matter fall into one of four categories: gravitation-

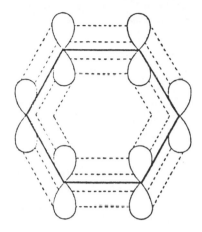

Figure 21. *Schematic representation of delocalized p-bonds in benzene.*

al, strong, weak, and electromagnetic.[134] Of those, the latter is almost entirely responsible for the formation and cohesiveness of chemical structures. Whether the number of participating electrons is two (as in single bonds) or four, six, and so on (as in multiple bonds), whether the number of implicated atoms is two (localized bonds) or more (delocalized bonds), it is the sharing of electrons from different atomic nuclei that gives rise to molecular skeletons. This sharing takes place by way of an overlap of the atomic orbitals involved or, equivalently, through the interpenetration of the associated electronic clouds. The method of valence bonds, which we have barely skimmed over, describes the phenomenon in terms of an exchange of electrons between the participating atoms, and of exchange forces resulting from the fact that these electrons are indistinguishable.

These results relegate the hypothesis of hooks and grapples, proposed by the ancients to explain the linkage of atoms, to the museum of the history of science—perhaps even of philosophy. But the break with traditional conceptions is much more far-reaching. While the discovery of the complex structure of atoms shattered the notion of their indivisibility, the elucidation of the nature of chemical bonds delivered the ultimate death blow to the dogma of their impenetrability and rigidity.

Far from being entities forever impregnable, frozen in the splendor of immutable particles, atoms turned out to have an inner structure and to be deformable and open to external perturbations and even penetrations. Such is the price to pay for the privilege of forging societies. As always in life, this implies the ability and even obligation both to give and to receive.

At this point, the rupture with classical conceptions appears total. The atom retains its name only through the mercy of a long tradition.[135] Does this mean that the historical, philosophical, and scientific impact of the Greek legacy should be trivialized? The question is often couched in this form: Is the ancient atomic theory the genuine precursor of its modern version? Answers vary greatly from one expert to another. Schrödinger observed: "This question has often been asked, and opinions on the matter are quite varied. Gomperz, Cournot, Bertrand Russell and J. Burnet all answer yes. Benjamin Farrington answers 'in a way,' pointing out that both theories have much in common. Charles Sherrington answers no, indicting the purely qualitative character of ancient atomism and charging that the very name 'atom,' which means undissociable or indivisible and embodies the theory's fundamental tenet, has become totally inappropriate." But, adds Schrödinger: "I do not know of a single negative answer returned by a Hellenist." Schrödinger's personal opinion is unambiguous: "All fundamental characteristics of the ancient doctrine have survived in modern theories up to this day. Obviously, these characteristics have been improved and reworked, but they have remained basically intact, at least in terms of the criteria of natural philosophers, rather than from the myopic perspective of specialists."

Others emphasize the still current validity of the concept of elementary particles. Heisenberg pointed out: "We may ask how our present views about atoms and quantum mechanics compare with ancient concepts. Historically, the term 'atom' has been applied in modern physics and chemistry to something different from what was meant during the scientific renaissance of the seventeenth century, as the smallest particles in what we call chemical elements are still rather complex systems formed of subentities. Today, the name 'elementary particle' is reserved for these subentities, and it is evident that, if there exists anything in modern physics that can be compared to Democritus's atoms, it is elementary particles such as protons, neutrons, electrons, mesons. . . . Certain assertions in ancient philosophy retain a close resemblance to those of modern science. This simply indicates how far one can go in combining our ordinary knowledge of Nature—without conducting actual experiments—with systematic efforts to introduce some sort of logic in that knowledge in order to understand things from general principles."

In this respect, what the Greek atomists failed to predict was the existence of a stage intermediate between what might be called fundamental elementary entities and derivative elementary entities. We now know that chemistry in its current form is founded on the existence of these derivative elementary particles called atoms. This reality had been foreseen with great prescience by the great nineteenth-century chemist Kekulé when he wrote: "As a chemist, I consider the hypothesis of atoms not only highly advisable but absolutely necessary to chemistry. Even if scientific progress were to lead someday to a theory about the constitution of chemical atoms, as important as such knowlege might be to a general philosophy of matter, it would make little difference to chemistry. The chemical atom will always remain the chemical building block."

With the caveat that Kekulé obviously could not have anticipated nuclear chemistry, his comment summarizes quite well the state of affairs at this point in history. The atom turned out to be divisible after all, but not in the way opponents of atomism had envisioned when they were passionately defending the unlimited and indefinite divisibility of matter.

Before closing this section, I would like to say a few words about a very special type of chemical bond. Three reasons prompt me to do this: (1) this bond is sufficiently different from those discussed previously to deserve a special mention; (2) it is of crucial importance in chemistry, and perhaps even more so in biology; and (3) it provides an answer to a question posed explicitly by an eighteenth-century philosopher who happened to be a proponent of the atomic theory, a question that might even have been on the mind of the one who started it all, namely, Thales of Miletus.

We start with the question itself, as it was phrased by Locke in a passage from his *Essay Concerning Human Understanding* (see chapter 12): "The little bodies that compose that fluid we call *water* are so extremely small that I have never heard of anyone who by a microscope (and I have heard of some that have magnified to 10,000; nay, to much above 100,000 times) pretend to perceive their distinct bulk, figure, or motion; and the particles of water are also so perfectly loose one from another that the least force sensibly separates them. Nay, if we consider their perpetual motion, we must allow them to have no cohesion one with another; and yet let but a sharp cold come and they unite, they consolidate, these little atoms cohere and are not, without great force, separable. He that could find the bonds that tie these heaps of loose little bodies together so firmly, he that could make known the cement that makes them stick so fast one to another, would discover a great and yet unknown secret." How could one not speculate that this question must also have preoccupied Thales, and that it was water's ability to exist as vapor, liquid, and solid that inspired him to consider it the primordial substance?

The answer to the question is found in a particular bond called the *hydrogen bond*, discovered in 1919 by M. L. Huggins. It has a very weak energy, typically 5 to 10 kcal/mol, which is considerably less than that of a conventional chemical bond. It consists of a hydrogen atom linked on one side to an electronegative atom in a molecule by means of an ordinary covalent bond, and loosely connected on the other side to another electronegative atom of the same or another molecule. We will now focus on the latter case. The symbolic representation of this type of bond is illustrated in Figure 22, which depicts the dimerization of water. The interactions involved are essentially electrostatic in nature. Because of the partly ionic character of all O-H bonds, each hydrogen atom carries a fractional positive charge and is therefore attracted, albeit loosely, toward the free electronic doublet present in the outer shell of the oxygen atom belonging to a second water molecule. The British chemist H. E. Armstrong gave this two-sided hydrogen link the whimsical name of "bigamous hydrogen," which provides additional justification to include the phenomenon in this section, devoted to "marriage" in the society of atoms.

The existence of such dimers has actually been observed in water vapor. The ability of a water molecule to form hydrogen bonds is, however, not limited to simple dimers. First, each water molecule has two hydrogen atoms, each of which can interact with oxygen atoms belonging to two other water molecules. Second, each oxygen atom possesses in its outer electronic shell four electrons distributed in two so-called free pairs, whose localization directions form with the two O-H bonds a more or less regular tetrahedron. Each of these free pairs can give rise to separate hydrogen bonds with another two water molecules. In total, each water molecule can thus be linked to four others in an approximately tetrahedral configuration, sketched in Figure 23. When a water molecule is completely surrounded by other similar molecules and its capacity for hydrogen bonds is fully saturated, its oxygen atom is at the center of a pyramid. The oxygen atoms of the four neighboring molecules occupy each vertex of the pyramid, each of which can itself be the center of other similar pyramids, and so on, leading to a veritable process of three-dimensional polymerization; that is precisely the structure adopted by water in the most stable crystalline form of ice. It is schematically illustrated in Figure 24.

In water vapor, the molecules have little chance to come into contact, and only dimers are observed. In ice, the other extreme, the tetrahedron has become rigid. But what happens in liquid water? The answer is easiest to understand by using ice as the starting point: When ice melts by absorbing heat, hydrogen bonds get distorted, some break apart, and a

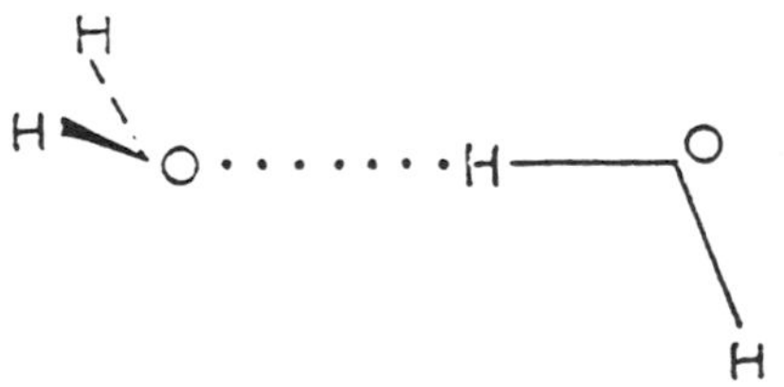

Figure 22. *Dimer of water, formed by one hydrogen bond.*

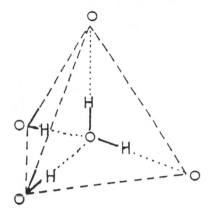

Figure 23. *Water molecules forming a three-dimensional array of hydrogen bonds.*

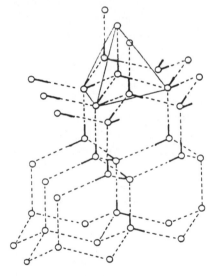

Figure 24. *The "hexagonal" structure of ice.*

certain disorder takes hold. Our current understanding of liquid water supports this picture of a three-dimensional network of hydrogen bonds, locally tetrahedral, constantly stretching, rupturing, and reforming. The relative weakness of the distended bonds impairs the stability of the lattice, enabling molecules to readily exchange partners in the fluid.

In short, this "great secret," the discovery of which Locke, and perhaps Thales as well, dreamed about, turns out to be the hydrogen bond.

Sadly, they will never learn of it, unless they are able to follow the latest scientific developments in the paradise of philosophers, in which case they can only marvel at the pivotal role this "great secret" also plays in biological phenomena. Indeed, hydrogen bonds are responsible for the structure of proteins and, to an even greater degree, nucleic acids, in which they control the linkage between complementary pairs of purines and pyrimidines; they are instrumental in storing and transmitting the genetic code.

Addendum: I have expressed the opinion that quantum mechanical studies of chemical bonds have relegated to museums the hypothesis of hooks that Leucippus and Democritus had proposed to explain how atoms link together. I cannot resist mentioning here an exotic result, reported in *La Recherche* 25 (1994), 450–451, which should somewhat temper my earlier judgment. Some time ago, the theorists E. Björkenskjöld and P. B. Calamares, and more recently H. A. Dok and K. Billol, predicted that in some highly excited states of certain atoms the electronic cloud could extend quite far away from the nucleus (up to three or four times the diameter of the atom), and that in the process of thinning out at such a distance its extremity would end up curving back. As it turns out, the existence of such "hooked" atoms has been confirmed quite recently by W. Loudmer in experiments involving the bombardment of phosphorus atoms with a beam of helium ions. Also in agreement with theoretical predictions, the experiments demonstrated the association of several (from two to four) "hooked" atoms linked together at the extremities of their electronic clouds. Obviously, the linkage is not a mechanical process, involving instead quantum mechanical interactions between electronic clouds. Nevertheless, it seems that we have come full circle. Leucippus and Democritus may have the last laugh after all and may even be partly justified in claiming that they were right all along!

THE NANOWORLD: THE VISIBLE AND TANGIBLE ATOM

Thomas (who was called "the twin") . . . said to them:
Unless I see the mark of the nails in his hands, and put
my finger in the mark of the nails and my hand in his
side, I will not believe.

—John 20:24–25

The lady in the audience was stubborn. "Have you ever
seen an atom?" she insisted. . . . My attempts to answer
this thorny question always begin with trying to
generalize the word "see.". . . Finally, in desperation,
I ask, "Have you ever seen the pope?" "Well, of course,"

*is the usual response. "I saw him on television." Oh,
really? What she saw was an electron beam striking
phosphorus painted on the inside of a glass screen. My
evidence for the atom, or the quark, is just as good.*

— Leon Lederman

The invisible nature of atoms, a sentence to which they seemed condemned for eternity, was an axiom which, until very recently, not even the most dedicated supporter of the atomic theory dared question. Many an opponent of the theory used this fact as ammunition to reject the hypothesis as lacking any scientific value, arguing that it could never be verified experimentally and thus belonged forever in the domain of metaphysics. Indeed, the presumed tiny size of atoms seemed to put them out of the reach of any detecting apparatus conceivable by man, even allowing for the inevitable technological development of more powerful probes designed to explore the structure of matter.

The future was to prove these diffident scientists and thinkers sorely mistaken. The first indications that atomic dimensions were becoming accessible came toward the middle of the twentieth century with the invention of the field ion microscope. But the decisive breakthrough in this area occurred only in the last decade with the advent of near-field microscopy and, more specifically, with the twin developments of the tunneling electron microscope and the atomic force microscope. Of these last two inventions, the former has produced the most plentiful and spectacular results, to the point that the acronym STM (for scanning tunneling microscope) has become an integral part of modern scientific jargon.

The STM was perfected by Gerd Binning and Heinrich Rohrer at the IBM Zurich Laboratory. In recognition of this remarkable achievement, both were awarded the 1986 Nobel prize in physics (which they shared with E. Ruska, the inventor of the conventional electron microscope a few decades earlier). Utilizing the general principle of near-field microscopy, the STM senses the surface of metals and semiconductors by means of a probe made of an exquisitely fine tungsten point, the tip of which often consists of a feature the size of only a few atoms (a tip made of a single atom would be ideal), scanning the surface at an incredibly short distance, typically less than a nanometer (10^{-9} m). When a small voltage is applied between the tip and the surface, a transfer of electrons across the gap turns up in one direction or the other, depending on the polarity of the voltage. That this current exists at all is likely to stupefy any physicist not well versed in quantum mechanics, since the transfer of electrons must take place across a potential barrier that would be considered insurmountable in classical physics. It is, however, readily accounted for by quantum mechanical "tunneling," a concept first introduced by George Gamow (1904–1968) to explain the emission of alpha particles by radioactive elements; it allows elementary particles to transit *through* potential barriers that would otherwise be impenetrable.

The current generated in an STM is sometimes called tunnel current. It turns out to be extremely sensitive to the distance separating the tip of the probe from the surface under study. Consequently, minute variations in the height of the surface scanned by the tip are sufficient to produce measurable changes in the magnitude of the current. An appropriate sensor records those changes (in practical instruments, an electronic feedback system continuously adjusts the distance between probe and surface so as to keep the current constant), and additional computer manipulation of the data generates a topographic image then displayed on a television screen. The pictures obtained show in remarkable detail things like lattices and atomic layers, and even individual atoms. The resolution of the instrument is determined largely by the accuracy with which the displacement of the tip can be controlled. At present, this accuracy appears to approach the astoundingly small value of 10^{-12} m. Toward the end of the 1980s, the pictures were even refined with the addition of colors. Obviously, these are not "true" colors of the sample, but arbitrary choices programmed into the computer by means of an instruction code matching specific colors with a range of topographic heights. One of the main advantages of color is to enable different kinds of atoms to be easily identified.

The atomic force microscope, known by its acronym AFM, was also developed under Gerd Binning's leadership. It replaces the tunnel current by a direct contact between probe and surface. In this instrument, the probe is made of a fine diamond tip mounted on a mobile cantilevered arm. The picture created displays the variations in intermolecular forces between probe and surface, which are reflected by the displacements of the mobile arm. The main advantage of this apparatus is its ability to handle insulating as well as conducting materials.

As important as the visualization of surfaces with atomic resolution may be for a definitive proof of the granular structure of matter, it was to be only a first step in an accelerating trend toward a genuine "taming" of atoms, to use H. C. von Baeyer's colorful expression.[136] The STM has been called upon not only to see atoms, but even to move them individually from one location to another. This delicate operation has become relatively routine with so-called adatoms, which are atoms adsorbed on a solid surface. The trick is to take advantage of the extremely high precision with which the tip can be positioned: By reducing the gap between it and the targeted atom, the force experienced by that atom can increase to the point where it can be gently dragged along by moving the tip laterally. When the desired destination has been reached, all that has to be done is to increase the distance between tip and surface, and the atom will remain pinned at its new spot. Not only have atoms been made to glide along surfaces, but they have even been transported by literally lifting them off a surface and depositing them at a new location. The most spectacular demonstrations of this new technique were achieved at the IBM Research Laboratory in Almaden, California, by Donald Eigler and his associates, who succeeded not only in "writing" their company's logo with thirty-five xenon atoms adsorbed on a nickel surface, but in building an electronic switch whose movable part was made of a single xenon atom alternating on command between two stable positions on the tip or the surface in response to an applied voltage of appropriate polarity.[137] More recently, they also succeeded in creating a "quantum corral" by building a circular barrier made of forty-eight iron adatoms on a copper surface, which, as a bonus, even enabled them to determine the local density of electronic states inside the circle.[138] Similar types of manipulations can be accomplished using individual molecules rather than atoms.[139]

In parallel with the development of techniques aimed at visualizing the corpuscular character of bulk matter, another avenue of research, perhaps even more ambitious, had as its objective to zero in on individual atoms in empty space.[140] While images generated by STM show clearly the atomic composition of surfaces, they still suggest an apparent continuity, which results both from the limited resolution of the instrument and from real interactions between atoms in a solid. Under these conditions, the only conclusive approach is to deal with isolated atoms surrounded by a vacuum that cuts them off from any parasitic interaction. Short of these extraordinary precautions, suspicions and doubts about the existence of atoms might linger on.

The unquestioned hero in this great enterprise is Hans Dehmelt (NP 1989). Perhaps surprisingly, the first particle to be isolated under such extreme conditions turned out to be an electron. Dehmelt managed in 1973 to keep a single electron in solitary confinement, so to speak, thus giving it a special identity, different from all the other electrons in the entire universe, at least in a symbolic sense. The prison that had the honor of housing this special inmate is known as Penning's trap: It is a small box made of copper and glass, in which the electron was confined between two negatively charged electrodes surrounded by a magnetic field. The experiment consisted in allowing the electrons within the box to escape one by one, until a single one was left. Its presence was detected by the radio frequency signal it emitted. It remained trapped for a period of six months, at which time it accidentally hit the wall of the trap and escaped forever from its guards.[141] Today, trapping electrons is performed in numerous laboratories, and techniques are even being developed to manipulate them.[142]

After electrons, it was the turn of individual atoms to be captured. The trap that made this feat possible was conceived and built by Wolfgang Paul (NP 1989, the same year as Dehmelt). The device relied on the same principles as Penning's, except that an oscillating electric field replaced the magnetic field.[143] The trapped particles were ions—charged atoms—and the box came to be known as an "ion trap." The initial success is due to Dehmelt and Toschek, who succeeded in 1978 in confining a cloud made of a few barium ions. A year later, they perfected their technique and managed to take a photograph of a single barium ion—the very first incontrovertible evidence of an individual atom.

The trapping of atoms happens to be considerably easier to achieve at low temperatures. In the last few years, the scientific world has witnessed fascinating developments relying on the use of lasers for that purpose.[144] By using several laser beams with suitably chosen geometries and wavelengths, it is possible to decrease the mean thermal velocity of gaseous atoms and chill them to temperatures as low as a few microkelvins. When sufficiently cold, atoms are apt to get caught in these beams of light, which act as a kind of optical molasses. These atoms can then be snared in magneto-optic traps created by a static magnetic field surrounding the "molasses."[145] With this technique, it has been possible to trap atoms of helium, sodium, rubidium, cesium, magnesium, and calcium. With a clever configuration of laser beams, scientists have even succeeded in using light to create periodic potential wells in which cesium atoms became stacked "much like eggs in a carton."[146] The authors of this exploit managed to obtain ordered structures in two- and, more recently, three-dimensional arrangements, in which "for the first time ever, atomic order was imposed by light."

All these examples illustrate the tremendous technological possibilities opened up by the study and mastery of matter on a very small scale— roughly of the order of a nanometer. Indeed, the term *nanotechnology* has been coined to designate the various techniques that allow the observation, fabrication, machining, manipulation, or control of objects of nanometer size. Strictly speaking, these objects should occupy a volume less than $10^3 \, nm^3$. In practice, though, the term is often applied to objects with just one dimension of the order of a nanometer. Important results have already been obtained with these cutting-edge techniques.[147] Many more are sure to follow. The conquest of the "nanoworld" has only begun.

At this point in our story, how could we avoid one last fond thought about the ancient authors of the atomic theory? Their dream, the atom, has now become reality. To be sure, the reality does not quite conform to the original dream. Today's composite atom is not the compact and undissociable atom that Democritus had boldly made the immutable stuff of the material world. The particles with a legitimate claim to the title of "elementary" are to be found ultimately beyond the atom itself. We will return to this point in the epilogue. Yet atoms do occupy a privileged position in the making of the world. Individualized in a limited number of types and defining distinct chemical elements, they constitute the fundamental entities out of which all existing substances are made. They are the uncontested intermediaries between subatomic particles and molecular or supermolecular structures, and they determine the composition and arrangement of all material objects. The special status of atoms in the physical world forever ensures a no less privileged place in the history of human thought to those who conceived of the idea. Atoms would surely have been discovered even without Democritus, but had it not been for his bold hypothesis, mankind would have been deprived of a thrilling intellectual epic. So many profound or erroneous reflections, amiable or hostile debates, passionate or acerbic comments, so many uplifting pronouncements would have been lost, were it not for the inspired imagination of a Greek philosopher whom tradition describes as joyful and for whom "any interaction with human beings was reason for laughter." Horace proclaimed that if Democritus ever returned to this earth, he would keep on laughing. Juvenal would see that as a sign of great wisdom.[148]

 **PROVISIONAL
EPILOGUE**

*Such are our myths and errors, which we devoted so
much effort contriving in order to supersede the ones
that preceded them! . . . Not everything was false in
what was discarded. Not everything is true in what
replaced it.*

—Paul Valéry

*The physical being . . . never reveals itself whole. That
is why physics will always remain open to question.*

—Bernard de Fontenelle

Atoms, of which we distinguish as many types
as there are chemical elements (not counting isotopes), are the building
blocks with which all substances in the material world are made. As we
know today, these building blocks are themselves made of even smaller
components, of which three have been identified: electrons, protons, and
neutrons. At this stage, it appears that the variety of qualitatively distinct
atoms can be traced to different combinations of a relatively small number of subatomic particles. That the unification of matter may be even
more fundamental than this looms as a distinct possibility. Could it be
that Democritus's dream has been simply pushed back from the atomic
to the subatomic level?

To put it more explicitly, should the three particles just mentioned be
considered truly "elementary," in the Democritean sense of particles that
are undissociable and irreducible, and might there be others that could
also lay claim to the title?

The short answer to the first of these two questions is that, within the
limits of our present knowledge, electrons are indeed elementary particles, while protons and neutrons are not. The equally short answer to the
second is that there are indeed other particles besides electrons that
are—for the moment at least—also considered elementary. The number
of such particles may be limited, but there have been plenty of potential
candidates. Owing to the extraordinary advances of nuclear and high-
energy physics, literally hundreds of particles have been discovered that
at one time or another made a bid to join this exclusive club. It took a

remarkable cooperation between experimenters and theorists to tell which were legitimate contenders (if only for the time being) and which, in all likelihood, were not.[1]

At the basis of this effort is the division of this bevy of particles into two general classes, epitomized by electrons, on the one hand, and neutrons and protons, on the other. These two categories are, it turns out, of very different sizes.

The electron is the prototypical particle in the family of *leptons*, a restricted group consisting of only six members. In addition to the electron, the group includes *muon* and *tau* particles, both resembling a kind of heavy electron in that they carry a negative unit electrical charge, and three neutrinos, which have a very small (or perhaps even zero) mass and are electrically neutral. All these particles have sizes far smaller than the experimentally accessible limit of 10^{-20} m. Some theorists estimate them to be as small as 10^{-31} m. One of the main properties of these leptons is to be affected by only one or two of the three fundamental interaction forces that, together with gravitation, govern the universe (gravitation has a negligible effect in the microscopic world). For instance, neutrinos are subject to weak interactions only. As a result, they can go virtually unhindered through the densest materials, while electrons and their heavier variants are affected by both weak and electromagnetic interaction forces. These six particles are believed to be without any internal structure, which qualifies them *de facto* as elementary. For completeness, we should mention that the lifetime of electrons is considered infinite, while that of muons and tau particles is quite short (10^{-6} and 10^{-12} second, respectively).[2]

Protons and neutrons belong in the group of *hadrons*, as does the vast majority of potentially elementary entities generated in accelerators, in which high-energy particles strike fixed targets or are made to collide head-on with each other. Most hadrons turn out to be simply excited states of protons and neutrons (which are themselves different energy states of one and the same particle called a nucleon) with very short lifetimes (ranging from 10^{-6} to 10^{-23} second).

When left to itself, a free neutron is unstable. Its lifetime is only about twelve minutes, at which time it decays into a proton, an electron, and an antineutrino. Its great stability in atomic nuclei is attributable to the strong interaction force that dominates within them. Free protons are not stable forever, either, but their "instability," if that is the proper word, is on a different time scale altogether—according to the most advanced theories (notably the grand unified theory, to be discussed below), its lifetime would be 10^{32} years—much longer than the age of the universe itself, which is estimated to be about 10^{10} years.

None of the hadrons qualifies as elementary. After discovering that their heavy masses could be organized into a few families with specific properties, Murray Gell-Mann (NP 1969) showed, in a remarkable display of mathematical insight, that the situation could be accounted for by postulating that they all consist of a limited number of more fundamental particles named *quarks*. Three quarks were initially proposed, denoted *up* (*u*), *down* (*d*), and *strange* (*s*). The number was eventually increased to six, with the addition of three other quarks called *charm* (*c*), *bottom* or *beauty* (*b*), and *top* or *truth* (*t*).[3] People commonly speak of six "flavors" of quarks, each of which is associated with three "colors," red, green, and blue. Needless to say, these terms are not to be understood in their literal sense, but are merely whimsical names designating specific quantum properties. One of these properties is especially startling: Quarks carry fractional electrical charges.

A particularly important result in this new framework is that protons and neutrons are no longer considered elementary particles. Instead, they are postulated to consist of three quarks each: two *u* quarks and one *d* quark for a proton, and two *d* quarks and one *u* quark for a neutron. This structure was experimentally confirmed by Richard Taylor, Jerome Friedman, and Henry Kendall, who studied the large-angle scattering of high-energy electron beams by protons. The discovery earned them the 1990 Nobel prize in physics. It should be emphasized, however, that, although the composite nature of protons and neutrons is no longer in doubt, no quark has ever been extracted individually from these particles and observed in isolation. Protons and neutrons do consist of parts, but these parts are inextricably entangled and remain confined to the interior of the particles they make up. They can move around freely within this interior, but with no possibility for escape. We have already mentioned the peculiar property that they individually carry fractional electrical charges. Yet they form stable associations that carry integral numbers of charges: 0, ±1, ±2. Since the charge of the *u* quark is $+^2/_3$ and that of the *d* quark is $-^1/_3$ (in units of electronic charge), it is easy to see how a *uud* grouping (characteristic of a proton) gives a total charge of +1, while *udd* (corresponding to a neutron) has a charge of 0.

The proposition that the structure of matter involves six leptons and six quarks is a part of what is known as the *standard model* of the universe. These corpuscles are only part of the picture, however, since a complete model must also include the particles corresponding to the forces regulating their interactions.

While matter and force are distinct concepts in classical physics, the advent of quantum theory forced a drastic reassessment of this point of view by requiring energy to be transmitted in discrete quanta. We have already encountered one such example when we discussed the energy of an electromagnetic field, for which the corresponding quanta are photons. Photons can thus be considered elementary particles associated with the field. The electromagnetic field that exists between two charged particles—say, two electrons, or an electron and a proton—exerts its action by exchanging photons, referred to as *intermediate* or *messenger* particles. As we will see toward the end of this epilogue, when we discuss the properties of vacuum, this exchange implies a complex ballet of *virtual* particles, which have exceedingly short lifetimes and are therefore undetectable for all practical purposes. The theory describing these interactions is known as *quantum electrodynamics*. The weak nuclear force, which has a range of 10^{-16} cm and was first recognized in 1930 by Enrico Fermi (1901–1954, NP 1938), is responsible for certain processes of radioactive decay and more generally for any reaction involving neutrinos. The carriers mediating this force are known as intermediate vector bosons and are designated W^-, W^+, and Z^0. The 1984 Nobel prize in physics went to Carlo Rubbia and Simon van der Meer for discovering them.[4] The strong nuclear force, which has a range of 10^{-13} cm, binds quarks together to form protons and neutrons and, in turn, binds those to form atomic nuclei. It has the unusual property of increasing monotonically with increasing distance (the shorter the distance, the weaker the force). Its mediating particles are called *gluons* (of which there are eight different kinds). The theory dealing with this interaction is called *quantum chromodynamics*. The name was chosen by analogy with quantum electrodynamics, in recognition of the fact that the "color" of quarks plays a role somewhat analogous to an electrical charge.[5] Another particle, named the *graviton*, corresponds to gravitational quanta. It remains hypothetical for now.[6]

Our present corpuscular view of the universe is based on twelve elementary particles making ordinary matter and twelve elementary particles transmitting forces, for a total of twenty-four. This number jumps to thirty-six if one takes into account that each quark "exists" with three different color properties, and to sixty when one includes antiparticles.

An important remark is in order, though. To the extent that a Democritean influence has shaped our conception of the world, there has been a tendency to stress the corpuscular angle of the standard model and to introduce a certain formal distinction between particles of matter and intermediary particles associated with force fields. As a result, we may have given the impression that this corpuscular aspect provides the most exact description of physical reality. Such a view would be unfortunate, as it might obscure what is considered today the most plausible picture of reality, which not only unifies the concepts of particles and fields, but even considers fields preeminent over particles.

This picture is the result of a remarkable effort to synthesize the two major theoretical contributions of the twentieth century, namely, the theory of relativity, with its fundamental law of equivalence between mass and energy, and quantum mechanics. A number of eminent scientists distinguished themselves in that effort: They include Pascual Jordan, Max Born, Werner Heisenberg, Wolfgang Pauli, Eugene Wigner, Paul Dirac, Enrico Fermi, Richard Feynman, and Julian Schwinger. What emerged came to be known as the *relativistic quantum theory of fields,* in which the fundamental and underlying reality of the world is embodied in the existence of a slew of fields and in their interactions. In accordance with the rules of quantum theory, the field intensity at a given point in space is a measure of the probability of finding in that location the quanta associated with the field—in other words, the energy packets corresponding to the elementary particles that are observed by experimenters or postulated by theorists. Just as photons are the quanta of electromagnetic fields, electrons are the quanta of electronic fields, and so on. The distinction, implied in our earlier discussion, between the fundamental nature of quanta of matter and that of field quanta turns out not to be entirely justified. Fields are the ultimate reality, and there are as many fundamental fields as there are elementary particles. In the framework of present-day knowledge, the elementary nature of entities such as leptons, quarks, and intermediary particles mediating interaction forces suggests just as many underlying fundamental fields. Reality boils down to an assortment of fields interacting with one another via their respective quanta particles. And so the dualistic description of the world recedes in favor of a more unified vision.

The world may be more unified, but is that sufficient to satisfy physicists (or, for that matter, philosophers)? As one could easily guess, the answer is no, because what they have all been looking for since the time of Thales is not homogenization of the world but unification of its vision—unification into a fundamental principle during antiquity, and into a single, all-encompassing mathematical equation in modern times. From that point of view, the standard model is hardly the answer. Lederman and Teresi observe: "As a compact summary of everything we know, the standard model has two major defects: one aesthetic, one concrete. Our aesthetic sense tells us there are too many particles, too many forces. . . . The concrete problem is one of inconsistency. When the force-field theories, in impressive agreement with all of the data, are asked to predict the results of experiments carried out at very high energies, they churn out physical absurdities."[7]

Numerous efforts are currently underway to try to improve the model. The most significant of those is aimed at the unification of the fundamental interactions into a common mathematical formalism. Success so far has been limited to the unification of the electromagnetic and weak nuclear interactions into what is called the *electroweak force.* In recognition of this remarkable achievement, the 1979 Nobel prize went to its authors:

Steven Weinberg, Abdus Salam, and Sheldon Glashow. The next stage would be the so-called grand unification theory, which would extend this formalism to include the strong interaction. A further extension, which would ultimately include the gravitational force, faces enormous difficulties that, at present, seem overwhelming.

Interestingly, the quest for grand unification goes hand in hand with exploring the very distant past of the cosmos. In a sense, the journey toward grand unification is also a journey back toward the big bang. All evidence suggests that the universe began in a state of extreme uniformity, and that the differentiation of force and matter, as they appear to us today, came about only later. A number of specialized books offer very vivid descriptions of these transformations, with a wealth of details so impressive as to make one believe one were witnessing the events live. Briefly and schematically, the process can be summarized as follows. The unification of forces requires extremely high temperatures (or energies, which is equivalent), so high, in point of fact, that they existed only very briefly following the birth of the universe. The 10^{-43}-second mark after the big bang is known as *Planck's time.* The limitations of our present theoretical models prevent us from understanding what went on before that time, which is referred to as the era of "theoretical obscurity."[8] At any rate, it is currently believed that 10^{-43} second after the big bang, when the universe had a diameter of only 10^{-33} cm, and its temperature reached as high as 10^{32} K, the electromagnetic, strong nuclear, and weak nuclear forces were effectively dissolved into a single force known as the *electronuclear force.* Particles of light and matter were undifferentiated then.[9] At 10^{-35} second the temperature had dropped to 10^{27} K and the strong nuclear force separated itself out, leaving the electromagnetic and weak nuclear forces still lumped together into the so-called electroweak force. By 10^{-12} second the electroweak force in turn split into its two components.

The splitting off of forces was accompanied by a gradual crystallization of the universe. At 10^{-43} second, the world existed in the form of a "quantum vacuum." The material world proper was born during what has been dubbed the "inflationary" phase—a brief spurt between 10^{-35} and 10^{-32} second—when the universe expanded by a factor of some 10^{50}. It saw the appearance of the elementary particles known today, such as quarks, electrons, and neutrinos, as well as their antiparticles. Baryons appeared at 10^{-6} second (when the temperature had cooled to 10^{13} K) and dominated the scene during the "hadronic era," which lasted until 10^{-4} second and was followed by the "leptonic era." This entire scenario unfolded against a continuous backdrop of energy turning to matter, annihilation of matter into energy, and competition between matter and radiation, to say nothing of the contest between matter and antimatter.[10] As it happens, matter ended up winning (otherwise we would not be here to talk about it).[11] This happened for two reasons. First, photons gradually lost their energy as the universe expanded and cooled, until they eventually could no longer prevent the organization of the material world. Second, nature turns out to have a very slight but definite preference for matter over anti-matter, much to our advantage. The famous Soviet physicist Andrei Sakharov (1921–1989) was the first to have recognized that preference. This "symmetry violation" is attributed in the grand unified theory to the effect of a hypothetical superheavy boson, the so-called X boson, which is able to transform quarks into either antiquarks or leptons, and vice versa. Be that as it may, the differentiation of fundamental forces and elementary particles, which resulted from the cooling and expansion of the universe and paved the way for the complexity of our present world, occurred in less than a second after the "creation." Everything else followed in due time. The first atomic nuclei (deuterium, or heavy hydrogen, and helium 4) were formed at about 100 seconds (at a temperature of 10^6 K). The first atoms (hydrogen and helium) materialized 700,000 years later (at a temperature of 4,000 K), when matter emerged from its plasma state, decoupling itself from radiation and making the universe transparent. Heavier atoms did not appear until considerably later. They occurred three or four billion years after the "beginning," in the interior of stars by fusion of helium nuclei. The process is known as stellar nucleosynthesis, the theory of which was developed largely by Hans Bethe (1907– , NP 1967); it produced increasingly heavy atoms, up to iron. Finally, the heaviest atoms came about through a process of neutron capture during gigantic explosions of very massive stars (supernovas).

We have talked about matter, which Democritus recognized, and we have talked about forces, which he ignored. What we have yet to discuss is the concept of void, or vacuum, which Democritus had made the second principle and fundamental constituent of the world, on a par with atoms. The two notions were completely separate in his view. Vacuum existed where atoms did not; it filled any and all space between atoms. The compact nature of atoms, on the other hand, precluded them from containing even the slightest amount of vacuum in their interior. These two fundamental realities were mutually exclusive.

Quantum mechanics changed all that and shattered our conception of vacuum.[12] Modern theory holds that vacuum, no matter how "empty," is never completely devoid of electromagnetic field. The random fluctuations of that field have the paradoxical effect of populating vacuum with—you guessed it!—particles. The key to that process is found in Heisenberg's fourth uncertainty relation, which stipulates restrictions on the simultaneous determination of the two "canonically conjugate" variables of energy and time. One relation states that the product of the uncertainties ΔE (energy) and Δt (time) must exceed Planck's constant h divided by 2π, according to the formula

$$\Delta E \cdot \Delta t > \frac{h}{2\pi}.$$

The implication is that the shorter the lifetime of a particle, the more uncertain its energy. An important consequence of this proposition is that it allows for particles to be created out of vacuum, as long as they exist for only a short duration. By virtue of the equivalence between mass and energy, that duration must be inversely proportional to the mass of the particle created. The required amount of energy is borrowed from the vacuum and must be returned to the vacuum. The greater the energy borrowed, the quicker the restitution has to be completed. The particles return their "borrowed" energy by annihilating themselves within the exceedingly short prescribed time. In light of such strict conditions, they are undetectable and are said to be *virtual.*

Thus, vacuum functions as a sort of permanent energy bank. As such, far from being empty, it is actually continually filled with a host of virtual particles of all types, and even, in a manner of speaking, of various degrees of virtuality, since each virtual pair can give rise to another pair with a higher order of virtuality. From this perspective, vacuum can be regarded as "the state of minimum energy of all fields stripped of any particle of real matter," or "space purged of all real particles," or even "an ocean of virtual particles."

Under these conditions, there is an obvious mechanism that allows a transition from virtuality to reality: It requires supplying enough energy to guarantee that particles will materialize. Estimating the amount of energy required is straightforward. It simply depends on the mass of the particle—more precisely, of the particle-antiparticle pair—we want to create.[13] It is convenient to express the mass of particles in units of energy rather than weight, on the basis of the formula $E=mc^2$. Using this conversion, the mass of the electron (conventionally quoted as 9.109×10^{-31} kg) is 0.511 MeV (1 MeV = 1 mega-electron volt = 10^6 eV), while that of the proton (1.673×10^{-27} kg) is 938 MeV, just short of 1 Gev (1 GeV = 1 giga-electron volt = 10^9 eV). Temperature, rather than energy, is also sometimes used, the conversion being defined by the relation $E=kT$, where k is Boltzmann's constant, equal to 8.67×10^{-5} eV/K. One eV corresponds to about 10,000 K, 1 MeV to 10^{10} K, and 1 GeV to 10^{13} K. In order to produce a particle-antiparticle pair from electromagnetic energy, the photons must have an energy at least equal to a threshold corresponding to the sum of the rest energies of both members of the pair. This amounts to 1 MeV for an electron-positron pair, and 2 GeV for a proton-antiproton (or neutron-antineutron) pair. Such energies are routinely produced nowadays in particle accelerators. They were also pervasive for a short time following the big bang, producing proton-antiproton pairs during the first 10^{-5} second, and electron-positron pairs for up to 5 seconds.

From this perspective, the barrier Democritus had erected between vacuum and matter vanishes. In the framework of vacuum, defined as the state of minimum energy of all fields, real particles become "eigenmodes of excited states of these fields." Vacuum is a latent state of reality, while matter, made of elementary particles, is its actualized state. The two "principles" fuse into a single one, of which they are merely different manifestations.

The vacuum-matter complementarity, or, equivalently, the virtual-material duality of particles, reveals itself in several different ways. To start with, according to the quantum theory of fields, virtual particles constitute the forces acting on material particles. All interactions occurring in nature between particles or material objects can be reduced to an exchange of virtual particles. Many are of the opinion today that the definition itself of material elementary particles ought to be generalized to include the cloud of virtual particles representing the forces they are subject to.

But that is not all. Classical vacuum, whether interatomic, as it had been since the time of Democritus until the end of the nineteenth century, or intra-atomic, as it became early in the twentieth century with the discovery of the planetary structure of atoms, was not supposed to exert any influence on the physical characteristics of material particles.[14] That picture changed dramatically with the advent of quantum vacuum; the theory predicts, and experiments confirm, interaction effects between the modern version of vacuum and matter. These effects turn out to be far more profound than a trivial transmission of forces. They constitute a genuine and direct participation of vacuum in the very properties of matter.

These effects are further manifested in an important way by a phenomenon known as *vacuum polarization*, which is the last ingredient required for a complete quantum description of the world. When, say, a virtual electron-positron pair appears in the electric field between an electron and a proton, as short-lived as the pair may be, the material electron will tend to attract toward itself the virtual positron and repel the virtual electron, which confers on the pair a preferred direction in space. In the process, vacuum becomes polarized. As weak as this effect may be, it is quite measurable. In a hydrogen atom, for instance, it manifests itself as a tiny, yet measurable, correction to the energy of the electron. Quantum electrodynamical calculations taking into account the combined effects of the fluctuations and polarization of vacuum have produced estimates of the magnitude of the associated perturbations on the hyperfine structure of certain spectral lines of the hydrogen atom. The predictions were quantitatively verified in an experiment devised by W. E. Lamb, which earned him the 1955 Nobel prize in physics. The effect came to be known as the Lamb shift.

Another consequence of vacuum polarization is to cause an apparent decrease in the charge of the electron. The virtual pairs popping up around an electron and oriented by the polarization cause a partial electrostatic screening. As a result, to an outside observer, the electron's charge appears smaller than it really is. The thicker the screen, which means the farther the distance between observer and electron, the greater the apparent decrease. In other words, the electronic charge is no longer a constant, and quantum vacuum is directly responsible for that.

These effects are not the only ones to confirm the interdependence of the properties of vacuum and matter. Another proof is provided by the so-called Casimir effect, named after the renowned Dutch physicist, who showed in 1948 that the quantum fluctuations of vacuum between the two plates of a capacitor produce a force of attraction inversely proportional to the fourth power of the distance separating the plates. Of more significance to us, because more directly related to atomic entities, are the results recently obtained by D. Kleppner. He showed in 1985 that what is called *spontaneous emission* of radiation by atoms, which had until then been considered an intrinsic and invariant property, could in fact be influenced and even completely inhibited by altering the quantum nature of vacuum, for instance by suitably changing the geometry of the container in which the measurement is carried out.[15]

In light of all these data, it is clear that the fundamental and unconditional distinction Democritus had introduced between atoms and vacuum can no longer stand. Not only has vacuum invaded atoms to the point that it occupies virtually their entire volume, and not only have atomic and subatomic particles been plunged into a veritable ocean of virtual particles with which they continually interact, but even the most intimate properties of atoms can no longer escape vacuum's dictates. Democritus's vacuum was an inert playground for atoms to frolic on. It has now been supplanted by an active vacuum participating in the making and evolution of the world. For better or worse, it even appears possible that it actually was the primordial source from which the universe sprang to life through "a gigantic original fluctuation," whatever is meant by that phrase.[16] Thanks to quantum randomness and Einstein's equivalence of matter and energy, Democritus's two principles end up merging into a unified concept. To be sure, on closer examination, the two principles retain a certain specificity, embodied for instance in the different properties of fermionic and bosonic fields, but, on a fundamental level, these are merely two facets of the same reality.

In the end, the theory of atoms, which should have been named from the outset the theory of atoms *and* vacuum, has covered a long trek indeed, which often resembled an obstacle course, and underwent numerous substantial transformations. It took twenty-five centuries to go from the invisible and indivisible atom to one that is divisible and visible, from the corpuscular atom to the wave-particle atom, from empty and inert void to a vacuum that is filled and active, to learn to annihilate matter and materialize things out of vacuum. In light of such a lesson from the past, who knows what surprises are in store for the next twenty-five centuries? Charles Coulson, who taught quantum chemistry at Oxford University from the 1940s to the 1960s, once received a letter addressed to him as "professor of phantom chemistry." The blooper rather appealed to him, inasmuch as it evoked the endless pursuit of an elusive truth. In that sense, the letter might appropriately have been addressed to all of us.

NOTES

Chapter One

1. B. Farrington, *Greek Science: Its Meaning for Us* (New York: Penguin Books, 1944). See chapter 2. Included in the chapter is a discussion of the many qualifications to apply to this statement.
2. Science and philosophy were indistinguishable at the time, which is evidently not the case today. See, for example, L. de Crescenzo, *Les Grands Philo-sophes de la Grèce antique* (The great philosophers of ancient Greece), vol. 1, *Les Présocratiques* (The pre-Socratics) (Paris: Julliard, 1985): "Man has reached the highest summits of civilization through two fundamental disciplines: science and religion. While science, relying on reason, studies natural phenomena, religion, which satisfies an inner need of the human soul, looks for something absolute, beyond the ability to know with the senses and the intellect. Philosophy, on the other hand, lies halfway between science and religion, closer to one or the other depending on whether it is promoted by the so-called rationalistic philosophers or by those favoring a more mystical view of things." For Bertrand Russell, the renowned British philosopher of the rationalist school, "Philosophy is a kind of no-man's-land, a bridge between science and theology, and subject to attack from both sides."
3. According to some sources (Diogenes Laertius, Simplicius, and St. Epiphanius), Leucippus, the presumed founder of atomism, also hailed from Miletus; Epicurus denied that Leucippus ever existed.
4. At about the same time, in Croton, Pythagoras proposed that the "prime reality" was not natural substances but numbers, which, according to him, constituted all "things." Pythagoras was born on the island of Samos, not far from Miletus. Hippocrates, the most renowned physician of antiquity, hailed from the island of Cos, also not far from Miletus.
5. Quoted by J. Voilquin in *Les Penseurs grecs avant Socrate. De Thalès de Milet à Prodicos* (Greek thinkers before Socrates: From Thales of Miletus to Prodicos) (Paris: Flammarion, 1964).
6. F. Nietzsche, *Philosophy in the Tragic Age of the Greeks* (Chicago: Henry Regnery Co., 1982).
7. This destruction is probably responsible for the extinction of the Milesian school after Anaximenes. Later, Miletus was again devastated by Alexander and, finally, laid to ruin by the Goths (A.D. 263).
8. Among them was Abdera, the hometown of Democritus, who was one of the founding fathers of atomism (according to some sources, Democritus would even have been born in Miletus). Elea, another famous colony to the south of Naples, was home to the important Eleatic school of philosophers (Parmenides, Zeno, Melissus). Elea had been established by the Phocaeans, who also came from Ionia (the founder of the Eleatic school, Xenophanes, originally came from Colophon, another city in Ionia). The string of colonies in modern-day Italy reminds us of the beautiful words of Jean d'Ormesson, who called Segesta and Agrigentum "pieces of Greece lost in Sicily."

9. In fact, almost all the philosophers of the time about whom information is available (Thales, Pythagoras, Heraclitus, Parmenides, Empedocles, Melissus, Democritus, Anaxagoras), belonged to prominent and wealthy families.

10. R. Lenoble, *Histoire de l'idée de la nature* (History of the concept of nature) (Paris: Albin Michel, 1968).

11. A. Pichot, *La Naissance de la science* (The birth of science), vol. 2, *Grèce présocratique* (Pre-Socratic Greece) (Paris: Gallimard, 1991).

12. E. Schrödinger, *Nature and the Greeks* (New York: Cambridge University Press, 1996).

13. This argument was revived a few centuries later by a number of Christian thinkers: Cyrillus Alexandrinus, Basil of Caesarea, John Chrysostum, St. Augustine, and others.

14. As we will see in chapter 2, this judgment is far too negative. At the very least, we owe to this school—and particularly to Thales—the very important notion of the unity of the universe, of the principle that "all is one."

15. D. Rops, *Histoire sainte. Le peuple de la Bible* (Holy history. The people of the Bible) (Paris: Fayard, 1943). Clement of Alexandria, a Christian philosopher born in Athens in the middle of the 2nd century A.D., who defended Hellenism in the midst of Christianity, captured this duality in the phrase "God gave the law to the Hebrews and philosophy to the Greeks."

16. G. Minois, *L'Église et la science. Histoire d'un malentendu* (The Church and science: History of a misunderstanding), vol. 1, *De saint Augustin à Galilée* (From St. Augustine to Galileo) (Paris: Fayard, 1990).

17. As Nietzsche said in *Philosophy in the Tragic Age of the Greeks*, "Other peoples have saints, the Greeks have sages."

18. E. Ludwig, *The Mediterranean: Saga of a Sea*, trans. June Barrows-Mussey (New York: McGraw-Hill, 1942).

Chapter Two

1. L. de Crescenzo, *Les Grands Philosophes de la Grèce antique* (The great philosophers of ancient Greece), vol. 1, *Les Présocratiques* (The pre-Socratics) (Paris: Julliard, 1985). Where de Crescenzo says "in the history of philosophy," we would prefer to say in the history of the philosophy of nature, which at the time represented a rudimentary form of science.

2. G. E. R. Lloyd, *Early Greek Science: Thales to Aristotle* (New York: W. W. Norton, 1971).

3. F. Nietzsche, *Philosophy in the Tragic Age of the Greeks* (Chicago: Henry Regnery Co., 1982).

4. Most of the doxographers' quotations used in this book come from *Les Présocratiques* (The pre-Socratics), ed. J.-P. Dumont (Paris: Gallimard, "Bibliothèque de la Pléiade," 1988).

5. J.-P. Vernant and P. Vidal-Naquet, *La Grèce ancienne* (Ancient Greece), vol. 1, *Du mythe à la raison* (From myth to reason) (Paris: Éd. du Seuil, 1990).

6. According to Simplicius and Hippolytus, Anaximander was the first to use the term *archè*, or "principle."

7. The importance of equilibrium in Anaximander's conception of the world is manifest in his proposition that "the earth remains at rest because of its equilibrium," staying where it is by virtue of its equal distance toward all things. In contrast, Thales assumed that earth rested on water, while Anaximenes, as we will see in the next section, thought it was supported by air.

8. A. Pichot, *La Naissance de la science* (The birth of science), vol. 2, *Grèce présocratique* (Pre-Socratic Greece) (Paris: Gallimard, 1991).

9. J. Voilquin, *Les Penseurs grecs avant Socrate. De Thalès de Milet à Prodicos* (Greek Thinkers before Socrates. From Thales of Miletus to Prodicos), (Paris: Flammarion, 1964).

10. J. R. Partington, *A History of Chemistry*, vol. 1 (London: Macmillan, 1961).

11. Pichot, *La Naissance de la science*.

12. Besides Parmenides, whom we will discuss in the next section, one can mention Oenopides of Chios (circa 425 B.C.) among the advocates of *two* primordial substances. Oenopides chose fire and air.

13. With perhaps a degree of preeminence of fire (or heat, as we will see later), particularly in living systems; for instance, sleep was considered a temporary cooling, and death a permanent cooling, caused by the separation of the igneous element contained in the body.

14. The word *element (stoïkhéion)*, which, incidentally, was not introduced until later by Plato, can evidently be used with a double meaning: a rigorous one, as it is understood in modern chemistry, and a more poetic one. It is clearly the latter, broader sense that is used by the ancient philosophers of nature.

15. Besides their infinite divisibility, the minimalist particles of Anaxagoras differ from the atoms of Leucippus and Democritus by their qualitatively diversified nature, while the atoms of ancient thinkers had merely quantitative differences in dimension and shape. It is an essential conceptual difference, which we will encounter repeatedly, and which will find a solution only in modern atomism. In the nineteenth century, the British physicist James Clerk Maxwell would describe Anaxagoras's doctrine as the most radically opposed to the atomic theory. He wrote: "To another very eminent philosopher, Anaxagoras, best known to the world as the teacher of Socrates, we are indebted for the most important service to the atomic theory, which, after its statement by Democritus, remained to be done. Anaxagoras, in fact, stated a theory which so exactly contradicts the atomic theory of Democritus that the truth or falsehood of the one theory implies the falsehood or truth of the other" (J. C. Maxwell, "Molecules," *Nature* 8 [1873], 437–441). One and a half centuries earlier, the French philosopher Pierre Bayle (1647–1706), better inclined to compromise, had adopted a more conciliatory stance by writing in his *Dictionnaire historique* (Dictionary of History), published in 1740: "It is a pity that Anaxagoras and Democritus were not acquainted, and that these two great minds did not collaborate: the flaws of one hypothesis might have been mitigated by the perfections of the other."

16. In *Dictionnaire des philosophes* (Dictionary of philosophers), ed. D. Huisman (Paris: Presses Universitaires de France, 1984), 81.

17. J.-P. Dumont, *La Philosophie antique* (Ancient philosophy) (Paris: Presses Universitaires de France, Collection "Que sais-je?", 1962).

18. Crescenzo, *Les grands philosophes de la Grèce antique*.

19. Pichot, *La Naissance de la science*.

Chapter Three

1. *Les Présocratiques* (The pre-Socratics), ed. J.-P. Dumont (Paris: Gallimard, "Bibliothèque de la Pléiade," 1988).

2. Plato would later also adopt the idea that similar things attract each other.

3. J.-F. Duvernoy, *L'Épicurism et sa tradition antique* (Epicureanism and its ancient tradition) (Paris: Bordas, 1990).

4. J. Brunschwig, in *Dictionnaire des philosophes*, (Dictionary of philosophers), ed. D. Huisman (Paris: Presses Universitaires de France, 1984).

5. J. d'Ormesson, *La Douane de mer* (The maritime custom house) (Paris: Gallimard, 1993).

6. F. Nietzsche, *Philosophy in the Tragic Age of the Greeks,* (Chicago: Henry Regnery Co., 1982).

7. G. Minois, *L'Église et la science. Histoire d'un malentendu* (The Church and science: History of a misunderstanding), vol. 1, *De saint Augustin à Galilée* (From St. Augustine to Galileo) (Paris: Fayard, 1990).

8. Historically, the period of Democritus and Epicurus coincides with the heydays of Plato and Aristotle, the two giants of Greek philosophy. These two philosophers formulated commentaries on and critiques of the atomic theory, some of which were to influence Epicurus's thinking. Nevertheless, in the interest of continuity in the presentation of ideas that have contributed to the overall development of this theory, we defer until later chapters the analysis of the commentaries of Plato and Aristotle.

9. It may be interesting, and amusing to some, to recall that "The Difference in the Philosophies of Nature of Democritus and Epicurus" was the subject of the doctoral thesis of the young Karl Marx. The thesis is reproduced in Karl Marx, *Œuvres* (Collected works), vol. 2, *Philosophie* (Philosophy) (Paris: Gallimard, "Bibliothèque de la Pléiade," 1982); see chapter 28 of that volume.

10. M. Conche, *Lucrèce* (Lucretius) (Paris: Éd. de Magrave, 1990).

11. The reader is encouraged to consult J. Perrin's book *Atoms*, trans. D. Hammick (Woodbridge, Conn.: Ox Bow Press, 1990).

12. M. Cariou, *L'Atomisme. Trois essais: Gassendi, Leibniz, Bergson et Lucrèce* (Atomism. Three essays: Gassendi, Leibniz, Bergson and Lucretius) (Paris: Aubier-Montaigne, 1978).

13. Conche, *Lucrèce.*

14. G. E. R. Lloyd, *Greek Science after Aristotle* (New York: W. W. Norton, 1973).

15. Conche, *Lucrèce.*

16. This argument is remarkably similar to the modern view of the quark structure of protons and neutrons in atomic nuclei (see Provisional Epilogue).

17. R. Lenoble, *Histoire de l'idée de la nature* (History of the concept of nature) (Paris: Albin Michel, 1968).

18. J.-P. Sartre, *Matérialisme et révolution* (Materialism and revolution), in *Interaction III* (Paris: 1949)

19. Cicero called the Garden of Epicurus a garden of pleasure. Credit for the rehabilitation of the ethical doctrine of Epicurus goes to Gassendi, at the threshold of modern times (see chapter 12).

20. Bread and water sufficed for Epicurus's happiness. One of his disciples and friends, Metrodorus of Lampsacus (330–277 B.C.) insisted, however, on adding wine (see M. Conche, in *Dictionnaire des philosophes*).

21. J. O. de La Mettrie, *Œuvres philosophiques* (Philosophical writings), vol. 1 (Paris: Fayard, 1984).

Chapter Four

1. Such a proposition naturally raises all kinds of questions. As an example, the following fictitious dialogue takes place between Antithenes (445–360 B.C.), the founder of the philosophy of Cynicism, which accepts only the existence of concrete and individual beings, and Plato. Antithenes: "I can easily see the horse, but I fail to see equinity." Plato: "Precisely. To see the horse, you need eyes, and you have those. To see equinity, you need intelligence, and of that you have none."

2. The term *creation* is fraught with ambiguity. With respect to cosmogonic issues, the two most common meanings of the word involve either the sudden organization of a world out of preexisting material chaos or the creation of the primordial matter itself. In fact, the vast majority of Greek thinkers rejected any notion of creation at all. Such was, for instance, the position of Democritus, who, as we have seen, conceived the world as organizing itself gradually, by way of collisions of atoms that have existed through eternity. By contrast, the Judeo-Christian tradition propounds the creation by God of both the original matter—out of nothing—and from it the organized world. The biblical account is probably rooted in certain Mesopotamian myths.

3. The problem of the possible limits to God's power will be a topic of reflection for a great many thinkers, as were many other questions raised by Plato. Paul Valéry said: "Thinkers are people who think again, and who think that what was thought before was never thought through enough." In general, those Greek philosophers who accepted the supremacy of a unique being never accorded him the unlimited powers of the Judeo-Christian God, whose only restriction was, according to the Jewish philosopher Maimonides of Cordoba (1135–1204), the inability to do evil (St. Augustine later stated that evil is not a part of creation, but "a lack of being"). This restriction is evident with Plato as well, for whom, as we saw, good is an integral part of the ultimate design of the craftsman. Even in modern times, some (including Einstein) will question whether God really had the option to create a universe other than the one he did create, the argument being that the existing universe could be the only one to possess a complete logical consistency.

4. *Les Présocratiques* (The pre-Socratics), ed. J.-P. Dumont (Paris: Gallimard, "Bibliothèque de la Pléiade," 1988).

5. This metempsychosis can take on quite baffling and unfortunate forms. One can read in the *Timaeus*: "He who has made good use of the time allotted him will be allowed to return to the star assigned to him and will live there in happiness; but he who has squandered this opportunity will be transformed into a woman upon his second birth, and, in that state, if he does not desist being wicked, he will, according to the nature of his wickedness, be transformed upon each new birth into the animal which he resembles most by his conduct." And also: "The species of land animals and wild beasts are descended from men who pay no heed to philosophy." Another passage states: "All animals pass from one to the other, according to whether they lose or gain in intelligence or stupidity."

6. Recall that the number of shapes of atoms was infinite in the Leucippus-Democritus-Epicurus theory.

7. Transmutations of the "roots" were forbidden in Empedocles's theory.

8. G. E. R. Lloyd, *Early Greek Science: Thales to Aristotle* (New York: W. W. Norton, 1971).

9. J. R. Partington, *A History of Chemistry*, vol. 1 (London: Macmillan, 1961).

10. Aristotle, *De Generatione et Corruptione*, trans. C. J. F. Williams (New York: Oxford University Press, 1982).

11. E. Chambry, in the preface to his edition of *Timée* (Timaeus) (Paris: Flammarion, 1969).

12. B. Russell, *Religion and Science* (New York: Oxford University Press, 1961). Plato's case is far from unique. The "esthetic" argument concerning the sphere and the circle in the history of astronomy and cosmography carried a great deal of weight; see, for example, A. Koyré, *Études d'histoire de pensées scientifiques* (Studies of the history of scientific thoughts) (Paris: Gallimard, 1966). The argument would prevent even Galileo and Newton from accepting Kepler's law of the elliptical motion of the planets.

13. L. de Crescenzo, *Les Grands Philosophes de la Grèce antique* (The great philosophers of ancient Greece), vol. 1, *Les Présocratiques* (The pre-Socratics) (Paris: Julliard, 1985).

14. His own "geometric" theory found few supporters during the course of history. One might still mention his defense by the heir apparent of the Neo-Platonic school in Athens, Proclus of Byzantium (412–485), shortly before the closing of the Academy by an edict from Justinian I in 529, after an existence of nearly a millennium. Some twenty centuries after Plato, Johannes Kepler (1571–1630) used a geometrical theory of atoms inspired by that of the Greek philosopher to explain the crystalline structure of snowflakes.

15. J.-F. Revel, *Histoire de la philosophie occidentale de Thalès à Kant* (History of Western philosophy from Thales to Kant) (Paris: Nil Éditions, 1994).

Chapter Five

1. Aristotle, *Leçons de physique* (Lessons in physics), trans. and commentary by J.-B. Saint-Hilaire (Paris: Presses Pocket, 1990).
2. In the appendix to Aristotle, *Leçons de physique*. For an interesting discussion of the rivalry between reason and senses among the Greek philosophers of nature, see E. Schrödinger, *Nature and the Greeks* (New York: Cambridge University Press, 1954).
3. R. Lenoble, *Origine de la pensée scientifique moderne. Histoire de la science* (The origin of modern scientific thought: A history of science) (Paris: Gallimard, "Encyclopédie de la Pléiade," 1957).
4. While distinguishing four causes may be useful in describing certain phenomena, it can at times lead Aristotle to some rather bizarre assertions. He believed, for instance, that in sexual reproduction the mother supplies the embryonic matter, while the father provides both the formal and the effective cause. The atomists, on the other hand (supported by Hippocrates and, later, by Galen), were proponents (perhaps even the authors) of the theory of pangenesis, or "dual seed," which holds that generation is the product of the mixing of seeds secreted by every part of two organisms, male and female.
5. Aristotle also rejected Plato's triangle theory. He wrote in *De Generatione et Corruptione*, trans. C. J. F. Williams (New York: Oxford University Press, 1982): "It is impossible for the planar figures in the *Timaeus* to be the nutriment and the primordial matter of things."
6. Note that Theophrastus of Eresa (372–285 B.C.), Aristotle's successor at the head of the Lyceum, asserted that fire could not be placed on the same level as the other three elements.
7. Aristotle, *De Generatione et Corruptione*.
8. See, for example, P. Duhem. *Le Mixte et la combinaison chimique* (The Mixed and chemical combinations) (Paris: Fayard, 1985 [1902]).
9. G. Bachelard, *Les Intuitions atomistiques* (Atomistic intuitions) (Paris: J. Vrin, 1975).
10. In anticipation of the great debates to come in modern times on the topic of vacuum, it is interesting to note the following excerpt from Aristotle's *Leçons de physique*, p. 162. After defining void as "what is not filled with a body perceptible to the touch, that which has weight or lightness being sensible to the touch," he posed the following question: "Hence a difficulty: what if the interval contains color or sound; is it void or not?" This is a remarkable premonition of modern wranglings over such issues.
11. B. Russell, *A History of Western Philosophy* (New York: Simon & Schuster, 1959). In this book, Russell states: "Any causal explanation must have an arbitrary beginning."
12. Sooner or later, philosophers will propose any possible conception of the world. For the atomists, the multiple worlds are different from one another. Some twenty-two centuries later, Leibniz, who also accepted a multiplicity of worlds, would prefer them all alike.
13. Koyré, in the appendix to Aristotle, *Leçons de physique*.
14. Russell, *A History of Western Philosophy*
15. A. Koestler, *The Sleepwalkers: A History of Man's Changing Vision of the Universe* (London: Arkana, 1959).
16. J.-M. Aubert, *Philosophie de la nature* (Philosophy of nature) (Paris: Beauchesne et fils, 1965), p. 69.
17. The citations reproduced in this section come from the volume *Les Stoïciens* (The Stoics), commentary by E. Bréhier, ed. P. M. Schuhl (Paris: Gallimard, "Bibliothèque de la Pléiade," 1962).
18. Carneades, a Stoic dialectician who lived in the third century B.C., distinguished the necessary from the inevitable: what happens is inevitable but not necessary, for the

opposite event could also have happened logically. He taught that, while it is quite true that every event has a cause, such a cause can be independent of other causes. It would be a fortuitous cause, possibly responsible for voluntary events.

19. See, for instance: Plotinus, *The Enneads*, 3rd ed., rev. by B. S. Page, trans. Stephen MacKenna (New York: Pantheon Books, 1961).

Chapter Seven

1. L. Mabilleau, *Histoire de la philosophie atomistique* (History of atomistic philosophy) (Paris: Félix Alcan, 1895).

2. J. R. Partington, *Annals of Science* 4 (1939), 245–283; Helmuth von Glasenapp, *Die Philosophie der Inder: eine Einführung in ihre Geshichte und ihre Lehren* (Philosophy of India: An introduction to its history and its teachings) (Stuttgart: A. Kroner, 1958). For the benefit of readers particularly interested in the question of chronology, let us mention a comment contained in the writings of Strabo, a Greek geographer (58 B.C.–A.D. 25), who, incidentally, referred on this topic to Posidonius, a Stoic (135-50 B.C.). The comment alludes to a Phoenician hailing from Sidon named Mochos, "who lived at the time of Troy and who was the author of the ancient opinion on atoms." Since the Trojan War, the most famous event in the history of that city, took place in the twelfth century before our era, the birth of the atomic theory would be pushed back to a much more distant past. At the dawn of modern times, some philosophers (including Henry More and Ralph Cudworth) suggested that Mochos was none other than Moses, which allowed them to consider an original atomism cleansed of the sin of atheism.

3. In addition to the sources cited in note 2, see M. Biardeau, "Philosophies de l'Inde" (Philosophies of India), in *Histoire de la philosophie* (A History of Philosophy), vol. 1 (Paris: Gallimard, "Encyclopédie de la Pléiade," 1990).

4. M. K. Gangopadhyaya, "The Atomic Hypothesis," in D. Chattopadhyaya, *History of Science and Technology in Ancient India*, vol. 2, *Formation of the Theoretical Fundamentals of Natural Science* (Calcutta: 1992); also from the same author, *Indian Atomism, History and Sources* (Calcutta: 1980).

5. This deficiency will be one of the arguments used by Vedantic philosophers, the most eminent of whom is Shankara, in the eighth century B.C., to refute atomism.

6. Glasenapp, *Die Philosophie der Inder*.

7. Mabilleau, *Histoire de la philosophie atomistique*.

8. Recall that the granular appearance of dust in a ray of sunshine had also been noted by Epicurus (see chapter 3).

9. G. Bachelard, *Les Intuitions atomistiques* (Atomistic intuitions) (Paris: J. Vrin, 1975).

10. Mabilleau, *Histoire de la philosophie atomistique*; Glasenapp, *Die Philosophie der Inder*.

11. Mabilleau, *Histoire de la philosophie atomistique*.

12. Glasenapp, *Die Philosophie der Inder*.

13. Ibid.

14. Furthermore, it is tempting to see an analogy between the rules of Indian atomism stipulating how dyads combine and certain modern propositions concerning the formation of chemical elements following the big bang. Modern theory holds that, after the production of hydrogen and helium nuclei, about one hundred seconds after the "beginning," the formation of elements came to a stop. The helium nucleus, being extremely stable, failed to associate with other nucleons or with themselves. It was not until stars were formed several billion years later that *three* helium nuclei, *but not two*, were willing to unite during thermonuclear reactions to form a nucleus of carbon 12. From that moment on, the formation of new elements resumed. See, for example, T. X. Thuan, *The Secret Melody* (New York: Oxford University Press, 1995), or J. R. Grib-

bin, *In Search of the Big Bang: Quantum Physics and Cosmology* (New York: Bantam Books, 1986).

15. Voltaire would surely use here his aphorism claiming that the substance of which the soul is made matters little, as long as it is virtuous.

Chapter Eight

1. St. Augustine, *The City of God* (New York: Modern Library, 1993).

2. St. Basil, *Exegetic Homilies*, trans. Agnes Clare Way (Washington, D.C.: Catholic University of America Press, 1963). See "Creation of the Heavens and the Earth," Homily 1 of the nine homilies on the Hexaemeron. St. Basil is also known as Basil the Great, or Basil of Caesarea, bishop of Cappadocia.

3. St. Augustine, *Dialogues philosophiques* (Philosophical dialogues) (Paris: Desclée de Brouwer, 1939).

4. Strictly speaking, St. Basil taught that God initially created only the earth, and left it up to the intelligence he granted to men to "introduce within his creation the other three elements," fire, water, and air.

5. St. Basil, *Exegetic Homilies*.

6. For a more complete account, see S. Swiezawski, *Redécouvrir Thomas d'Aquin* (Rediscovering Thomas Aquinas) (Paris: Nouvelle Cité, 1989).

7. To add a lighter touch to this discussion, we quote an excerpt from St. Augustine's *Confessions* (Book XI, chap. 12) trans. F. J. Sheed (Indianapolis, Ind.: Hackett Pub. Co., 1992): "I come now to answer the man who says: 'What was God doing before He made Heaven and earth?' I do not give the jesting answer—said to have been given by one who sought to evade the force of the question—'He was getting Hell ready for people who pry too deep.' . . . I make bold to reply: Before God made Heaven and earth, He did not make anything."

8. Lactantius, "Divinae Institutiones," in Hans F. von Campenhausen, *The Fathers of the Latin Church*, trans. Manfred Hoffmann (Stanford, Calif.: Stanford University Press, 1969).

9. St. Basil, *Exegetic Homilies*.

10. E. Schrödinger, *Nature and the Greeks* (New York: Cambridge University Press, 1996).

11. In spite of their differences, if not outright opposition, Christians and Epicureans have at times been the target of a common animosity. One example is the following story told by Fontenelle (*Histoire des oracles, Œuvres complètes*, vol. 2 [History of oracles: Complete works] [Paris: Fayard, 1991]): "The custom of excluding Epicureans from all mysteries was so widespread and so necessary to the safety of sacred institutions that it was adopted by this great fiend, whose life Lucian describes so delightfully. . . . He even added the Christians to the Epicureans, because, in his opinion, they were not worth much more than the latter; and before the start of his ceremonies, he would exclaim: 'Let us chase all Christians away from here'; to which the People would reply, as in a choir: 'Let us chase away all the Epicureans.'"

12. The declaration of the Council of Trent specifies that: "Through the consecration of bread and wine, all the substance of the bread is changed into the substance of the body of Christ, our Lord, and all the substance of the wine into the substance of his blood."

13. P. Redondi, *Galileo Heretic*, trans. Raymond Rosenthal (Princeton, N.J.: Princeton University Press, 1987).

14. A compromise solution could be envisioned—and it came to pass—in the form of a cosubstantiation in the consecrated wafer of the body of Christ and of bread. Rejected by the Catholic Church, this solution was adopted by Luther.

15. Duns Scotus was nonetheless opposed to atomism. To refute the doctrine, he used the

argument of concentric circles: "Concentric circles are all intersected by any radius emanating from the center; they must, therefore, all contain the same number of atoms, and, consequently, they must all be equal to each other." (Cited by G. Minois in *L'Église et la science. Histoire d'un malentendu* [The Church and science: History of a misunderstanding], vol. 1, *De saint Augustin à Galilée* [From St. Augustine to Galileo] [Paris: Fayard, 1990].) This excellent treatise is a valuable source of information on the topic covered in this chapter. William of Ockham, on the other hand, was rather in favor of the atomic theory (see chapter 9).

Chapter Nine

1. E. Bréhier, *La Philosophie du Moyen Âge* (Philosophy of the Middle Ages) (Paris: Albin Michel, 1937).

2. J. Jolivet, *La philosophie médiéviale en Occident* (Western medieval philosophy), in *Histoire de la philosophie* (A History of philosophy), vol. 1 (Paris: Gallimard, "Encyclopédie de la Pléiade," 1969) For lack of a better opportunity, one might mention here the names of two physicians of antiquity, long before medieval Christianity, who could properly wear the label of "sympathizers" of atomism. They are Asclepiades of Prusa in Bithynia, (124–40 B.C.), and Hero of Alexandria, who lived during the second century A.D. Some authors (notably J. R. Partington, *A History of Chemistry*, [London: Macmillan, 1961], and Marie Boas, "The Establishment of the Mechanical Philosophy," *Osiris* 10, [1952] 412–541) considered them "atomists" because they accepted that matter and the human body are composed of particles and of pores in which these particles lodge themselves. However, it is difficult to consider them proponents of the Democritean or Epicurean doctrine, since particles were not necessarily indivisible in their system. Asclepiades, in particular, recognized their ability to subdivide indefinitely under the effect of collisions.

3. Quoted by E. Bréhier, *The History of Philosophy*, vol. 1, trans. Joseph Thomas (Chicago: University of Chicago Press, 1963).

4. In *L'Église et la science. Histoire d'un malentendu* (The Church and science: History of a misunderstanding), vol. 1, *De saint Augustin à Galilée* (From St. Augustine to Galileo) (Paris: Fayard, 1990), G. Minois mentions another Franciscan—actually the general of the Order—Gérard of Odon, who aligned himself firmly on the side of atomism. He quotes an excerpt from Odon's *Traité de prédicaments* (Treatise on predicaments): "Let us then conclude with Democritus that any continuum is, in the final analysis, composed of indivisibles, and that it can ultimately be resolved into these indivisibles; it is not composed of an infinity of indivisibles, but of indivisibles in infinite number. Adhere to the Peripatetics or the Platonics who will; still, Democritus seems to speak more reasonably." Note that the Jesuits, on the other hand, will be almost unanimously and profoundly opposed to atomism. The book by Minois contains an excellent account of these conflicting positions.

5. Bachelard routinely referred to the Christian atomists in the Middle Ages as "non-Aristotelian." See G. Bachelard, *Le Materialism rationel* (Rational materialism) (Paris: Presses Universitaires de France, 1953).

6. Cited in Bréhier, *La Philosophie du Moyen Âge*.

7. Ibid.

Chapter Ten

1. The vast majority of Jewish philosophers were either proponents of the Neo-Platonic philosophy, like Isaac ben Solomon Israeli (850–932), Solomon ibn Gabirol (also known as Avicebron, 1021–1050), and Abraham ibn Ezra (1089–1164), or Aristotelian supporters, like Maimonides, Abraham ibn Daud (1110–1180), and Levi ben Ger-

shom, also called Gersonides (1288–1344). They either adopted the four elements, which they sometimes derived from an even more basic primordial matter created by God, or envisioned their own set of "simple substances." For additional details, see C. Sirat, *A History of Jewish Philosophy in the Middle Ages* (New York: Cambridge University Press, 1985); M. R. Hayoun, *L'Éxégèse philosophique dans le judaïsm médiéval* (Philosophical exegesis in medieval Judaism) (Paris: J. C. B. Moher, 1992); and I. Husik, *A History of Mediaeval Jewish Philosophy* (New York: Meridian Books, 1958).

2. Sirat, *A History of Jewish Philosophy in the Middle Ages*. This excellent reference mentions two Karaitic writers: Jacob al-Qirqisani and Yefet ben Ali, both of whom did accept the four elements and rejected void.

3. A letter, dated 1625, from an Italian Jew residing in Jerusalem documents the presence in Jerusalem of no more than twenty Karaites. (Quoted in *The Jews in their Land*, ed. Ben Gourion [New York: Doubleday, 1966]).

4. Ibn Ezra, too, strongly criticized the Karaitic exegesis, although his objections centered mostly on problems specific to Jewish religious life, such as the calendar, the process of fixing holy days, and ancestral laws; see M. R. Hayoun, *La Philosophie médiévale juive* (Jewish medieval philosophy) (Paris: Presses Universitaires de France, 1991).

5. See G. Vajda, *Introduction à la pensée juive du Moyen Âge* (Introduction to Jewish thought in the Middle Ages) (Paris: Librairie Philosophique J. Vrin, 1947).

6. Though they were viewed as "different" Jews by the Rabbinites, the Karaites were not considered heretical in spite of their very unconventional position vis-à-vis certain aspects of cult, the observance of holy days, the rules of purity and dietary laws (*kashruth*). For instance, the Karaites softened the scope of the biblical prescription "thou shall not cook a lamb in its mother's milk" (Exodus 23:19). They adopted its literal meaning and interpreted it as banning the consumption of the meat of a mammal cooked in the milk *of an animal of the same species*. That is a considerable relaxation of the rule of orthodox Judaism, which forbids mixing meats with any dairy product. See, for example, E. Trevisan-Semi, *Les Caraïtes. Un autre judaïsm* (The Karaites: A different Judaism) (Paris: Albin Michel, 1992).

7. S. Munk, the eminent expert in medieval Jewish and Arab philosophy, renders the following verdict on the historical role of the Karaites in the development of Jewish theology: "Although it is true that the Karaites, who lacked any fixed principle and recognized only the individual opinions of their own scholars, ended up surrounding themselves in a maze of contradictions and endless arguments far more difficult to unravel than the Talmudic debates, no one can deny that Karaitism, by its principle, gave Jewish scholars a beneficial nudge by resorting to weapons of reasoning to combat rabbinism, and by forcing the rabbis to use those same weapons to defend themselves." See S. Munk, *Mélanges de philosophie juive et arabe* (Mixture of Jewish and Arab philosophy) (Paris: J. Vrin, 1988 [1857]).

8. For an excellent biography, consult A. Heschel, *Maimonide* (Paris: Fondation Sefer, 1982).

9. In a delightful romanticized biography of Maimonides (*The Doctor from Cordova*, trans. Barbara Wright [Garden City, N.Y.: Doubleday, 1979]), Herbert Le Porrier writes the following imaginary exchange. In it, ibn-Rushd (more commonly known as Averroës, who, like Maimonides, was a native of Cordova, Spain) claims that "in the three great religions issued from a common trunk of primitive monotheism, the revelation starts from an impossible assumption [the creation *ex nihilo*], which takes all credibility away from them." Maimonides responds thus: "I objected, as I was already in disagreement with him, that it was inappropriate to take the revealed word literally, and that it should be accepted, instead, in a purely allegorical sense. Could it be that metaphysical truth is a bad thing to say? Not at all. The prophet had the wisdom not to formu-

late it in a confusing manner, and to devise gradual stages of initiation. I could only compare this to someone forcing a newborn to eat wheat bread and meat, and to drink wine; it would undoubtedly kill him; not because these foods are intrinsically bad and contrary to man's nature; but because a newborn trying to absorb them would be in no position to digest them and derive any nourishment from them."

10. Maimonides wrote: "If the novelty [of the world] were proven, all that has been said by the philosophers to refute us would fall apart; and likewise, had they succeeded in demonstrating the eternity [of the world], in accordance with Aristotle's view, all of religion would collapse and we would be driven toward other opinions." This adherence to a truth regarded as the cornerstone of a dogma reminds us the words of St. Paul to the Corinthians: "If Christ was not resurrected, then we should eat and drink, for our faith would be in vain." As a comparison, ibn Daud, an Aristotelian like Maimonides, did not reject the Aristotelian doctrine of eternity, though he did believe in creation *ex nihilo.*

11. A. Neher, *La Philosophie juive médiévale* (Medieval Jewish philosophy), in *Histoire de la philosophie* (A History of philosophy) (Paris: Gallimard, "Encyclopédie de la Pléiade," 1969).

12. M. Maimonides, *The Guide of the Perplexed*, trans. Chaim Rabin (Indianapolis, Ind.: Hackett Pub. Co., 1995).

13. Ibid.

Chapter Eleven

1. E. Bréhier, *La Philosophie du Moyen Âge* (Philosophy of the Middle Ages) (Paris: Albin Michel, 1971).

2. Sometimes quite different spellings are used to describe Arab thinkers and their teachings. For instance, the atomic doctrine is written variously as Kalam, Kalâm, or Calam. Its proponents are called Mutakallemîn, Mutacallemin, or Mutakallimun, and their sects are designated Motazites, Mutazilites, Mu'tazilites, Motazales, and Ascarite, Ash'arites, Asch'arites, etc. Faced with so many choices, we have adopted the spellings that seem to us the most consistent with modern usage, and, to the extent possible, the simplest.

3. See D. Gimaret, *La Doctrine d'Al-Ashari* (The Doctrine of al-Ashari) (Paris: Cerf, 1990).

4. Gerard Heym, "Al-Razi and Alchemy" *Ambix* 1 (1938), no. 1, 184–191.

5. The Karaitic thinker Aaron ben Elijah, already mentioned in our discussion of Jewish thought concerning atoms, argued in his book *The Tree of Life* that, by adopting the corpuscular doctrine, the Mutakallmun were motivated not just by a desire to exploit arguments in favor of the noneternity of the world, but also for its inherent philosophical value. After all, as ben Elijah remarked, the Greek thinkers who propounded this doctrine were all opponents of the Creation.

6. L. Mabilleau, *Histoire de la philosophie atomistique* (A History of atomistic philosophy) (Paris: F. Alcan, 1895).

7. Moses Maimonides, *The Guide of the Perplexed*, trans. Chaim Rabin (Indianapolis, Ind.: Hackett Pub. Co., 1995).

8. It is difficult to miss the parallel between the central idea in the sixth proposition of the Kalam and the concept of transcreation developed centuries later by Leibniz in his treatise *Pacidius Philatethi*, published in 1676, in which he advanced the thesis that the continuous existence of material objects demands their constant re-creation by God. In particular, to avoid certain difficulties in the interpretation of continuous motion, he decomposed it into a series of stops; he asserted that when a material body is in motion, it ceases to exist in one position and is re-created in another.

9. Mabilleau, *Histoire de la philosophie atomistique.*

10. Maimonides, *The Guide of the Perplexed.*
11. E. Gilson, *La Philosophie au Moyen Âge* (Philosophy in the Middle Ages) (Paris: Payot, 1988).
12. There is a striking analogy between this philosophical system and Occasionalism, a metaphysical theory developed several centuries later by Malebranche (1638–1715). He asserted that material movements and the action of soul on body are not "primitive causes," but only occasional and circumstancial causes. The only efficient cause was to be found in God. He wrote in *Recherche de la vérité* (The search after truth): "The efficacy of God's immutable decrees on actions is determined only by the circumstances of causes commonly called natural; I feel I should call them occasional to avoid reinforcing the dangerous misconcepption of confusing nature and efficacy with God's will and omnipotence." Nevertheless, Malebranche saw the structure of matter as continuous: "If men could consider the true notion of things with any attention, they would quickly realize that, all material bodies having extent, their nature or essence has nothing in common with numbers and that they cannot be indivisible."
13. A. Heschel, *Maimonide* (Paris: Fondation Sefer, 1982).
14. Speculations concerning the connection between God and atoms, or God and the laws of nature, are inevitable in any theology that includes a corpuscular concept of the structure of matter. In Part Three of this book, we will hear echoes of a controversy between Newton and Leibniz, some aspects of which will hark back to the preoccupations of the Mutakallimun.
15. Maimonides, *The Guide of the Perplexed.*
16. Ibid.
17. See E. Renan, *Averroës et l'averroïsm* (Averroës and Averroism) (Paris: Calmann-Lévy, 1856).
18. Note that certain Nominalists, in particular Jean Buridan (c. 1300–1358) and Albert of Saxony (1316–1390), were also proponents of minimalist particles. However, contrary to the Averroists who, as mentioned earlier, considered these particles requisite for the specific action of material bodies, they held that the very existence of these bodies demanded that they possess a minimum dimension below which they become unstable.
19. For more details on the various contributions of these proponents of minimal particles, see A. G. van Melsen, *From Atomos to Atoms: The History of the Concept of Atom* (Pittsburgh: Duquesne University Press, 1952).

Chapter Twelve

1. B. Brundell, *Pierre Gassendi: From Aristotelianism to a New Natural Philosophy* (Dordrecht: Reidel Publishing Co., 1987).
2. Likewise, the idea of the existence of objects smaller than what our vision can behold was made possible by the invention of the microscope in 1590. Pascal's position toward the atomic theory is peculiar in that, while he himself contributed evidence of the reality of void, he appears not to have placed a great deal of faith in the significance of atoms. Although he did use the term in a pictorial manner—"I see only infinities in all directions which confine me like an atom"—he also wrote: "Thus, since all things are both caused or causing, assisted and assisting, mediate and immediate, providing mutual support in a chain linking together naturally and imperceptibly the most distant and different things, I consider it as impossible to know the parts without knowing the whole as to know the whole without knowing the individual parts." This celebrated phrase, taken from Thought 199 (see Pascal, *Pensées*, trans. A. J. Krailsheimer [New York: Penguin Books, 1966]) articulates a global, "holistic" worldview (the whole is more than the sum of the parts). Bréhier (*The History of Philosophy,*

vol. 2, trans. Joseph Thomas [Chicago: University of Chicago Press, 1963]) interpreted it as a rejection of the atomic philosophy, consistent with Pascal's conviction that
human knowledge is incapable of grasping the first "principles."

3. A. Koyré, *Études d'histoire de la pensée scientifique* (Studies in the history of scientific
thought) (Paris: Gallimard, 1973).

4. Nevertheless, the atom is not necessarily "without parts." It could be made up of parts,
themselves obviously infinitesimal, but *inseparable*, which Lucretius expressed in a
paradoxical phrase: "The atom is one in its parts" (see M. Cariou, *L'Atomisme. Trois
essais: Gassendi, Leibniz, Bergson et Lucrèce* [Atomism. Three essays: Gassendi, Leibniz,
Bergson and Lucretius] [Paris: Aubier-Montaigne, 1978]).

5. The importance of this point will become particularly significant when we discuss (in
chapter 13) Descartes's views and, more specifically, his fundamental thesis concerning the identification of physical matter with geometrical extent, a notion that was to
be vigorously opposed by Gassendi.

6. Two centuries later, J.-B. Dumas stigmatized in his *Leçons sur la philosophie chimique*
(Lectures on chemical philosophy) (Paris: Gauthiers-Villars, 1837) Gassendi's excessive imagination and even rebuked all proponents of the atomic theory: "Gassendi,
perfecting the then-prevailing image of atoms and their mutual interactions, brought
it closer to our modern concept by accepting forces holding atoms in equilibrium as
well as spaces between them that are wider than the atoms themselves. While up to
this point Gassendi is correct, or at least remains in the realm of the plausible, he
soon strays from reasonable hypotheses into pitfalls that have oftentimes and rightfully exposed the proponents of atoms to the disdain of all exact and positive minds. For
instance, he ascribes light to round atoms, while other specific atoms are responsible
for heat, cold, odors, flavors; sound itself is ascribed to atoms. All these errors, recognized or condemned by subsequent physicists, brought about the indiscriminate sink-
ing of what could be useful and correct in his basic ideas."

7. Attempts to reconcile the ancient theory of "elements" with the atomic hypothesis can
also be found in the writings of other proponents of the corpuscular doctrine during
this period. They, however, remain lesser-known than Gassendi. They include: Daniel
Sennert, who appeared to recognize void; Jean Magnien, who recognized only three
elements, water, fire, and earth, air replacing, in his view, "void" which he rejected; and
Sébastien Basso, who also rejected void and replaced it with a "subtle spirit." One century later, Hermann Boerhaave, an eminent Dutch chemist, tried again to promote the
coexistence of the atomic theory with that of the four elements.

8. It might be useful to add that, from Gassendi's point of view, the motion of atoms
constitutes the internal cause of motion at all levels: When atoms aggregate to form
an extended body, the micro-movements of the former combine to determine the
macro-movements of the latter. Thus, for Gassendi, the motion of celestial bodies is
entirely determined by the motion of their atoms. This marks a return to the original
doctrine of the classical atomists, as well as a reaffirmation of the physical unity of the
celestial and terrestrial worlds, in further contradiction with the teachings of Aristotle
(see chapter 5).

9. Aristotle (see chapter 5) also rejected the concept of creation, although this did not
prevent his philosophy from becoming intimately associated with Christian teachings, at
least from the thirteenth century on. From this perspective, Epicureanism was, in
Gassendi's opinion, eminently suitable to replace Aristotelianism in such an association.

10. L. Mabilleau, *Histoire de la philosophie atomistique* (History of atomistic philosophy)
(Paris: F. Alcan, 1895). For more details, see F. Pillon, "L'Évolution historique de
l'atomisme" (Historical evolution of atomism), *Année philo-sophique* (1891), part 1,
chap. 2, 67–112.

11. The ecclesiastical tribunal before which Galileo appeared was composed of ten cardinals. Only seven of them voted for condemnation. There are therefore three judges whose abstention partly saved the honor of the Church. The names of the prelates in question deserve to be mentioned: Gasparo Borgia; Francesco Barberini, nephew of Pope Urban VIII; and Laudivio Zacchia. For more details, see Mario d'Addio, *Considerazioni sui processi a Galileo* (Considerations about Galileo's process) (Rome: Herder, 1985).

12. Galileo Galilei's *Sidereus nuncius*, published in March 1610, can be found in English translation as *The Sidereal Messenger* (Chicago: University of Chicago Press, 1989).

13. A Dutch physicist, Christian Huygens (1629–1695), was to make this identification about half a century later.

14. Galileo was born in Pisa to a Florentine family. He kept a special fondness for Florence during his entire life.

15. P. Redondi, *Galileo Heretic* (Princeton, N.J.: Princeton University Press, 1987).

16. 1623 was the year when cardinal Maffeo Barberini, who had always publicly expressed his admiration for Galileo, was elected Pope under the name Urban VIII. Galileo dedicated his book to him but, unfortunately, his relations with the pontiff deteriorated a few years later.

17. Galileo Galilei, *Dialogue Concerning the Two Chief World Systems: Ptolemaic and Copernican,* trans. Stillman Drake (Berkeley, Calif.: University of California Press, 1967).

18. For a lucid and comprehensive account of the vagaries of the relations between atomism and the Jesuits, see G. Minois, *L'Église et la science. Histoire d'un malentendu* (The Church and science: History of a misunderstanding), vol. 1, *De saint Augustin à Galilée* (From St. Augustine to Galileo) (Paris: Fayard, 1990).

19. Quoted in P. Redondi, *Galileo Heretic, op. cit.*

20. Ibid.

21. Ibid.

22. Galileo Galilei, "Documenti del Processo di Galileo Galilei" (Documents of the Trial of Galileo Galilei) (Vatican City: Pontificia Academia Scientiarum, 1984).

23. To be sure, there was the example of Giordano Bruno, burned alive at the stake thirty-three years earlier, in the year 1600. But, as we will see, his case was rather different.

24. Egidio Festa, "La querelle de l'atomisme. Galilée, Cavalieri et les jésuites" ("The quarrel about atomism: Galileo, Cavalieri, and the Jesuits), *La Recherche* 21, no. 224, (1990), 1038.

25. A. Koestler, *The Sleepwalkers: A History of Man's Changing Vision of the Universe* (London: Arkana, 1959).

26. The recently issued *Nouveau Catéchisme de l'Église catholique* (The new catechism of the Catholic Church) (Paris: Mame/Plon, 1992) reaffirms the nature of the Eucharist. Its article 1374 reads: "Christ is present in the Eucharistic species in a unique way. . . . The very Holy Sacrament of the Eucharist contains, truly, genuinely, and substantially, the Body and the Blood, together with the soul and divinity of our Lord Jesus Christ, and, consequently, Christ as a whole." Article 1377 further states: "The Eucharistic presence of Christ begins at the moment of the consecration and lasts as long as the Eucharistic species subsist. Christ is entirely within each species and within each of their parts, so that the breaking of the bread does not divide Christ." Article 1375 reads: "It is not man who transforms the offerings into the Body and the Blood of Christ, but Christ Himself." One is reminded in this article of the words of St. Ambrose: "The strength of the benediction takes precedence over that of nature."

27. P. N. Mayaud, "Science et foi: comment comprendre la transsubstantation" (Science and faith: How to understand transsubstantiation), *La Recherche* 20 (1984), 522. Article 1381 of the *Nouveau Catéchisme de l'Église catholique* is reminiscent of the words of St.

Thomas: "The presence of the genuine Body of Christ and of the genuine Blood of Christ in this sacrament cannot be apprehended by the senses, but only by faith, which rests on the authority of God."

28. G. Minois, *L'Église et la science. Histoire d'un malentendu* (The Church and science: History of a misunderstanding), vol. 2, *De Galilée à Jean-Paul II* (From Galileo to John Paul II) (Paris: Fayard, 1991).

29. Koyré, *Études d'histoire de la pensée scientifique.*

30. Although this was the basis of Galileo's thinking, he could not escape the "esthetic prejudice" in favor of the circle, which prevented him from accepting Kepler's ellipses as trajectories of planets around the sun.

31. A. C. Crombie, *Robert Grosseteste and the Origins of Experimental Science* (Oxford: Clarendon Press, 1953).

32. The intimate link between the physical theory of atoms and the mathematization of the universe in Galileo's time found an additional stimulus in the development of the "geometry by means of indivisibles," of which one of the most important pioneers was Bonaventura Cavalieri (1598–1647). The objective of this effort was to search for ways to conceive of the division of continuous media into ever smaller parts. In other words, the problem was to determine if one could legitimately represent a surface and a volume by means of "all the lines" and "all the planes," respectively, that these figures could contain.

33. Koyré, *Études d'histoire de la pensée scientifique.*

34. *Einstein, Galileo: Commemoration of Albert Einstein* (Vatican City: Libreria Editrice Vaticana, 1980).

35. P. Poupard, ed., *Après Galilée. Science et foi: nouveau dialogue* (After Galileo: Science and faith; a new dialog) (Paris: Desclée de Brouwer, 1994).

36. Giordano Bruno, *De minima* (1591), quoted in Redondi, *Galileo Heretic.*

37. Minois, *L'Église et la science. Histoire d'un malentendu*, vol. 1.

38. A. Koyré, *From the Closed World to the Infinite Universe* (Baltimore: Johns Hopkins University Press, 1994). Galileo did not take a clear position vis-à-vis the infiniteness of the universe. It is likely that the fate of Giordano Bruno, whose name he actually never mentioned in his writings, contributed to this great prudence. We have already seen that Gassendi's world is finite.

39. An English translation of G. Bruno's *De l'infinito universo et mundi* (1584) (On the infinite universe and worlds) can be found in Dorothea Waley Singer, *Giordano Bruno: His Life and Thought* (New York: Greenwood Press, 1968).

40. Ibid.

41. Singer, *Giordano Bruno: His Life and Thought.* This passage is from the first of Bruno's five dialogues.

42. Giordano Bruno was born in Nola, near Naples.

43. Bruno's rejection of the dogma of transubstantiation does not appear to have played any decisive role in his condemnation, casting further doubts on the plausibility of the claim alluded to in the preceding section concerning a "hidden agenda" in Galileo's trial.

44. P. Thuillier, "Martyre de la science ou illuminé? Le cas de Giordano Bruno" (Martyr of science or seer? The case of Giordano Bruno), *La Recherche* 19, no. 198 (1988), 510. See also Mabilleau, *Histoire de la philosophie atomistique.*

45. We should point out, however, that Newton, a strict adherent to monotheism, rejected the dogma of the Trinity, which he regarded as a "corruption" of the original Christian doctrine by Platonic ideas. He denied the divinity of Jesus Christ whom he saw simply as mediator between God and man. See, for example, A. R. Hall, *Isaac Newton: Adventures in Thought* (Oxford: Blackwell, 1992).

46. I. Newton, *Mathematical Principles of Natural Philosophy* (New York: Greenwood Press, 1969).

47. It will, however, be endorsed by Voltaire (1694–1778), who was an avowed deist (see his *Lettres philosophiques* [Philosophical letters], published in 1743). Voltaire also happened to be an enthusiastic popularizer of Newton's physics, as evidenced by his book *Les éléments de la physique de Newton* (Elements of Newton's physics), published in 1738. A French translation of Newton's *Principia* appeared as early as 1756. It was prepared by the Marquise du Châtelet, a dear friend of Voltaire's.

48. Koyré, *From the Closed World to the Infinite Universe.*

49. H. Metzger, *Newton, Stahl, Boerhaave et la doctrine chimique* (Newton, Stahl, Boerhaave, and the chemical doctrine) (Paris: Librairie Albert Blanchard, 1974). This book contains a detailed discussion of the way chemists reacted to the law of attraction.

50. J. Merleau-Ponty, "Laplace: un héros de la science 'normale'" (Laplace: Hero of "normal" science), *La Recherche* 10 (1979): 251–258.

51. Laplace, an ardent proponent of Newton's theory, but resolutely deterministic, excluded divine intervention in any form. As a parallel with Newton's celebrated words *Hypotheses non fingo*, readers might enjoy the following exchange between Laplace and Napoleon: after Laplace finished explaining the broad outlines of his system of the world (as narrated in his famous work *La Mécanique céleste* [Celestial mechanics]), Napoleon asked what became of God in all this. "I do not need this hypothesis," was Laplace's reply. We might add that Laplace had serious reservations about the atomic theory of Dalton. See P. Thuillier, "La résistible ascension de la théorie atomique" (The resistible rise of the atomic theory), *La Recherche* 4, no. 36 (1973), 705. The distinction between molecules and atoms was far from clear in those days (see chapter 19).

52. Although Boyle did not necessarily reject alchemy as a whole, he was deeply suspicious of Paracelsus and his followers. He wrote about them: "It seems to me that their writings, like their furnaces, produce as much smoke as light." It should be pointed out that alchemists were, for the most part, proponents of the four elements. Even Paracelsus, whose real name was Theophrastus von Hohenheim (1493–1541) and who was the most famous of the alchemists, called the four elements "mothers of all things," although he considered that God extracted them from some primordial matter. He claimed that, together, they formed a single substance, which God subdivided in "three principles" (mercury, sulfur, and salt) endowed with "all the force and power of perishable things." Agrippa von Nettesheim (1486– 1535), a contemporary of Paracelsus's, pushed the cult of the four elements so far as to profess that they existed "in exalted forms in the heavens, the stars, the angels, and even in God himself." Nearly two centuries later, an English philosopher named John Hutchinson (1674–1737) would go even further in trying to deify the elements; he asserted in his book *Moses' Principia*, published in 1724, that the world is based on three elements—fire, air, and light—which he assimilated to the three entities of the Trinity.

53. In agreement with Democritus, but unlike Epicurus, Boyle did not consider weight a primary property of atoms.

54. See, for example, P. Thuillier, "Sciences, religion et politique: le cas de Newton" (Science, religion, and politics: The case of Newton), *La Recherche* 11 (1980), 1340–1346.

55. In the words of G. Minois commenting on Boyle, this was done "to offset somewhat the effects of the original sin." See *L'Église et la science. Histoire d'un malentendu*, vol. 2.

56. This quotation, and other below, come from portions of Bentley's sermons reproduced in A. Koyré, *From the Closed World to the Infinite Universe.*

57. Quoted in Koyré, *From the Closed World to the Infinite Universe.*

58. Ibid.

59. In the sense of the modern "anthropic principle" of Brandon Carter, who holds that the universe was designed precisely so as to lead inexorably, at a particular time in its existence, to the emergence of conscious beings. On this topic, it is worthwhile to consult, for example, P. C. W. Davies, *God and the New Physics* (New York: Simon & Schuster, 1983), and Trinh Xuan Thuan, *The Secret Melody* (New York: Oxford University Press, 1995).

60. Quoted in Koyré, *From the Closed World to the Infinite Universe.*

61. Protagoras, a Sophist from Abdera who lived in the fifth century before our era (485–410 b.c.), had already expressed similar views in his particularly compact and incisive statement: "Man is the measure of all things." Diderot would later say: "Without man, the universe remains silent."

62. A century and a half later, Auguste Comte, the father of positivism, will do far worse: His aversion toward atoms will prompt him to eschew the use of a microscope! This attitude is reminiscent of the refusal of Cremonini and other colleagues of Galileo's in Padua to look at the sky through a telescope, for fear of discovering phenomena contradicting Aristotle's teachings. At the other end of the spectrum, the astronomer and philosopher Robert Hooke (1635–1703) recognized the potential benefits of the microscope, even though he did not believe in atoms. He essentially adopted Descartes's view that matter may have a corpuscular structure in fact, but that it is infinitely divisible in practice. He expressed in his *Micrographia* the belief that, by revealing the microstructure of chemical compounds, the microscope would expose "the true essence" of things, a knowledge that man had been deprived of, in Hooke's opinion, since the original sin. Locke, on the other hand, believed that only angels could have such a direct comprehension of things: they alone could reach a "true" (angelic) science of nature, while human science was condemned to remain forever hypothetical.

63. Jean Deprun, "Locke et la philosophie corpusculaire" (Locke and corpuscular philosophy), in vol. 2 of *Histoire de la philosophie* (A History of philosophy) (Paris: Gallimard, "Encyclopédie de la Pléiade," 1972).

64. From a more mundane point of view, chemists faced the same problem. For a long time, indeed until the beginning of the nineteenth century, they did not even consider the possibility of synthesizing organic compounds in the laboratory, being convinced that these could be produced only by vegetal or animal organisms, which were thought to be endowed, for this purpose, with some obscure "vital force." It would not be until 1828 and the synthesis of urea, achieved by Friedrich Wöhler (1800–1882), from CO_2 gas and ammonia, that the link between the mineral world and the organic and biological world would become evident, as no one could question the animal quality of the synthesized product. This initial laboratory achievement was followed by many others, but it was not until the middle of the century that numerous "organic" syntheses, realized notably by Hermann Kölbe (1818–1884) and Marcellin Berthelot (1827–1907), delivered the final blow to the theory of a "vital force."

65. E. Cassirer, *The Philosophy of the Enlightenment*, trans. Fritz C. A. Koelln and James P. Pettegrove (Princeton, N.J.: Princeton University Press, 1979).

66. P. L. Maupertuis, *Réponse aux objections de M. Diderot* (Reply to the objections of Mr. Diderot) (Paris: J. Vrin, 1984).

67. P. L. Maupertuis, *Système de la nature* (System of nature) (Paris: J. Vrin, 1984).

68. Among the great scientists of that period, Georges-Louis Leclerc de Buffon (1707–1788), an adherent of the atomic theory, is one of the few who considered gravitational forces capable of playing a determining role in the formation and transformation of chemical substances, as well as in biological processes—particularly in "the combination of the seminal fluids of the two sexes." From the second volume (published in 1749) of his voluminous *Histoire naturelle* (Natural history), which features a

total of thirty-six volumes, we quote: "The laws of affinity governing the way the parts composing different substances separate and reunite to form homogeneous matter are the same as the general law governing the action of all the celestial bodies on one another." Yet, apparently concerned by the isotropic character of the gravitational forces, Buffon completed his proposition on their presumed participation in the association of atoms by emphasizing the influence of the shape of these corpuscles. He postulated "that, at a very small distance, the shape of atoms attracting one another is as important, and perhaps more so, than their mass in the expression of the law, this shape becoming then the primary factor in the element of distance." Laplace shared this dual point of view. Furthermore, given Buffon's stature and importance in the sciences during this Age of Enlightenment, it is worth mentioning the logical, if not philosophical, argument with which he defended the very idea of atomicity: "Some will undoubtedly claim that matter must be regarded as infinitely divisible, since an atom, no matter how small we imagine it to be, can always be mentally divided into several parts. My reply is that the notion of such divisibility ad infinitum is as illusory as any other type of geometrical and arithmetical infinity: it is simply an abstraction of our mind, and does not exist in the nature of things; and should one insist on considering the infinite divisibility of matter as an absolute infinite, it becomes even easier to prove that it cannot exist in this sense: for, the moment we assume the smallest possible atom, our very assumption makes this atom indivisible since, if it were divisible, it could not be the smallest possible atom, which would be contrary to the assumption. It then seems to me that any hypothesis that is based on a process proceeding to infinity must be rejected not only as false, but also as devoid of any sense."

69. Maupertuis, *Système de la nature*.
70. Cassirer, *The Philosophy of the Enlightenment*.
71. Maupertuis, *Système de la nature*.
72. Ibid.
73. Ibid.
74. Maupertuis, *Réponse aux objections de M. Diderot*.
75. P. L. Maupertuis, *Essai de cosmologie* (Essay on cosmology) (Paris: J. Vrin, 1981[1750]).
76. D. Diderot, *Entretiens entre d'Alembert et Diderot* (Conversations between d'Alembert and Diderot) (Paris: Flammarion, 1965). The opinions of Maupertuis and Diderot were shared by other famous philosophers of the day, including Jean-Jacques Rousseau (1712–1778), who was certain that "what is incapable of thought could not have produced thinking beings," and Charles-Louis de Secondat, Baron de Montesquieu (1689–1755), who exclaimed in his *Esprit des lois* (Spirit of laws): "What greater absurdity is there than a blind fatality supposedly producing intelligent beings?" An anonymous work titled *L'Âme matérielle* (The material soul), written sometime between 1692 and 1704, also credits atoms with a rudimentary conscience. According to its author, reason consists of a propitious ordering of the delicate atoms of the soul; the effect of alcohol in its journey through the body is to dislodge some of these particles from their proper places.
77. D. Diderot, *Le Rêve de d'Alembert* (D'Alembert's dream) (Paris: Garnier-Flammarion, 1965); D. Diderot, *Pensées philosophiques* (Philosophical thoughts), (Paris: Garnier-Flammarion, 1972).
78. Diderot, *Entretiens entre d'Alembert et Diderot*.
79. See, for example, C. Guyot, *Diderot* (Paris: Éd. du Seuil, 1953).
80. Diderot, *Le Rêve de d'Alembert*.
81. Diderot, *Pensées philosophiques*.
82. D. Diderot, *Lettre sur les aveugles* (Letter on the blind) (Paris: Garnier-Flammarion, 1972).

83. Diderot, *Le Rêve de d'Alembert*.
84. Guyot, *Diderot*.
85. It is worth mentioning here the attempt by Savinien de Cyrano de Bergerac (1619–1655), hero of the well-known drama by Rostand (1897)—although the character bears little resemblance to him—to prove by means of a primitive probability calculation that atoms, driven by chance in the infiniteness of time, could actually produce, through their innumerable combinations, the organized world as we know it. See *Les Œuvres libertines de Cyrano de Bergerac* (The libertine works of Cyrano de Bergerac) (Paris: La Chèvre, 1921).
86. D. Diderot, "Letter to Sophie Volland," dated October 15, 1759, reproduced in C. Guyot, *Diderot*.
87. In a more prosaic vein, William Petty, a somewhat less romantically inclined English author, would in 1674 attribute sexual characteristics to atoms. His motivation was to explain elasticity (!) [cited in G. Bachelard, *Les Intuitions atomistiques* (Atomistic Intuitions) (Paris: J. Vrin, 1975)].
88. Paul Henri Thiry, baron d'Holbach, *The System of Nature: Laws of the Moral and Physical World* (New York: Garland Pub., 1984).
89. This conclusion is based on some of de La Mettrie's comments. For instance, he wrote in his *Abrégé des systèmes pour faciliter l'intelligence du Traité de l'âme* (Synopsis of systems to assist in the comprehension of the treatise on the soul): "It is not the nature of the principles of solid bodies that is responsible for their great variety, but the varied configurations of their atoms." See J. O. de La Mettrie, *Œuvres philosophiques* (Philosophical writings), vol. 1 (Paris: Fayard, 1984).
90. J. O. de La Mettrie, *Système d'Épicure*, in *Œuvres philosophiques*.
91. J. O. de La Mettrie, *L'Histoire naturelle de l'âme*, in *Œuvres philosophiques*.
92. J. O. La Mettrie, *L'Homme-machine*, in *Œuvres philosophiques*.
93. Herschel wrote in his work *A Preliminary Discourse on the Study of Natural Philosophy*: "We are certain that atoms can be grouped in a small number of classes, each of which is composed of entities with properties similar in all respects. When we observe a large number of objects that are completely alike, we are inclined to believe that this similarity is due to a common principle independent of the observation. . . . This conclusion, which would not be inconsequential, even if it were to apply to only two individuals perfectly identical in all respects and at all times, acquires overwhelming force when their number multiplies beyond what the imagination can grasp. It seems to me that the discoveries of modern chemistry dispel the notion of an eternal matter existing by itself, by attributing to each of these atoms the essential characteristics both of a manufactured object and of a subordinate agent." See John F. W. Herschel, *A Preliminary Discourse on the Study of Natural Philosophy* (Chicago: University of Chicago Press, 1987).
94. J. C. Maxwell, "Molecules," *Nature* 8 (1873), 437–441. This paper is a transcript of an address Maxwell gave before the British Association for the Advancement of Science.

Chapter Thirteen

1. A complete listing of Descartes's scientific errors would be quite lengthy. First among the most glaring ones, there was his "memorable mistake" (Leibniz's words) in his formulation of the principle of conservation of movement (instead of energy) as a fundamental law of collisions. Next came the assertion of vanishing gravity in void, followed by erroneous theories on the instantaneous transmission of light, on the circulation of blood, on color, which he considered a "modification" of white light, on magnetism, on the "state of rest" of earth, supposedly "transported" around the sun by "vortexes," on the importance of the epiphysis (or pineal gland) in spiritual life,

and many more. See, for example, J.-F. Revel, *Histoire de la philosophie occidentale. De Thalès à Kant* (History of Western philosophy: From Thales to Kant) (Paris: J. Vrin, 1989).

2. R. Descartes, *Principles of Philosophy*, trans. Valentine Roger Miller and Reese P. Miller (Dordrecht: Reidel, 1983). The celebrated phrase *"Cogito, ergo sum,"* actually only echoed the words uttered two millennia earlier by Parmenides: "Thought and being are one and the same." Paul Valéry composed a number of variants on Descartes's maxim. Two of these are: "Man thinks, therefore I am, said the universe," and "Sometimes I think and sometimes I am." Voltaire, for his part, declared: "I am body, and I think."

3. Descartes, *Principles of Philosophy*.

4. Ibid.

5. Ibid.

6. In this respect, he and Gassendi have similar conceptions of the shape of what amounts to atoms. Nicolas Lemery (1645–1715) was another notable proponent of the atomic structure of matter, who also adopted this visual picture.

7. Nicolas Malebranche (1638–1715), a post-Cartesian who later developed unmistakably anti-Cartesian theses, nevertheless, made subtle matter the centerpiece of his own physics. See his *Entretiens sur la métaphysique* (Conversations on metaphysics) (1688). "Vortexes" would find an ardent defender in Bernard de Fontenelle (1657–1757), another opponent of the Newtonian system. For instance, he wrote in his *Histoire de l'Académie des sciences* (History of the academy of sciences) (1734): "When one has conceived of this immense matter, divided and subdivided into vortexes, where an infinitude of movements prevail which neither interfere with nor hinder one another,... where nothing exerts an action except by causes whose existence is quite constant and whose idea is quite familiar, it seems no longer possible to consider Newtonian void without doing oneself violence." From another passage: "To invoke attraction is to permit the reintroduction of sympathies, abhorrences, of everything that was the disgrace of the ancient Scholastic philosophy." Among his preoccupations was also the question, "Why does attraction follow the square of distances, rather than any other power?"

8. J.-J. Rousseau, *Discours sur les sciences et lettres* (Discourse on science and literature) (Paris: Garnier-Flammarion, 1971).

9. For more details, see R. Lenoble, *Mersenne, ou la Naissance du mécanisme* (Mersenne: the birth of mechanism) (Paris: J. Vrin, 1971), or, by the same author, *Esquisse d'une histoire de l'idée de nature* (Outline of the history of the idea of nature) (Paris: Albin Michel, 1968).

10. R. Descartes, *Conversation with Burman*, trans. and commentary by John Cottingham (New York: Oxford University Press, 1976).

11. Quoted by G. Minois, *L'Église et la science. Histoire d'un malentendu* (The Church and science: History of a misunderstanding), vol. 1, *De saint Augustin à Galilée* (From St. Augustine to Galileo) (Paris: Fayard, 1990).

12. Hermeticism is a doctrine of the occult, embraced by alchemists; its name comes from Hermes Trismegistus, the legendary king of Egypt and author of books on magic, astrology, and alchemy. The cabala is a Jewish doctrine about God and the universe, according to which the text of the Bible contains a hidden meaning beyond its literal sense. The hidden message can be deciphered by attributing to each letter an esoteric and divine meaning.

13. See, for instance, A. Koyré, *From the Closed World to the Infinite Universe,* (Baltimore: Johns Hopkins University Press, 1957). The quotations from Henry More reproduced below all come from this source.

14. Although, according to More, both material and spiritual substances have extent, there are, nevertheless, differences between the two: The "body" is divisible and impenetrable, while the "spirit" is indivisible and penetrable. More's body-spirit duality differs from the unitary conception of the world embraced by the atomists of antiquity.

15. Quoted in Koyré, *From the Closed World to the Infinite Universe.*

16. Ibid. See also H. More, *The Antidote Against Atheism*, in *Philosophical Writings of Henry More*, ed. Flora Isabel MacKinnon (New York: AMS Press, 1969).

17. It is worthwhile to consult A. R. Hall, *Henry More. Magic, Religion and Experiment* (Cambridge, Mass.: Blackwell, 1990).

18. A. Hannequin, *Essai critique sur l'hypothèse des atomes dans la science contemporaine* (Critical essay on the hypothesis of atoms in contemporary science) (New York: Arno Press, 1981).

19. R. Lenoble, "Origines de la pensée scientifique moderne" (Origins of modern scientific thought), in *Histoire de la science* (History of science) (Paris: Gallimard, "Encyclopédie de la Pléiade," 1957).

20. "Isaac Newton was not a pleasant man." Such was the opinion of Stephen Hawking, his distant successor to the chair of mathematics at Cambridge University. See S. Hawking, *A Brief History of Time* (Toronto: Bantam Books, 1988).

21. In the context of this book, it is interesting to add that Spinoza rejected the two fundamental propositions of the atomic theory: the existence of atoms, and that of void. This is hardly surprising on the part of someone who recognized only a single substance in the world. On the subject of atoms, Spinoza evidently embraced Descartes's views, although he warned in *The Principles of Descartes's Philosophy* (La Salle, Ill.: Open Court Pub. Co., 1961) that he was not always in agreement with his French colleague. For instance, he wrote: "The topic of atoms has always been important and confused. Some argue that atoms exist because an infinite cannot be larger than another infinite; but if a quantity, say A, and a quantity twice as large as A, are infinitely divisible, they can also, through the power of God—who understands their infinite parts at first glance—be actually divided in infinite parts. Therefore, since one infinite cannot be, as we have stated, larger than another, the quantity A would have to be equal to its double, which is absurd. It is also debated whether half an infinite number is infinite too, whether it is even or odd, and other similar questions. To all this, Descartes replies that we should not reject those things which fall within our understanding and are thus conceived clearly and distinctly, just because of other things which are beyond our understanding and grasp, and can thus be perceived only inadequately. But the infinite and its properties are beyond human understanding which, by its nature, is finite; it would then be foolish to reject as false what we do not conceive clearly and distinctly about space, or to consider it doubtful, because we do not understand the infinite. That is why Descartes regards as indefinite or unlimited those things about which we discern no bounds, such as the extent of the universe, the divisibility of the parts of matter, etc." Spinoza exchanged a correspondence, through third parties, with Boyle on the issue of void; see the letters of Spinoza to Henri Oldenburg of July 17, 1663, and that to Louis Meyer of August 3, 1663, in *Spinoza. Œuvres complètes* (Spinoza: Complete works) (Paris: Gallimard, "Bibliothèque de la Pléiade," 1954). Boyle himself accepted the concept of void.

22. G. W. Leibniz, *La Monadologie* (Monadology), ed. E. Boutroux (Paris: LGF, 1991 [1714]).

23. J. Rivelaygue, in the introduction to Leibniz, *La Monadologie.*

24. E. Boutroux, in the foreword to Leibniz, *La Monadologie.*

25. "The perfection of matter is to that of void, as something to nothing," wrote Leibniz in

his "Letter to the Princess of Wales" of May 12, 1716, reproduced in S. Clarke, *The Leib-niz-Clarke Correspondence* (New York: Philosophical Library, 1956). Leibniz's statement would later puzzle Bertrand Russell (*A History of Western Philosophy* [New York: Simon & Schuster, 1959]), who could not understand why to exist is better than not to exist.

26. Clarke, *The Leibniz-Clarke Correspondence.*
27. Clarke, "Letter to the Princess of Wales," May 12, 1716. The letter was appended by Clarke as a postscript to Leibniz's fourth paper.
28. Ibid.
29. Koyré, *From the Closed World to the Infinite Universe.*
30. Clarke, *The Leibniz-Clarke Correspondence* (Leibniz's fifth paper).
31. Ibid (Clarke's fourth reply).
32. Ibid (Leibniz's first paper).
33. Koyré, *From the Closed World to the Infinite Universe.*
34. Clarke, *The Leibniz-Clarke Correspondence* (Clarke's first reply).
35. M. Serres, in *Histoire de la philosophie* (A History of Philosophy), vol. 2, ed. Yvon Belaval (Paris: Gallimard, "Encyclopédie de la Pléiade," 1973).
36. Plotinus, *The Enneads*, 3rd ed., rev. by B. S. Page (New York: Pantheon Books, 1961).
37. Russell, *A History of Western Philosophy.*
38. A. Schopenhauer, *The World as Will and Representation*, trans. E. F. J. Payne (New York: Dover Publications, 1958).

Chapter Fourteen

1. See, for example, L. Lederman and D. Teresi, *The God Particle* (New York: Dell Publishing, 1993).
2. F. Pillon, "L'Évolution historique de l'atomisme" (Historical evolution of atomism"), *Année philosophique* (1891), chap.2, part 1, 67–112.
3. See A. Cauchy, *Sept Leçons de physique générale* (Seven lectures in general physics) (Paris: Gauthiers-Villars, 1868).
4. Lederman and Teresi, *The God Particle.*
5. F. Pillon, "L'Évolution historique de l'atomisme."

Chapter Fifteen

1. Christian Wolff (1679–1754) stated that "philosophy is the science of all possible things," while G. W. F. Hegel (1770–1831) proclaimed: "All philosophies are both true and false."
2. G. Berkeley, *Principles of Human Knowledge* and *Three Dialogues between Hylas and Philonous* (New York: Penguin Books, 1988).
3. The allusion to Locke is particularly pointed.
4. D. Diderot, *Lettre sur les aveugles* (Letter on the blind) (Paris: Garnier-Flammarion, 1972).
5. F. W. J. Schelling, *The Ages of the World*, trans. F. de Wolf Bolman (New York: Columbia University Press, 1942).

Chapter Sixteen

1. I. Kant, *Universal Natural History and Theory of the Heavens*, trans. W. Hastie (Ann Arbor, Mich.: University of Michigan Press, 1969).
2. I. Kant, *Metaphysical Foundations of Natural Science*, trans. J. Ellington (Indianapolis: Bobbs-Merrill, 1970).
3. E. Bréhier, *Histoire de la philosophie* (A History of Philosophy), vol. 2, (Paris: Gallimard, "Encyclopédie de la Pléiade," 1972).

4. I. Kant, *Critique of Pure Reason*, trans. W. S. Pluhar (Indianapolis, Ind.: Hackett Publishing. Co., 1996).

5. Kant is not the only philosopher to have defected from supporter to opponent of the atomic theory. Francis Bacon (1561–1626), who is often considered the founder of the English school of empiricism, went through a similar evolution one and a half centuries earlier. An atomist in 1605, as seen in his *Cogitationes de natura rerum*, he abandoned that doctrine in 1620 in his *Novum organum*. Before switching, he asserted that the atomic theory, even though it was often branded as atheism, was better able to demonstrate the value of religion than did the theory of the four elements. In fact, in his estimation, it was far less necessary to invoke God for the purpose of constructing an ordered and beautiful world if one accepted the four elements—actually five, because he also admitted the existence of ether—"correctly and eternally distributed," than it would be with an infinite number "of tiny particles or seeds without specific locations." This argument is quoted in J. H. Brooke, *Science and Religion: Some Historical Perspectives* (Cambridge: Cambridge University Press, 1991). As an aside, Francis Bacon should not be confused with his namesake Roger Bacon (1214–1294), who was an English philosopher in the Middle Ages. Roger Bacon was a solid antiatomist; he went so far as to assert that atomism was one of the greatest obstacles to the development of philosophy!

6. I. Kant, *Prolegomena to Any Future Metaphysics that Will Be Able to Come Forward as Science*, trans. Gary Hatfield (New York: Cambridge University Press, 1997).

Chapter Seventeen

1. J. R. Partington, *A History of Chemistry*, vol. 2 (London: Macmillan, 1961); H. Schepers, "La philosophie allemande au XVIIe siècle" (German philosophy in the seventeenth century), in *Histoire de la philosophie* (A history of philosophy), vol. 2 (Paris: Gallimard, "Encyclopédie de la Pléiade," 1972); E. Bréhier, *Histoire de la philosophie*, vol. 2; B. Bensaude-Vincent and I. Stengers, *A History of Chemistry*, trans. Deborah van Dam (Cambridge, Mass.: Harvard University Press, 1996).

2. A. G. Van Melsen, *From Atomos to Atoms: The History of the Concept of Atom* (Pittsburgh: Duquesne University Press, 1952).

3. See, for instance, G. B. Stones, "The Atomic View of Matter in the 15th, 16th, and 17th Centuries," *Isis* 10 (1928), 445–465.

Chapter Eighteen

1. P. Thuillier, "La résistible ascension de la théorie atomique" (The resistible rise of the atomic theory), *La Recherche* 4, no. 36 (1973), 705.

2. B. Vidal, *Histoire de la chimie* (History of Chemistry) (Paris: Presses Univeritaires de France, collection "Que sais-je?," 1985); J. Rosmorduc, *Une histoire de la physique et de la chimie. De Thalès à Einstein* (A history of physics and chemistry: From Thales to Einstein) (Paris: Éd. du Seuil, 1985); B. Wojtkowiak, *Histoire de la chimie. Technique et documentation* (History of chemistry: Technology and documentation) (Paris: Lavoisier, 1988); B. Bensaude-Vincent and I. Stengers, *A History of Chemistry*, trans. Deborah van Dam (Cambridge, Mass.: Harvard University Press, 1996); R. Mierzecki, *The Historical Development of Chemical Concepts* (Dordrecht: Kluwer Academic Publishers, 1991); M. J. Nye, *Molecular Reality: A Perspective on the Scientific Work of Jean Perrin* (New York: American Elsevier, 1972); J. R. Partington, *A History of Chemistry*, vol. 2–5 (London: Macmillan, 1961–1964).

Chapter Nineteen

1. The last famous proponent of the theory of primordial elements may well have been

none other than Napoleon. He is said to have quipped during one of his military campaigns: "God created a fifth element especially for Poland—mud." (Cited by A. Castelot in *Talleyrand* [Paris: Librairie Académique Perrin, 1980].)

2. Jöns Jakob Berzelius would be even more explicit in his *Lehrbuch der Chemie* (Textbook of chemistry). Among substances to be found on earth, he made a distinction between *simple* (those that are believed with certainty to be undissociable), *nondissociated* (which are presumed not to be simple, but have yet to be decomposed), and *composite* (those that can be dissociated by chemical reactions into their constitutive elements).

3. We might mention in passing that Joseph-Louis Proust was aware that several combinations of the same elements were possible. He failed to discover the law of multiple proportions probably because he concerned himself with the relative weights involved in these combinations solely in terms of percentages of the total weight. Such an approach does not reveal directly the law in question.

4. In the hope that the atomic weight of all elements would turn out to be exact multiples of the atomic weight of hydrogen, the English physicist William Prout (1785–1850) postulated that hydrogen might be the fundamental building block of the atoms of all elements, which would vindicate the notion of a single, universal constituent of all bodies. This hypothesis had to be abandoned as soon as precise measurements of atomic weights revealed that fractional numbers were in fact the rule rather than the exception. That did not prevent the idea from resurfacing later on (see chapter 20).

5. This number, known as Avogadro's number, is one of the fundamental constants of modern atomic theory. Its value is 6.029×10^{23} molecules in 22.4 liters of *any* gas. Bachelard commented in his *Intuitions atomistiques* (Atomistic intuitions) (Paris: J. Vrin, 1975): "The path leading from Avogadro's hypothesis to Avogadro's law, and from Avogadro's law to Avogadro's number, retraces the full scientific history of an entire century."

6. A. Avogadro, "Essai d'une manière de déterminer les masses relatives des molécules élémentaires des corps, et les proportions selon lesquelles elles entrent dans ces combinaisons" (Essay on a procedure for determining the relative masses of elementary molecules in substances and the ratios in which they combine), *Journal de physique, de chimie et d'histoire naturelle* 73 (1811), 58–76.

7. As an amusing diversion, one might quote here an anecdote contributed by Alfred Naquet (1834–1916), and quoted by J. Jacques in his book about Berthelot (*Berthelot, autopsie d'un mythe* [Berthelot: Autopsy of a myth] [Paris: Belin, 1987]). Berthelot, as we will see, was a fervent "equivalentist" and a bitter opponent of atomism. Half a century after Avogadro, "he still refused to write water H_2O and to assign an atomic weight of 16 to oxygen. He insisted on writing H_2O_2 for water and accepted only formulas where oxygen entered with an even exponent. In doing so, he implicitly recognized that the smallest quantity of oxygen that could transfer from one compound to another was not 8 but 16." When Naquet asked him why he subjected himself to such contortions, he answered: "It is because I do not want to see chemistry degenerate into a religion. I do not want people to believe in the real existence of atoms the way Christians believe in the real presence of Jesus Christ in the consecrated wafer." It is unclear whether Berthelot realized the danger of using the words atoms and wafer in the same sentence.

8. This statement should not encourage the reader to believe that these pioneers acted in a concerted fashion, or even that they inspired each other. That was, unfortunately, not the case at all. Dalton, for instance, never recognized the validity of Gay-Lussac's laws, which he described as "the French disease of volumes." In retaliation, Gay-Lussac never endorsed Dalton's theory.

9. Auguste Kekulé (1829–1896), another famous chemist whom we will mention again

later, even distinguished three "different molecular and atomistic units"—physical molecules, chemical molecules, and atoms.

10. One century before Dumas, Bernard de Fontenelle (1657-1757) wrote in his *Histoire de l'Académie royale des sciences* (History of the Royal Academy of Sciences): "Chemistry, through visible operations, resolves bodies into crude and palpable principles, like salts, sulfur, and so on. But physics, through delicate speculations, acts on these principles, the way chemistry does on bodies, and in turn resolves them into other, even simpler, principles, into small bodies moved and conceived in an infinitude of ways: such is the main difference between physics and chemistry."

11. The phrase is from M. Daumas and J. Jacques, in *Histoire générale des sciences* (General history of science), ed. R. Taton, vol. 3 (Paris: Presses Universitaires de France, 1966).

12. Ibid.

13. J. Pétrel, "La Négation de l'atome dans la chimie du XIXe siècle. Cas de J.-B. Dumas" (The denial of atoms in the nineteenth century: The case of J.-B. Dumas), *Cahiers d'histoire et de philosophie des sciences*, 13 (1979), 1-135. Pétrel is convinced that Dumas did see the correct solution but simply refused to accept it, because "the fundamental requirement of experimental science proved, once again, to be a parasite nourished by an excess of logic," and therefore that "it is his philosophy and not his science that led Dumas astray."

14. Faraday described the atomic theory as clumsy and awkward. He proposed to use "electrochemical equivalents," of which he even devised a table. He pushed the degree of speculation so far as to suggest that the equivalent weights of material bodies might be their masses containing similar quantities of electricity. Moreover, in a famous 1844 article entitled "A Speculation Concerning Electrical Conductivity and the Nature of Matter," he sided with Boscovitch in refusing to make a distinction between the atom as a material point and the force field surrounding it. He argued that the atom is simply the center of the field. That led him to conclude that matter can be considered formed as much by forces as by atoms, a point of view that is remarkably similar to modern concepts (see Provisional Epilogue).

15. Some of the most important excerpts of the minutes of the debates of this Congress can be found in B. Bensaude-Vincent and C. Konnelis, ed., *Les Atomes. Une anthologie historique* (Atoms: A historical anthology) (Paris: Presses Pocket, 1991).

16. H. E. Roscoe, *Lessons in Elementary Chemistry: Inorganic and Organic* (New York: Macmillan, 1878).

17. B. Russell, *A History of Western Philosophy* (New York: Simon & Schuster, 1959).

18. D. Collinson, *Fifty Major Philosophers: A Reference Guide* (London: Croom Helm, 1987).

19. G. W. F. Hegel, *Science of Logic*, trans. A. V. Miller (London: Allen and Unwin, 1969).

20. G. W. F. Hegel, *Encyclopedia of the Philosophical Sciences in Outline*, trans. Steven A. Taubeneck (New York: Continuum, 1990).

21. Ibid.

22. A. Lécrivain, private communication. See also J. Biard, et al., *Introduction à la lecture de la science de la logique de Hegel* (Introduction to the reading of Hegel's science of logic) (Paris: Aubier-Montaigne, 1981).

23. G. W. F. Hegel, *Lectures on the History of Philosophy*, vol. 1, trans. E. S. Haldane (Lincoln, Neb.: University of Nebraska Press, 1995).

24. G. W. F. Hegel, *Lectures on the History of Philosophy*, vol. 3, trans. E. S. Haldane and Frances H. Simson (Lincoln, Neb.: University of Nebraska Press, 1995).

25. G. W. F. Hegel, *La science de la logique. La Théorie de la mesure* (The science of logic: Theory of measure), trans. and commentary by A. Doz (Paris: Presses Universitaires de France, 1970).

26. A. Schopenhauer, *The World as Will and as Representation*, trans. E. F. J. Payne (New York: Dover Publications, 1958).

27. All the citations in this section come from ibid.

28. The parallel between Condillac and Locke is perhaps unexpected. While Locke was an enthusiastic supporter of the corpuscular theory of matter, the abbot of Condillac (Étienne Bonot, 1714–1780) was an avowed anti-Epicurean (see his *Traité des animaux* [Treatise on animals] [Paris: J. Vrin, 1987]). He advanced a specific argument against the Greek form of atomism. He believed he had demonstrated "that a cause is ineffective on a being it had not given birth to." From this he concluded that the Epicurean system is an impossibility, "since it assumes that substances existing independently [namely, atoms] can act on one another." As for the reference to Hegel, Schopenhauer had a very low regard for this philosopher, whom he branded a "scribbler of absurdities and wrecker of brains," as quoted in D. Raymond, *Schopenhauer* (Paris: Éd. du Seuil, "Écrivains de toujours," 1979).

29. A. Comte, *The Positive Philosophy*, trans. H. Martineau (New York: AMS Press, 1974).

30. B. Russell, *History of Western Philosophy*.

31. W. Dampier, *A History of Science and its Relations with Philosophy and Religion* (New York: Cambridge University Press, 1966).

32. For a particularly lucid account, see E. Bréhier, *Histoire de la philosophie* (History of philosophy), vol. 3 (Paris: Presses Universitaires de France, "Quadrige," 1964), 751–758.

33. Michel Serres, *Auguste Comte et le positivisme* (Auguste Comte and positivism), in vol. 3 of *Histoire de la philosophie* (A history of philosophy) (Paris: Gallimard, "Encyclopédie de la Pléiade," 1974), 151–180.

34. G. Bachelard, *Les Intuitions atomistiques* (Atomistic intuitions) (Paris: J. Vrin, 1975).

35. Serres, *Auguste Comte et le positivisme*.

36. Comte, *The Positive Philosophy*.

37. Ibid.

38. B. d'Espagnat and E. Klein, *Regards sur la matière. Des quanta et des choses* (Considerations of matter: On quanta and objects) (Paris: Fayard, 1993).

39. Serres, *Auguste Comte et le positivisme*.

40. J. Ullmo, *La Pensée scientifique moderne* (Modern scientific thought) (Paris: Flammarion, 1969).

41. R. Mierzecki, *The Historical Development of Chemical Concepts* (Dordrecht: Kluwer Academic Publishers, 1991); B. Bensaude-Vincent and I. Stengers, *A History of Chemistry*, trans. Deborah van Dam (Cambridge, Mass.: Harvard University Press, 1996); M. J. Nye, *Molecular Reality: A Perspective on the Scientific Work of Jean Perrin* (New York: American Elsevier, 1972); J. R. Partington, *A History of Chemistry*, vol. 4 (London: Macmillan, 1964).

42. For more details on this topic, see C. A. Russell's extensive monograph *The History of Valency* (Leicester: Leicester University Press, 1971).

43. As a related footnote, we should mention here Alexander Wilhelm Williamson, one of Gerhardt's followers in the development of the theory of types. He believed that all chemical compounds could be derived by successive substitution of a single fundamental type: water and "polycondensed water." This was a throwback via an original path to the ideas of Thales of Miletus, and equally unsuccessful, we might add.

44. Quoted in Nye, *Molecular Reality: A Perspective on the Scientific Work of Jean Perrin*.

45. Bachelard, *Les Intuitions atomistiques*.

46. H. Sainte-Claire Deville, *Leçons de la Société chimique* (Lectures to the Chemical Society), February 28 and March 6, 1867.

47. M. Berthelot, "Réponse à la Note de M. Wurtz, relative à la loi d'Avogadro et à la théorie atomique" (Reply to the note by Mr. Wurtz concerning Avogadro's law and the atomic theory), *Comptes Rendus de l'Académie des sciences*, 84 (1877) 1189–1195

48. An account of this protracted debate was presented by P. Colmant, "Querelle à l'Institut entre équivalentistes et atomistes" (The dispute at the Institute [of France] between equivalentists and atomists), *Revue des questions scientifiques*, 143 (1973), 493–519.

49. C. A. Wurtz, *La Théorie atomique* (The atomic theory) (Paris: Germer Baillère et Cie, 1879).

50. See the excellent biography by J. Jacques, *Berthelot, autopsie d'un mythe* (Berthelot: Autopsy of a myth) (Paris: Belin, 1987).

51. Bachelard, *Les Intuitions atomistiques.*

52. P. Duhem, *Traité d'énergétique et de thermodynamique générale* (Treatise on energeticism and general thermodynamics) (Paris: Gauthiers-Villars, 1911).

53. W. Ostwald, "La déroute de l'atomisme contemporain" (The disarray of contemporary atomism) *Revue générale des sciences pures et appliquées*, no. 21 (1895), 953–958.

54. See for example E. Mach, *The Science of Mechanics: A Critical and Historical Account of its Development*, trans. Thomas J. McCormack (La Salle, Ill.: Open Court Pub. Co., 1960), and *Analysis of Sensations: The Relation of the Physical to the Psychical*, trans. C. M. Williams (New York: Dover, 1959).

55. A. Brenner, *Duhem. Science, réalité et appearance* (Duhem: Science, reality and appearance) (Paris: J. Vrin, 1990).

56. P. Brouzeng, *Duhem. Science et providence* (Duhem: Science and providence) (Paris: Belin, 1987).

57. P. Duhem, *L'Évolution de la mécanique* (The evolution of mechanics) (Paris: A. Joanin, 1903).

58. P. Duhem, *Le Mixte et la combinaison chimique* (Mixed substances and chemical combinations) (Paris: Fayard, 1902).

59. R. Maiocchi, "Duhem et l'atomisme" (Duhem and atomism), *Revue internationale de philosophie* 46 (1992), 376–389.

60. Duhem, *Le Mixte et la combinaison chimique.*

61. Kekulé adopted a similar position. He believed that since valency is a fundamental property of elements, it is just as invariant as the atomic weight (see Russell, *The History of Valency*).

62. D. Naquet, "Sur l'atomicité de l'oxygène, du soufre, du sélenium et du tellure" (On the atomicity of oxygen, sulfur, selenium, and tellurium), *Comptes rendus de l'Académie des sciences*, vol. 58 (1864), 381.

63. W. Heisenberg, *Physics and Beyond: Encounters and Conversations* (New York: Harper & Row, 1971). Heisenberg also voiced his own opinion on the subject: "The positivists have a simple solution: the world must be divided into that which we can say clearly and the rest, which we had better pass over in silence. But can anyone conceive of a more pointless philosophy, seeing that what we can say clearly amounts to next to nothing? If we omitted all that is unclear, we would probably be left with completely uninteresting and trivial tautologies."

64. Partington, *A History of Chemistry*, vol. 4.

65. Brouzeng, *Duhem. Science et providence.*

66. Partington, *A History of Chemistry*, vol. 4.

67. J.-M. Aubert, *Philosophie de la nature* (Natural philosophy) (Paris: Beauchesne, 1965).

68. Nye, *Molecular Reality: A Perspective on the Scientific Work of Jean Perrin.*

69. *Textes philosophiques de Lénine* (Lenin's philosophical writings) (Paris: Éd. Sociales, 1978). See also L. Sève, *Une introduction à la philosophie marxiste* (An introduction to Marxist philosophy) (Paris: Éd. Sociales, 1980).

70. F. Nietzsche, *Philosophy in the Tragic Age of the Greeks* (Chicago: Henry Regnery Co., 1982).

71. F. Nietzsche, *The Gay Science* (New York: Random House, 1974).

72. Ibid. The reference is to Alexander von Humboldt (1769–1859), a German scientist of some fame. He is known primarily as the author of the book *Kosmos*, a noted essay containing a detailed physical description of the world.

73. Both works appear in Karl Marx, *Œuvres* (Collected works), vol. 3, "Philosophie" (Paris: Gallimard, "Encyclopédie de la Pléiade," 1982).

74. Karl Marx and Friedrich Engels, *Lettres sur les sciences de la nature* (Letters about the natural sciences), trans. and commentary by J.-P. Lefebure (Paris: Éd. Sociales, 1973).

75. F. Engels, *Dialectics of Nature*, trans. J. B. S. Haldane (New York: International Publishers, 1973).

76. M. Cariou, *L'Atomisme. Trois essais: Gassendi, Leibniz, Bergson et Lucrèce* (Atomism. Three essays: Gassendi, Leibniz, Bergson and Lucretius) (Paris: Aubier-Montaigne, 1978).

77. A. Hannequin, *Essai critique sur l'hypothèse des atomes dans la science contemporaine* (Critical essay on the atomic hypothesis in contemporary science) (New York: Arno Press, 1981).

78. H. Bergson, *Creative Evolution*, trans. A. Mitchell (Westport, Conn.: Greenwood Press, 1975).

79. See the excellent monograph by M. Barthélémy-Madaule, *Bergson* (Paris: Éd. du Seuil, collection "Écrivains de toujours," 1967).

80. H. Bergson, *Essay on the Immediate Data of Consciousness*, trans. F. L. Pogson (New York: Humanities Press, 1971).

81. Bergson, *Creative Evolution*.

82. Bergson, *Essay on the Immediate Data of Consciousness*.

83. H. Bergson, *Matter and Memory*, trans. Nancy M. Paul and W. Scott Palmer (New York: Zone Books, 1988).

84. Ibid.

Chapter Twenty

1. The thirteen values of Avogadro's number proposed by Jean Perrin ranged from 6.2 to 7.5×10^{23} (see J. Perrin, *Atoms*, trans. D. Hammick [Woodbridge, Conn.: Ox Bow Press, 1990]). The value accepted today is 6.029×10^{23}.

2. This does not mean its universal acceptance. Even in France, some prominent chemists such as Henry Le Châtelier (1850–1936) and Georges Urbain (1872–1938) would remain distrustful of the atomic theory long after the publication of the book.

3. J. Tannery, *Science et philosophie* (Science and philosophy) (Paris: Félix Alcan, 1912).

4. Today's accepted values are 0.5687×10^{-11} kg/C and 1.045×10^{-8} kg/C, respectively.

5. See S. Weinberg (NP 1979), *The Discovery of Subatomic Particles* (New York: W. H. Freeman, 1990).

6. J. J. Thomson, *The Atomic Theory* (Oxford: Clarendon Press, 1914). In connection with positivism's hostile attitude toward atomism discussed in the previous chapter, it is instructive to relate the misadventure experienced by a scientist from Berlin named Walter Kaufmann, as told by Weinberg in *The Discovery of Subatomic Particles*. Kaufmann determined the m/e ratio virtually at the same time as Thomson, and apparently even obtained a more accurate value. But he never claimed to have discovered an elementary particle. The reason, according to Weinberg, was that, strongly influenced by Mach, he believed that a man of science had no business being concerned with hypothetical particles such as atoms. In Weinberg's opinion, Thomson discovered the electron because, unlike Mach and Kaufmann, he was perfectly willing to accept that the search for fundamental particles was an integral part of a physicist's mission.

7. In an ill-fated later effort to "improve" things, Lord Kelvin proposed in 1905 a more complicated model in which the atom would consist of alternating layers of "vitreous" (positive) and "resinous" (negative) electricity, the first being predominant for some

strange reason. The terms "vitreous" and "resinous" came from a nomenclature proposed in 1733 by Charles-François de Fay (1698–1739).

8. A similar model was proposed a few years earlier by a Japanese scientist named Hantaro Nagaoka (1865–1950). But he contributed hardly anything of value to prove its validity.

9. Leon Lederman (NP 1988), the renowned expert in high-energy physics, stressed in his book (with Dick Teresi) *The God Particle* (London: Dell Publishing, 1993) how lucky Rutherford had been to use X rays with an energy of only about 5 million eV, sufficiently low to be scattered by an atomic nucleus without penetrating it. A higher energy would have created all kinds of strange effects that would have considerably complicated the interpretation of the results. To give a sense of the order of magnitude of the phenomena involved here, this energy is still a million times greater than what is released in a typical chemical reaction.

10. We know today that, while the dimensions of the atom are conveniently expressed in units of Angströms (10^{-10} meter), those of the nucleus are better expressed in units of Fermis (10^{-15} meter). By way of comparison, the electron has a radius less than 10^{-20} meter, which makes it look practically like a point. The proof that the proton has a finite volume is due to Robert Hofstadter (NP 1961).

11. There have been numerous speculations about the possible existence of worlds composed of antimatter and about their likely properties. The great physicist Richard Feynman (1918–1988, NP 1965) mused that an extraterrestrial made of antimatter might have his heart "on the wrong side." He gives the following sound advice to an earthling about to meet a Martian: "You walk up to him and put out your right hand to shake hands. If he puts out his right hand, O.K., but if he puts out his left hand, then watch out . . . the two of you will annihilate with each other!" See R. Feynman, *The Character of Physical Law* (Cambridge, Mass.: M.I.T. Press, 1965), 107.

12. A. Pais, *Niels Bohr's Times in Physics, Philosophy, and Polity* (New York: Oxford University Press, 1991).

13. Gravitational forces are completely negligible in atomic systems. The electrostatic force between an electron and a proton exceeds the gravitational force by a factor of 10^{41}!

14. See A. George's introduction to M. Planck, *Autobiographie scientifique et derniers écrits* (Scientific autobiography and later writings) (Paris: Albin-Michel, 1960).

15. M. J. Nye, *Molecular Reality: A Perspective on the Scientific Work of Jean Perrin* (New York: American Elsevier, 1972).

16. As L. de Broglie stated in *The Revolution in Physics: A Non-Mathematical Survey of Quanta*, trans. Ralph W. Niemeyer (New York: Greenwood Press, 1969): "The way Einstein defines a quantum of light involves a non-corpuscular entity, namely, a frequency. Therefore, the definition of the photon energy as the product of the frequency and Planck's constant cannot serve as the basis for a purely corpuscular conception of radiation. Actually, it is rather like a bridge connecting the wave nature of light, well known since Fresnel's time, and its particle nature revealed by the discovery of the photoelectric effect." This comment should be kept in mind when we discuss Louis de Broglie's proposition concerning the wavelike characteristics of the electron.

17. W. Heisenberg, *Physics and Beyond: Encounters and Conversations* (New York: Harper & Row, 1971).

18. Niels Bohr, *Atomic Theory and the Description of Nature* (Cambridge: University Press, 1934, rev. ed. Woodbridge, Conn.: Ox Bow Press, 1987).

19. H. A. Boorse, L. Motz, and J. H. Weaver, *The Atomic Scientists: A Biographical History* (New York: John Wiley, 1989).

20. Bohr, *Atomic Theory and the Description of Nature.*

21. Ibid.

22. De Broglie, *The Revolution in Physics: A Non-Mathematical Survey of Quanta.*

23. Heisenberg, *Physics and Beyond: Encounters and Conversations.*

24. See an excellent discussion of this topic in C. Chevalley's introduction and glossary in N. Bohr, *Physique atomique et connaissance humaine* (Atomic physics and human knowledge) (Paris: Gallimard, 1991).

25. Pais, *Niels Bohr's Times in Physics, Philosophy, and Polity.*

26. Ibid.

27. Heisenberg, *Physics and Beyond: Encounters and Conversations.*

28. Inert gases were unknown in Mendeleev's time. They were added to a new column of the periodic table only after their discovery by William Ramsey (1857–1916, NP 1904) and his students in the waning years of the nineteenth century.

29. Pais, *Niels Bohr's Times in Physics, Philosophy, and Polity.*

30. Today, protons and neutrons are viewed as two different energy states of the same particle called a *nucleon*. The mass of a neutron (1.6749×10^{-27} kilogram) is slightly larger than that of a proton (1.6726×10^{-27} kilogram). A neutron corresponds to a higher energy state. When left to itself, a neutron decays into a proton, together with an electron and another particle called the antineutrino. The half-life of the decay process (the time necessary for half the neutrons originally present to disappear) is 12 minutes. Under normal conditions, such as those prevailing inside a nucleus, nuclear forces (corresponding to the so-called strong interactions) prevent this transformation from taking place, although it does occur when the number of neutrons is very much larger than the number of protons. The electrons emitted during the decay reaction constitute the β radiation.

31. John Dalton, *A New System of Chemical Philosophy* (New York: Philosophical Library, 1964).

32. Boorse, Motz, and Weaver, *The Atomic Scientists: A Biographical History.*

33. Thomson, *The Atomic Theory.*

34. The holy grail of alchemists, which was to make gold starting from other elements, finally became reality in 1949, when the American physicist Arthur J. Dempster achieved the transmutation of mercury into gold.

35. Bohr, *Atomic Theory and the Description of Nature.*

36. P. S. Laplace, *Essai philosophique sur les probabilités* (Philosophical essay on probabilities) (Paris: 1814).

37. The thesis committee was presided over by Jean Perrin and included Élie Cartan and Charles Mauguin. Paul Langevin joined the proceedings.

38. De Broglie, *The Revolution in Physics: A Non-Mathematical Survey of Quanta.*

39. It has been said in jest that Bohr devised his theory of the hydrogen atom by using classical mechanics on Mondays, Wednesdays, and Fridays, and quantum mechanics on Tuesdays, Thursdays, and Saturdays.

40. G. Lochak, *Louis de Broglie: Un prince de la science* (Louis de Broglie: A prince of science) (Paris: Flammarion, 1992).

41. Ibid.

42. De Broglie, *The Revolution in Physics: A Non-Mathematical Survey of Quanta.*

43. When speaking of Thomson in the context of atomic theory, it is important to carefully keep track of the first names: G. P. Thomson is J. J. Thomson's son. Interestingly, while the father first discovered the electron as a particle, the son completed the work by demonstrating that it was also a wave. As a result, the family can boast of having discovered both aspects of the electron's "personality."

44. See D. Carnal and J. Mlyneck, "L'Optique atomique" (Atomic optics), *La Recherche* 23 (1992), 1134–1142.

45. E. Schrödinger, *Collected Papers on Wave Mechanics*, trans. J. F. Shearer and W. M. Deans (New York: Chelsea Pub., 1978).

46. For more details, see, for instance, the excellent book by Walter Moore, *Schrödinger: Life and Thought* (Cambridge: Cambridge University Press, 1989).

47. For more details on Schrödinger's views and how they evolved over the years, see Michel Bitbol, "La Clôture de la représentation" (The closure of description), an extensive introduction to Schrödinger's *La Nature et les Grecs* (Nature and the Greeks) (Paris: Éd. du Seuil, 1992), and, by the same author, "L'Élision: essai sur la philosophie d'E. Schrödinger" (The elision: Essay on E. Schrödinger's philosophy), an equally extensive introduction to another of Schrödinger's works, *L'Esprit et la matière* (Mind and matter) (Paris: Éd. du Seuil, 1990). As Bitbol wrote: "Later on, despite all the difficulties inherent to the notion of wave packets, Schrödinger persisted in identifying particles with excitation states of a wave continuum, and their localization in space to regions of enhanced intensity in this continuum."

48. Lederman and Teresi, *The God Particle*.

49. A complex function involves the imaginary quantity i defined by $i^2 = -1$. The complex conjugate of a function f, designated f^*, is obtained by substituting $-i$ for i in the expression for f. The square of the module of the function f is the product $f.f^*$. It is written $|ff^*|$ and is always a real number.

50. A. Einstein, *Out of My Later Years* (Westport, Conn.: Greenwood Press, 1970).

51. A. George, ed., *Louis de Broglie, physicien et penseur* (Louis de Broglie: Physicist and thinker) (Paris: Albin Michel, 1953).

52. The solution to Schrödinger's equation can be found, in various levels of detail, in numerous textbooks on quantum mechanics. As a lighter footnote to this extraordinary intellectual adventure, Schrödinger was unable to solve by himself his own equation in the case of the hydrogen atom. He had great difficulties with the radial part of the solution expressed in a spherical coordinate system. That explains his exclamation: "If only I knew more mathematics!" It was the famous mathematician Herman Weyl (1885–1955) who steered him toward the correct solution (Moore, *Schrödinger: Life and Thought*).

53. The electron spin is a relativistic property. Therefore, it cannot be deduced from Schrödinger's equation, which does not include such effects. A relativistic theory of the electron, devised by Paul Adrian Maurice Dirac (1902–1984, NP 1933), remedied this deficiency.

54. For more details on the development of Heisenberg's ideas, see the excellent biography by David C. Cassidy, *Uncertainty: The Life and Science of Werner Heisenberg* (New York: W. H. Freeman, 1992).

55. S. Weinberg, *Dreams of a Final Theory* (New York: Pantheon Books, 1992).

56. Quoted in Albert Einstein, *Correspondance 1903–1955* (Paris: Hermann, 1972).

57. Cassidy, *Uncertainty: The Life and Science of Werner Heisenberg*.

58. W. Heisenberg, *Physics and Philosophy: The Revolution in Modern Science* (New York: Harper, 1958).

59. Quoted in S. Petruccioli, *Atoms, Metaphors and Paradoxes: Niels Bohr and the Construction of a New Physics* (New York: Cambridge University Press, 1993).

60. J. L. Heilbron, "The Earliest Missionaries of the Copenhagen Spirit," *Revue de l'histoire des sciences* 35 (1985), 195–230.

61. It is related to the fact, already mentioned, that matrix algebra does not obey the same rules as scalar algebra. For example, while multiplication of conventional numbers is always commutative (meaning a × b = b × a), that property does not necessarily hold for matrix multiplication. If q denotes the position of an electron and p its momentum, the product pq, which represents a measurement of the position *followed*

by a measurement of the momentum, is not equal to the product qp, which corresponds to the same measurements but performed in reverse order. The fact that the two results are different implies that measuring one quantity perturbs the measurement of the other. It follows that it is not possible to determine with precision these two quantities simultaneously. That is the essence of the uncertainty principle. G. Lochak in *Louis de Broglie: Un prince de la science* quotes a joking remark made by Dirac: "Quantum mechanics boils down to classical mechanics written in a noncommutative algebra," to which he added sarcastically: "That did not do much to clarify ideas."

62. G. Bachelard, *The New Scientific Spirit*, trans. Arthur Goldhammer (Boston: Beacon Press, 1984).
63. George, introduction to Planck, *Autobiographie scientifique et derniers écrits*.
64. Laplace, *Essai philosophique sur les probabilités*.
65. David C. Cassidy, "Heisenberg, Uncertainty and the Quantum Revolution," *Scientific American* 266 (May 1992), 106–112.
66. De Broglie, *The Revolution in Physics: A Non-Mathematical Survey of Quanta*.
67. B. d'Espagnat and E. Klein, *Regards sur la matière. Des quanta et des choses* (Considerations of matter: On quanta and objects) (Paris: Fayard, 1993).
68. George, ed., *Louis de Broglie, physicien et penseur*.
69. Ibid.
70. Bachelard, *The New Scientific Spirit*.
71. Einstein, *Out of My Later Years*.
72. E. Schrödinger, *My View of the World*, trans. Cecily Hastings (Cambridge, UK: Cambridge University Press, 1964).
73. See, for example, Pais, *Niels Bohr's Times in Physics, Philosophy, and Polity*; George, introduction to Planck, *Autobiographie scientifique et derniers écrits*; Chevalley, introduction to Bohr, *Physique atomique et connaissance humaine*; Lochak, *Louis de Broglie: Un prince de la science*; Moore, *Schrödinger: Life and Thought*; Cassidy, *Uncertainty: The Life and Science of Werner Heisenberg*.
74. A good example is M. Jammer, *The Philosophy of Quantum Mechanics: The Interpretations of Quantum Mechanics in Historical Perspective* (New York: Wiley Interscience, 1974). This book contains an extensive bibliography on this topic.
75. D'Espagnat and Klein, *Regards sur la matière. Des quanta et des choses*; J. E. Baggott, *The Meaning of Quantum Theory* (New York: Oxford University Press, 1992); F. Selleri, *Le Grand Débat de la théorie quantique* (The great debate about quantum theory) (Paris: Flammarion, 1986); H. R. Pagels, *The Cosmic Code: Quantum Physics as the Language of Nature* (New York: Simon & Schuster, 1982); Michel Blay, ed., "Bohr et la complémentarité" (Bohr and complementarity), special issue of *Revue d'histoire des sciences*, 38 (1985), no. 3–4, 195–351.
76. J. L. Heilbron summarized the atmosphere at the Fifth Solvay Congress, held in Brussels in October 1927, in those words: "Lorentz began a general discussion by rejecting Heisenberg's uncertainty principle. . . . Bohr responded with a reference to his Como presentation. Pauli declared himself in agreement with Bohr. Dirac attacked Schrödinger's realism . . . while Pauli attacked de Broglie's. Heisenberg insisted that it is the observer, and not nature, that chooses which type of phenomenon is revealed. Einstein objected that Heisenberg's interpretation violated the theory of relativity. Schrödinger and de Broglie jumped at the chance to restate their position, etc." (Heilbron, "The Earliest Missionaries of the Copenhagen Spirit").
77. See P. C. W. Davies and J. R. Brown, *The Ghost in the Atom: A Discussion of the Mysteries of Quantum Physics* (New York: Cambridge University Press, 1986).
78. C. Chevalley, "Complémentarité et langage dans l'interprétation de Copenhague"

(Complementarity and language in the Copenhagen interpretation), in Michel Blay, ed., "Bohr et la complémentarité," 251–292.

79. Bohr, *Atomic Theory and the Description of Nature.*
80. Chevalley, "Complémentarité et langage dans l'interprétation de Copen-hague."
81. De Broglie, *The Revolution in Physics: A Non-Mathematical Survey of Quanta.*
82. Paul Valéry, *Vues* (Views) (Paris: La Table ronde, 1948).
83. George, ed., *Louis de Broglie, physicien et penseur.*
84. This is an appropriate place to point out a paradoxical aspect of this issue. Scrödinger's equation—the fundamental equation of quantum mechanics—is completely deterministic. The energy levels of atoms and molecules are perfectly fixed. Likewise, if the wave function is known at any given time, it remains known at any subsequent time. As S. Ortoli pointed out in *L'Homme face à la science* (Man face-to-face with science) (Paris: Criterion, 1992): "The only problem is that it ceases to be deterministic the moment there is a measurement, and that is quite obviously the only time an observation can be made."
85. N. Bohr, *Essays 1932–1957 on Atomic Physics and Human Knowledge* (Woodbridge, Conn.: Ox Bow Press, 1987).
86. Heisenberg: *Physics and Beyond: Encounters and Conversations.*
87. Long before Einstein, Cardinal Melchior de Polignac (1661–1742), who was resolutely anti-Enlightenment, had expressed a similar distaste for games of dice. In his poem "Anti-Lucretius," he repudiates Epicureanism because, among other things, the constancy of physical laws seemed to him incompatible with the many possible outcomes of the "throws of a dice." He wrote the poem in order to "redress the damage Lucretius inflicted on poetry by prostituting it to atheism." The cardinal used a profusion of arguments (the poem is more than seven hundred pages long!) to denounce every single premise of ancient atomism. For example, he exclaimed: "If all these delusions which you, Epicurus, have spoken had been promoted by religion, how much scorn would you heap on them?" He supported Descartes's physics, particularly his distinction between matter and spirit. Of course, neither Einstein nor, *a fortiori*, Cardinal de Polignac was inclined to consider that God might have used loaded dice the way Holbach had suggested (see chapter 12).
88. D'Espagnat and Klein, *Regards sur la matière. Des quanta et des choses.*
89. George, ed., *Louis de Broglie, physicien et penseur.*
90. A. Pais, *Subtle Is the Lord: The Science and the Life of Albert Einstein* (New York: Oxford University Press, 1982).
91. Heisenberg, *Physics and Beyond: Encounters and Conversations.*
92. Chevalley, "Complémentarité et langage dans l'interprétation de Copen-hague."
93. Among the most famous of these thought experiments or paradoxes are "Young's double slits," the "suspended screen," Bohr's "photon box," and "Schrödinger's cat."
94. One is reminded of the words of Jean d'Ormesson (in *La Douane de mer*, [Paris: Gallimard, 1993]): "The world, in which logic and necessity reign with implacable rigor, is but an immense paradox. When we get some answers—and science gets many of them—we usually forget the right question. And when we do ask the right question, we get no answer."
95. D'Espagnat and Klein, *Regards sur la matière. Des quanta et des choses.*
96. Ibid.
97. For a simple account, see, for example, J. M. Lévy-Leblond, "Théorie quantique: un débat toujours actuel" (Quantum theory: The debate goes on), *La Recherche* 17 (1986), 394.
98. George, ed., *Louis de Broglie, physicien et penseur.*

99. Selleri, *Le Grand Débat de la théorie quantique.*

100. Ibid.

101. Heisenberg, *Physics and Philosophy.*

102. P. A. M. Dirac, in *Directions in Physics,* ed. H. Hora and J. R. Shepanski (New York: Wiley, 1978).

103. Quoted in Lochak, *Louis de Broglie: Un prince de la science.*

104. Moore, *Schrödinger: Life and Thought.*

105. Selleri, *Le Grand Débat de la théorie quantique.*

106. Lochak, *Louis de Broglie: Un prince de la science.*

107. L. de Broglie, *Recherches sur la théorie des quanta* (Research on the theory of quanta) (Paris: Masson, 1963 [1924]).

108. An account that is fairly easy to follow can be found in George, ed., *Louis de Broglie, physicien et penseur.*

109. At the time of his death in 1992, D. Bohm left us a book, written in collaboration with B. J. Hiley, *The Undivided Universe: An Ontological Interpretation of Quantum Theory* (London: Routledge, 1993), which summarizes his main ideas.

110. B. Jarroson, *Invitation à la philosophie des sciences* (Invitation to the philosophy of science) (Paris: Éd. du Seuil, 1992); J.-P. Longchamp, *Science and Belief* (Slough, UK: St. Pauls, 1993).

111. Valéry, *Vues.*

112. M. Planck, *Scientific Autobiography and Other Papers,* trans. Frank Gaynor, (Westport, Conn.: Greenwood Press, 1968); E. Schrödinger, *Mind and Matter* (Cambridge: Cambridge University Press, 1959); Heisenberg, *Physics and Beyond: Encounters and Conversations*; A. Einstein, *The World as I See it,* trans. Alan Harris (New York: Philosophical Library, 1949).

113. A. S. Eddington, *The Nature of the Physical World* (Cambridge: Cambridge University Press, 1928).

114. Heisenberg, *Physics and Beyond: Encounters and Conversations.*

115. Moore, *Schrödinger: Life and Thought.*

116. Ibid.

117. Bohr, *Atomic Theory and the Description of Nature.*

118. Pais, *Subtle Is the Lord: The Science and the Life of Albert Einstein.*

119. Einstein, *The World as I See it.*

120. Pais, *Subtle Is the Lord: The Science and the Life of Albert Einstein.*

121. Einstein, *The World as I See it.*

122. See note 21 to chapter 13.

123. Planck, *Scientific Autobiography and Other Papers.*

124. Quoted in George, introduction to Planck, *Autobiographie scientifique et derniers écrits.*

125. Heisenberg, *Physics and Beyond: Encounters and Conversations.*

126. The Dutch physician and chemist Herman Boerhaave (1668-1738) wondered during the Age of Enlightenment: "Why is it that Christianity made such rapid progress when it was preached by ignorants, and came to a virtual crawl when preached by learned men?"

127. G. Minois, *L'Église et la science. Histoire d'un malentendu* (The Church and science. History of a misunderstanding), vol. 2, *De Galilée à Jean-Paul II* (From Galileo to John Paul II) (Paris: Fayard, 1991).

128. *Speeches Delivered by Popes Pius XI, Pius XII, John XXIII, Paul VI, and John-Paul II to the Pontifical Academy of Sciences from 1936 to 1986* (Vatican: Pontificia Academia Scientiarium, 1986).

129. Paul Valéry (*Vues*) saw in Volta's discoveries "the beginning of a new era" in the history of mankind: "If history were truly a rational discipline systematically concerning

itself with the relative importance of events it chooses to record, I think it would teach us that the most important development—the one with the most far-reaching and visible impact on our present day-to-day life—that occurred between 1789 and 1815 may not be one of the great traditional historical dramas such as the Revolution or the Empire. . . . The most important event of the period from 1789 and 1815 is the invention of the galvanic battery and the discovery of electric currents by Volta in 1800."

130. For more details, see, for example, B. Vidal, *La Liaison chimique: Le concept et son histoire* (The chemical bond: The concept and its history) (Paris: J. Vrin, 1989).

131. G. N. Lewis, "The Atom and the Molecule," *Journal of the American Chemical Society* 38 (1916), 762.

132. The method of valence bonds is technically difficult to apply in quantitative research on molecular systems of even moderate size. But it is highly prized by specialists in its qualitative form, known as "resonance theory," because it enables the description of complex molecular structures by means of ordinary formulas familiar to any chemist. Although many researchers contributed to it, it owes much to Linus Pauling (1901–1994, NP 1954), whose celebrated book *The Nature of the Chemical Bond* (Ithaca: Cornell University Press, 1939) is considered the bible of the method and can be found in every chemistry laboratory throughout the world.

133. More in-depth accounts of the method of atomic orbitals can be found in a number of sources, for example: A. Pullman and B. Pullman, *Les Théories électroniques de la chimie organique* (Electronic theories in organic chemistry) (Paris: Masson, 1952); C. A. Coulson, *Valence* (Oxford: Oxford University Press, 1952); B. Pullman and A. Pullman, *Quantum Biochemistry* (New York: Wiley Interscience, 1963); R. S. Mulliken and W. C. Ermler, *Diatomic Molecules: Results ab initio Calculations* (New York: Academic Press, 1977).

134. See, for example, A. Rousset, *Les Nouvelles frontières de la connaissance* (New frontiers of knowledge) (Paris: Ellipses, 1983).

135. L. B. Guyton du Morveau would probably have vigorously objected to such leniency. He wrote in 1782: "To the extent possible, names must fit the essence of things. . . . When we do not have a sufficient understanding of a quality to be used as the basis for a name, a name that conveys nothing is preferable to one that expresses an erroneous concept."

136. For an excellent discussion of this subject, see H. C. von Baeyer, *Taming the Atom: The Emergence of the Visible Microworld* (New York: Random House, 1992).

137. For a review of the topic, see P. Zeppenfeld, D. M. Eigler, and E. K. Schweitzer, "On manipule même les atomes" (One can manipulate even atoms), *La Recherche* 23 (1992), 360–362. See also D. M. Eigler, C. P. Lutz, and W. E. Rudge, "An Atomic Switch Realized with Scanning Tunnelling Microscope," *Nature* 352 (1991), 600–603.

138. M. F. Crommie, C. P. Lutz, and D. M. Eigler, "Confinement of Electrons to Quantum Corrals on a Metal Surface," *Science* 262 (1993), 218–220.

139. J. S. Foster, J. E. Frommer, and P. C. Arnett, "Molecular Manipulation Using a Tunnelling Microscope," *Nature* 331 (1988), 324–326; G. Dujardin, R. E. Walkup, and P. Avouris, "Association of Individual Molecules with Electrons from the Tip of a Scanning Tunnelling Microscope," *Science* 255 (1992), 1232–1235.

140. Eigler, Lutz, and Rudge, "An Atomic Switch Realized with Scanning Tunnelling Microscope."

141. R. S. Van Dyck, P. B. Schwinger, and H. G. Dehmelt, "Electron Magnetic Moment from Geonium Spectra: Early Experiments and Background Concepts," *Physical Review D* 34 (1986), 722.

142. For a review of this topic, see, for example, M. Devoret, D. Estève, and C. Urbina, "Single Electron Transfer in Metallic Nanostructures," *Nature* 360 (1992), 547–553.

143. For a review, see W. Quint, W. Schleich, and H. Walther, "Le piègeage des ions" (Trapping ions), *La Recherche* 20 (1989), 1194–1205.

144. A. Aspect and J. Dalibard, "Le refroidissement des atomes par laser" (Cooling atoms with lasers), *La Recherche* 25 (1994), 30–37; G. Grynberg, "Une matrice de lumière pour ranger les atomes" (A matrix of light to arrange atoms), *La Recherche* 24 (1993), 896–897.

145. Translator's note: The 1997 Nobel prize in physics was awarded jointly to Stephen Chu, Claude Cohen-Tannoudli, and William D. Phillips for their pioneering work in the field of laser cooling technology.

146. Grynberg, "Une matrice de lumière pour ranger les atomes."

147. For an overall review and extensive bibliography, see "Observatoire français des techniques avancées," *Nanotechnologies et micromachines* (Nanotechnology and micromachines) (Paris: Masson, 1992).

148. Cicero believed, on the contrary, that Epicurus "was not the joking type and did not have the traits traditionally associated with his birthplace" (an allusion to the fact that his father was from Attica, whose citizens had a reputation for joviality).

Provisional Epilogue

1. We can afford only a very superficial summary of such a vast topic. Interested readers may want to consult several recent books intended for nonspecialists. See, for instance: L. Lederman and D. Teresi, *The God Particle* (New York: Dell Publishing, 1993); H. C. von Baeyer, *Taming the Atom* (New York: Random House, 1992); S. Weinberg, *Dreams of a Final Theory* (New York: Pantheon Books, 1992); Stephen W. Hawking, *A Brief History of Time* (New York: Bantam Books, 1988); A. Rousset, *Les Nouvelles frontières de la connaissance* (New frontiers of knowledge) (Paris: Ellipses, 1993); Heinz. R. Pagels, *The Cosmic Code* (New York: Simon & Schuster, 1982); S. Weinberg, *The First Three Minutes: A Modern View of the Origin of the Universe* (New York: Basic Books, 1977); T. X. Thuan, *The Secret Melody* (New York: Oxford University Press, 1995); P. C. W. Davies, *God and the New Physics* (London: Penguin Books, 1983); J. R. Gribbin, *In Search of the Big Bang* (New York: Bantam Books, 1986).

2. The estimate of the lifetime of electrons (or of any other particle) assumes that they never encounter positrons, their antimatter counterparts, in which case mutual annihilation would occur instantaneously.

3. At the time I was writing these lines, the top quark was the only one of the six quarks to remain strictly a hypothetical entity. A few days later, on April 26, 1994, the *New York Times* published an article, carried by the French newspaper *Le Monde* on April 28, announcing the discovery of the sixth and last quark by researchers at Fermilab near Chicago. Between the French and English versions of this book, another article in the February 9, 1996, issue of the journal *Science* reported results of experiments conducted at the same laboratory hinting at the possibility that quarks might themselves have a composite structure.

4. The term *boson* actually has a broader generic meaning and applies to any force-mediating particle. By contrast, particles of matter are called *fermions*. It is useful to remember that bosons have spins that are integers (including 0), while fermions always have half-integer spins. Another important difference is that fermions can be neither created nor destroyed (except by annihilation with their antiparticle counterparts), while no such restriction applies to bosons.

5. Before quarks were proposed, the interaction responsible for the cohesion of nuclei was called the "nuclear interaction." As early as 1934, Hideki Yukawa (1907–1981, NP 1949) had attributed it to an exchange of massive particles between protons and neutrons. This remarkable prediction was confirmed by the discovery in 1947 of the π-

meson (or pion) by C. F. Powell (1903–1969, NP 1950). Today, *mesons* constitute one of two subgroups of hadrons: just like protons and neutrons (which belong in the other subgroup, under the heading of *baryons*), mesons are also "composed" of quarks. But unlike baryons, they are made of only two quarks (in fact, one quark and one anti-quark of different types).

6. The existence of gravitational waves (as opposed to "gravitons," which are the bosons mediating the gravitational force) was confirmed quite recently by Russell Hulse and Joseph Taylor. The achievement earned them the 1993 Nobel Prize in physics.

7. Lederman and Teresi, *The God Particle*. The "esthetic" objection is sometimes expressed in the form of a practical argument. The existence of such a plethora of particles is especially troubling to some since there exists in nature only a very small number—four, to be exact—of stable particles: electrons, protons, neutrons (inside nuclei), and neutrinos. All existing matter is ostensibly made of the first three (neutrinos are apparently involved only in certain reactions, such as the radioactive decay of nuclei, but they play no significant role in the basic structure of matter and seem content to roam freely throughout the universe). Moreover, since protons and neutrons are made of only two types of quarks—up and down—united in undissociable groups of three, it could be said that virtually everything in today's universe is made up of these two quarks and of electrons. Indeed, some authors have not hesitated to assert that "as far as we know, if all the other particles suddenly ceased to exist, the universe would hardly be affected." This claim, however, may be a bit premature, and experiments aimed at proving it might well produce a few surprises.

8. This "theoretical obscurity" is due in large part to our inability to comprehend the role of gravitation. Given the extreme conditions of pressure and density prevailing before the 10^{-43}-second mark, gravitation must have been as important as the other three fundamental interactions. But physics as we know it is incapable of describing the behavior of matter and light when gravity is that strong.

9. George Lemaître (1894–1966), a Belgian priest who was one of the founders of modern cosmology and originated the concept of big bang—although he never actually used that term, which was introduced in 1950 by the British astronomer Fred Hoyle—was the first to propose, in 1927, the idea that the universe grew out of an originally superdense entity, which he named "primordial atom." However, he believed this "atom" to be a sphere thirty times larger than the sun.

10. This "provisional" epilogue must keep the door open to the future. As such, it is an appropriate forum to mention some fascinating recent developments on the topic of "constructive" interactions between matter and antimatter. Enrico Fermi and Edward Teller were the first to envision, as early as 1947, entities made of oppositely charged particles other than the usual protons and electrons. Such "exotic atoms" have in fact been created in the laboratory. The "antiprotonic helium atom," a helium atom in which one electron has been replaced by an antiproton, is a particularly interesting example. This new structure, denoted by the symbol pHe$^+$, has a lifetime of a few microseconds, which is sufficiently long for spectroscopic characterization. This metastable atom can also be considered an exotic diatomic molecule in which one nucleus has a charge of +2, while the other has the conventional charge of -1. Such entities have been given the whimsical name of "atomcules." Researchers are actively trying to create "antihydrogen," which is a "doubly exotic" atom composed of an antiproton and a positron (see, for instance, R. J. Hughes, "New Light on Antiprotonic Atoms," *Nature* 368 [1994], 813–814, and J. Eades, R. Hughes, and C. Zimmermann, "Antihydrogen," *Physics World* [July 1993], 44–48). The synthesis of antihydrogen was achieved in the interval between the French and English versions of this book. It occurred in September 1995. This success is credited to a team headed by the German

physicist Walter Oelert working at CERN (the acronym stands for Centre Européen de Recherches Nucléaires, or European Nuclear Research Center), in Geneva, Switzerland. *Nine* atoms of antihydrogen were produced (and survived for forty billionths of a second) by using antiprotons hurtling at very high speed around the giant LEP cyclotron and by injecting into their path a fine stream of xenon gas. On rare occasions, part of the resulting collision energy materialized into electron-positron pairs. On even rarer occasions, since the positrons and antiprotons had very nearly the same speed, these two particles briefly united to form an ephemeral antihydrogen atom.

11. It must be specified that the dominance of matter over radiation in today's universe is based on its energy content according to Einstein's formula $E = mc^2$. At the present time, matter has four thousand times more total energy than radiation. The balance swings the other way if one considers the number of particles involved: There are about one billion photons for every particle of matter. We might add that this ratio was the same during the one million years when radiation dominated over matter. The only difference is that the energy content of particles of matter has remained constant, while that of photons has continually decreased during the gradual expansion of cooling of the universe.

12. An excellent treatment of this topic can be found in a recent book by M. Cassé, *Du vide et de la création* (On vacuum and creation) (Paris: Odile Jacob, 1995).

13. We often take the mass of a particle for granted. In fact, the question of why elementary particles have different—indeed extremely different—masses and how they acquire that mass is one of the most vexing unresolved problems of modern physics. This is a good opportunity to at least mention the name of Peter Higgs, the author of the most advanced theory on this issue. It envisions that an as yet undiscovered particle, called appropriately enough the Higgs particle (or the Higgs boson, produced by the Higgs field), with very unusual properties, would be responsible for the mass of different elementary particles. The search for the Higgs boson constitutes, in Carlo Rubbia's words, "one of the most exciting detective stories of the last few years." It is to this very particle that Leon Lederman and Dick Teresi refer to in the title of their book *The God Particle.*

14. "The atom is just as porous as the solar system," observed A. Eddington.

15. For more details, see von Baeyer, *Taming the Atom.*

16. This possibility is currently prompting a great many discussions on its potential implications in terms of a religious view of the world's creation. This topic is addressed in several of the works cited in note 1 of this chapter.

INDEX